Mathematic

Teacher's Guide

CONTENTS

Revision Editor: Alan Christopherson, M.S.

Alpha Omega Publications ®

300 North McKemy Avenue, Chandler, Arizona 85226-2618
© MCMXCVI by Alpha Omega Publications, Inc.   All rights reserved.
LIFEPAC is a registered trademark of Alpha Omega Publications, Inc.

2

OVERVIEW

MATHEMATICS

Curriculum Overview
Grades K-12

Kindergarten

Lessons

1-40	41-80	81-120	121-160
Directions-right, left, high,low,etc.	**Directions**-right,left, high,low,etc.	**Directions**-right,left, high,low,etc.	**Directions**-right,left, high,low,etc.
Comparisons-big, little,alike,different	**Comparisons**-big, little,alike,different	**Comparisons**-big, little,alike,different	**Comparisons**-big, little,alike,different
Matching	**Matching**	**Matching**	**Matching**
Cardinal Numbers- to 9	**Cardinal Numbers**- to 12	**Cardinal Numbers**- to 19	**Cardinal Numbers**- to 100
Colors-red,blue,green, yellow, brown,purple	**Colors**-orange	**Colors**-black,white	**Colors**-pink
Shapes-circle,square, rectangle,triangle	**Shapes**-circle,square, rectangle,triangle	**Shapes**-circle square, rectangle,triangle	**Shapes**-circle,square, rectangle,triangle
Number Order	**Number Order**	**Number Order**	**Number Order**
Before and After	**Before and After**	**Before and After**	**Before and After**
Ordinal Numbers- to 9th	**Ordinal Numbers**- to 9th	**Ordinal Numbers**- to 9th	**Ordinal Numbers**- to 9th
Problem Solving	**Problem Solving**	**Problem Solving**	**Problem Solving**
	Number Words- to nine	**Number Words**- to nine	**Number Words**- to nine
	Addition-to 9	**Addition**-to 10 and multiples of 10	**Addition**-to 10 and multiples of 10
		Subtraction-to 9	**Subtraction**-to 10
		Place Value	**Place Value**
		Time/Calendar	**Time/Calendar**
			Money
			Skip Counting-2's, 5's, 10's
			Greater/ Less than

	Grade 1	Grade 2	Grade 3
LIFEPAC 1	**NUMBERS TO 99** • Number order, skip-count • Add, subtract to 9 • Story problems • Measurements, shapes	**NUMBERS TO 100** • Numbers and words to 100 • Operation symbols +, −, =, >, < • Add, subtract, story problems • Place value, fact families	**NUMBERS TO 999** • Digits, place value to 999 • Add, subtract, time • LInear measurements, dozen • Operation symbols +, −, =, ≠, >, <
LIFEPAC 2	**NUMBERS TO 99** • Add, subtract to 10 • Number words • Place value, shapes • Patterns, sequencing, estimation	**NUMBERS TO 200** • Numbers and words to 200 • Add, subtract, even and odd • Skip-count 2's, 5's, 10's, shapes • Ordinal numbers, fractions, money	**NUMBERS TO 999** • Fact families, patterns, fractions • Add, subtract - carry, borrow • Skip count 2's, 5's, 10's • Money, shapes, lines, even, odd
LIFEPAC 3	**NUMBERS TO 100** • Number sentences, • Fractions, oral directions • Story problems • Time, symbols =, ≠	**NUMBERS TO 200** • Add w/ carry to 10's place • Subtract, standard measurements • Flat shapes, money, AM/PM • Rounding to 10's place	**NUMBERS TO 999** • Add 3 numbers w/ carry • Coins, weight, volume, AM/PM • Fractions, oral instructions • Skip count 3's, subtract w/ borrow
LIFEPAC 4	**NUMBERS TO 100** • Add to 18, place value • Skip-count, even and odd • Money • Shapes, measurement	**NUMBERS TO 999** • Numbers and words to 999 • Add, subtract, place value • Calendar, making change • Measurements, solid shapes	**NUMBERS TO 9,999** • Place value to 9,999 • Rounding to 10's, estimation • Add and subtract fractions • Roman numerals, 1/4 inch
LIFEPAC 5	**NUMBERS TO 100** • Add 3 numbers - 1 digit • Ordinal numbers, fractions • Time, number line • Estimation, charts	**NUMBERS TO 999** • Data and bar graphs, shapes • Add, subtract to 100's • Skip-count 3's, place value to 100's • Add fractions, temperature	**NUMBERS TO 9,999** • Number sentences, temperature • Rounding to 100's, estimation • Perimeter, square inch • Bar graph, symmetry, even/odd rules
LIFEPAC 6	**NUMBERS TO 100** • Number words to 99 • Add 2 numbers - 2 digit • Symbols >, < • Fractions, shapes	**NUMBERS TO 999** • Measurements, perimeter • Time, money • Subtract w/ borrow from 10's place • Add, subtract fractions	**NUMBERS TO 9,999** • Add, subtract to 9,999 • Multiples, times facts for 2 • Area, equivalent fractions, money • Line graph, segments, angles
LIFEPAC 7	**NUMBERS TO 200** • Number order, place value • Subtract to 12 • Operation signs • Estimation, graphs, time	**NUMBERS TO 999** • Add w/ carry to 100's place • Fractions as words • Number order in books • Rounding and estimation	**NUMBERS TO 9,999** • Times facts for 5, missing numbers • Mixed numbers - add, subtract • Subtract with 0's in minuend • Circle graph, probability
LIFEPAC 8	**NUMBERS TO 200** • Addition, subtract to 18 • Group counting • Fractions, shapes • Time, measurements	**NUMBERS TO 999** • Add, subtract, measurements • Group count, 'think' answers • Convert coins, length, width • Directions-N, S, E, W	**NUMBERS TO 9,999** • Times facts for 3, 10 - multiples of 4 • Convert units of measurement • Decimals, directions, length, width • Picture graph, missing addend
LIFEPAC 9	**NUMBERS TO 200** • Add 3 numbers - 2 digit • Fact families • Sensible answers • Subtract 2 numbers - 2 digit	**NUMBERS TO 999** • Area and square measurement • Add 3 numbers - 20 digit w/ carry • Add coins and convert to cents • Fractions, quarter-inch	**NUMBERS TO 9,999** • Add, subtract whole numbers, fractions, mixed numbers • Standard measurements, metrics • Operation symbols, times facts for 4
LIFEPAC 10	**NUMBERS TO 200** • Add, subtract, place value • Directions - N, S, E, W • Fractions • Patterns	**NUMBERS TO 999** • Rules for even and odd • Round numbers to 100's place • Time - digital, sensible answers • Add 3 numbers - 3 digit	**NUMBERS TO 9,999** • Add, subtract, times facts 2,3,4,5,10 • Rounding to 1,000's, estimation • Probability, equations, parentheses • Perimeter, area

Grade 4	Grade 5	Grade 6	
WHOLE NUMBERS & FRACTIONS • Naming whole numbers • Naming Fractions • Sequencing patterns • Numbers to 1,000	**WHOLE NUMBERS & FRACTIONS** • Operations & symbols • Fraction language • Grouping, patterns, sequencing • Rounding & estimation	**FRACTIONS & DECIMALS** • Number to billions' place • Add & subtract fractions • Add & subtract decimals • Read and write Fractions	LIFEPAC 1
WHOLE NUMBERS & FRACTIONS • Operation symbols • Multiplication - 1 digit multiplier • Fractions - addition & subtraction • Numbers to 10,000	**WHOLE NUMBERS & FRACTIONS** • Multiplication & division • Fractions - +, –, simplify • Plane & solid shapes • Symbol language	**FINDING COMMON DENOMINATORS** • Prime factors • Fractions with unlike denominators • Exponential notation • Add & subtract mixed numbers	LIFEPAC 2
WHOLE NUMBERS & FRACTIONS • Multiplication with carrying • Rounding & estimation • Sequencing fractions • Numbers to 100,000	**WHOLE NUMBERS & FRACTIONS** • Short division • Lowest common multiple • Perimeter & area • Properties of addition	**MULTIPLYING MIXED NUMBERS** • Multiply mixed numbers • Divide decimals • Bar and line graphs • Converting fractions & decimals	LIFEPAC 3
LINES & SHAPES • Plane & solid shapes • Lines & line segments • Addition & subtraction • Multiplication with carrying	**WHOLE NUMBERS** • Lines - shapes - circles • Symmetric - congruent - similar • Decimal place value • Properties of multiplication	**DIVIDING MIXED NUMBERS** • Divide mixed numbers • Area and perimeter • Standard measurements	LIFEPAC 4
WHOLE NUMBERS • Division - 1 digit divisor • Families of facts • Standard measurements • Number grouping	**WHOLE NUMBERS & FRACTIONS** • Multiply & divide by 10, 100, 1,000 • Standard measurements • Rate problems • Whole number & fraction operations	**METRIC MEASURE** • Metric measures • Plane & solid shapes • Multi-operation problems • Roman Numerals	LIFEPAC 5
WHOLE NUMBERS & FRACTIONS • Division - 1 digit with remainder • Factors & multiples • Fractions - improper & mixed • Equivalent fractions	**FRACTIONS & DECIMALS** • Multiplication of fractions • Reading decimal numbers • Adding & subtracting decimals • Multiplication - decimals	**LCM & GCF** • LCM, GCF • Fraction and decimal equivalents • Percent • Variables, functions & formulas	LIFEPAC 6
WHOLE NUMBERS & FRACTIONS • Multiplication - 2 digit multiplier • Simplifying fractions • Averages • Decimals in money problems	**WHOLE NUMBERS & FRACTIONS** • Division - 2-digit divisor • Metric units • Multiplication - mixed numbers • Multiplication - decimals	**INTEGERS, RATIO & PROPORTION** • Positive and negative integers • Ratio & proportion • Fractions, decimals & percents • Statistics	LIFEPAC 7
WHOLE NUMBERS & FRACTIONS • Division 1 digit divisor • Fractions - unlike denominators • Metric units • Whole numbers - +, –, x, ÷	**WHOLE NUMBERS** • Calculators & whole numbers • Calculators & decimals • Estimation • Prime factors	**PROBABILITY & GRAPHING** • Probability • Graphs • Metric and standard units • Square root	LIFEPAC 8
DECIMALS & FRACTIONS • Reading and writing decimals • Mixed numbers - +, – • Cross multiplication • Estimation	**FRACTIONS & DECIMALS** • Division - fractions • Division - decimals • Ratios & ordered pairs • Converting fractions to decimals	**CALCULATORS & ESTIMATION** • Calculators • Estimation • Geometric symbols & shapes • Missing number problems	LIFEPAC 9
PROBLEM SOLVING • Estimation & data gathering • Charts & Graphs • Review numbers to 100,000 • Whole numbers - +, –, x, ÷	**PROBLEM SOLVING** • Probability & data gathering • Charts & graphs • Review numbers to 100 million • Fractions & decimals - +, –, x, ÷	**INTEGERS & OPERATIONS** • Mental arithmetic • Fraction operations • Variables & properties • Number lines	LIFEPAC 10

Mathematics LIFEPAC Overview

	Grade 7	Grade 8	Grade 9
LIFEPAC 1	**WHOLE NUMBERS** • Number concepts • Addition • Subtraction • Applications	**WHOLE NUMBERS** • The set of whole numbers • Graphs • Operations with whole numbers • Applications with whole numbers	**VARIABLES AND NUMBERS** • Variables • Distributive Property • Definition of signed numbers • Signed number operations
LIFEPAC 2	**MULTIPLICATION AND DIVISION** • Basic facts • Procedures • Practice • Applications	**NUMBERS AND FACTORS** • Numbers and bases • Sets • Factors and multiples • Least common multiples	**SOLVING EQUATIONS** • Sentences and formulas • Properties • Solving equations • Solving inequalities
LIFEPAC 3	**GEOMETRY** • Segments, lines, and angles • Triangles • Quadrilaterals • Circles and hexagons	**RATIONAL NUMBERS** • Proper and improper fractions • Mixed numbers • Decimal fractions • Per cent	**PROBLEM ANALYSIS AND SOLUTION** • Words and symbols • Simple verbal problems • Medium verbal problems • Challenging verbal problems
LIFEPAC 4	**RATIONAL NUMBERS** • Common fractions • Improper fractions • Mixed numbers • Decimal fractions	**FRACTIONS AND ROUNDING** • Common fraction addition • Common fraction subtraction • Decimal fractions • Rounding numbers	**POLYNOMIALS** • Addition of polynomials • Subtraction of polynomials • Multiplication of polynomials • Division of polynomials
LIFEPAC 5	**SETS AND NUMBERS** • Set concepts and operations • Early number systems • Decimal number system • Factors and multiples	**FRACTIONS AND PER CENT** • Multiplication of fractions • Division of fractions • Fractions as per cents • Per cent exercises	**ALGEBRAIC FACTORS** • Greatest common factor • Binomial factors • Complete factorization • Word problems
LIFEPAC 6	**FRACTIONS** • Like denominators • Unlike denominators • Decimal fractions • Equivalents	**STATISTICS, GRAPHS, & PROBABILITY** • Statistical measures • Types of graphs • Simple probability • And–Or statements	**ALGEBRAIC FRACTIONS** • Operations with fractions • Solving equations • Solving inequalities • Solving word problems
LIFEPAC 7	**FRACTIONS** • Common fractions • Decimal fractions • Per cent • Word problems	**INTEGERS** • Basic concepts • Addition and subtraction • Multiplication and division • Expressions and sentences	**RADICAL EXPRESSIONS** • Rational and irrational numbers • Operations with radicals • Irrational roots • Radical equations
LIFEPAC 8	**FORMULAS AND RATIOS** • Writing formulas • A function machine • Equations • Ratios and proportions	**FORMULAS AND GEOMETRY** • Square root • Perimeter, circumference, and area • Rectangular solid • Cylinder, cone, and sphere	**GRAPHING** • Equations of two variables • Graphing lines • Graphing inequalities • Equations of lines
LIFEPAC 9	**DATA, STATISTICS AND GRAPHS** • Gathering and organizing data • Central tendency and dispersion • Graphs of statistics • Graphs of points	**ALGEBRAIC EQUATIONS** • Variables in formulas • Addition and subtraction • Multiplication and division • Problem solving	**SYSTEMS** • Graphical solution • Algebraic solutions • Determinants • Word problems
LIFEPAC 10	**MATHEMATICS IN SPORTS** • Whole numbers • Geometry, sets, and systems • Fractions • Formulas, ratios, and statistics	**NUMBERS, FRACTIONS, ALGEBRA** • Whole numbers and fractions • Fractions and per cent • Statistics, graphs and probability • Integers and algebra	**QUADRATIC EQUATIONS AND REVIEW** • Solving quadratic equations • Equations and inequalities • Polynomials and factors • Radicals and graphing

Grade 10	Grade 11	Grade 12	
A MATHEMATICAL SYSTEM • Points, lines, and planes • Definition of definitions • Geometric terms • Postulates and theorems	**SETS, STRUCTURE, AND FUNCTION** • Properties and operations of sets • Axioms and applications • Relations and functions • Algebraic expressions	**RELATIONS AND FUNCTIONS** • Relations and functions • Rules of correspondence • Notation of functions • Types of functions	LIFEPAC 1
PROOFS • Logic • Reasoning • Two-column proof • Paragraph proof	**NUMBERS, SENTENCES, & PROBLEMS** • Order and absolute value • Sums and products • Algebraic sentences • Number and motion problems	**SPECIAL FUNCTIONS** • Linear functions • Second-degree functions • Polynomial functions • Other functions	LIFEPAC 2
ANGLES AND PARALLELS • Definitions and measurement • Relationships and theorems • Properties of parallels • Parallels and polygons	**LINEAR EQUATIONS & INEQUALITIES** • Graphs • Equations • Systems of equations • Inequalities	**TRIGONOMETRIC FUNCTIONS** • Definition • Evaluation of functions • Trigonometric tables • Special angles	LIFEPAC 3
CONGRUENCY • Congruent triangles • Corresponding parts • Inequalities • Quadrilaterals	**POLYNOMIALS** • Multiplying polynomials • Factoring • Operations with polynomials • Variations	**CIRCULAR FUNCTIONS & GRAPHS** • Circular functions & special angles • Graphs of sin and cos • Amplitude and period • Phase shifts	LIFEPAC 4
SIMILAR POLYGONS • Ratios and proportions • Definition of similarity • Similar polygons and triangles • Right triangle geometry	**RADICAL EXPRESSIONS** • Multiplying and dividing fractions • Adding and subtracting fractions • Equations with fractions • Applications of fractions	**IDENTITIES AND FUNCTIONS** • Reciprocal relations • Pythagorean relations • Trigonometric identities • Sum and difference formulas	LIFEPAC 5
CIRCLES • Circles and spheres • Tangents, arcs, and chords • Special angles in circles • Special segments in circles	**REAL NUMBERS** • Rational and irrational numbers • Laws of Radicals • Quadratic equations • Quadratic formula	**TRIGONOMETRIC FUNCTIONS** • Trigonometric functions • Law of cosines • Law of sines • Applied problems	LIFEPAC 6
CONSTRUCTION AND LOCUS • Basic constructions • Triangles and circles • Polygons • Locus meaning and use	**QUADRATIC RELATIONS & SYSTEMS** • Distance formulas • Conic sections • Systems of equations • Application of conic sections	**TRIGONOMETRIC FUNCTIONS** • Inverse functions • Graphing polar coordinates • Converting polar coordinates • Graphing polar equations	LIFEPAC 7
AREA AND VOLUME • Area of polygons • Area of circles • Surface area of solids • Volume of solids	**EXPONENTIAL FUNCTIONS** • Exponents • Exponential equations • Logarithmic functions • Matrices	**QUADRATIC EQUATIONS** • Conic sections • Circle and ellipse • Parabola and hyperbola • Transformations	LIFEPAC 8
COORDINATE GEOMETRY • Ordered pairs • Distance • Lines • Coordinate proofs	**COUNTING PRINCIPLES** • Progressions • Permutations • Combinations • Probability	**PROBABILITY** • Random experiments & probability • Permutations • Combinations • Applied problems	LIFEPAC 9
REVIEW • Proof and angles • Polygons and circles • Construction and measurement • Coordinate geometry	**REVIEW** • Integers and open sentences • Graphs and polynomials • Fractions and quadratics • Exponential functions	**CALCULUS** • Mathematical induction • Functions and limits • Slopes of functions • Review of 1200 mathematics	LIFEPAC 10

LIFEPAC

M
A
N
A
G
E
M
E
N
T

11

STRUCTURE OF THE LIFEPAC CURRICULUM

The LIFEPAC curriculum is conveniently structured to provide one teacher handbook containing teacher support material with answer keys and ten student worktexts for each subject at grade levels two through twelve. The worktext format of the LIFEPACs allows the student to read the textual information and complete workbook activities all in the same booklet. The easy to follow LIFEPAC numbering system lists the grade as the first number(s) and the last two digits as the number of the series. For example, the Language Arts LIFEPAC at the 6th grade level, 5th book in the series would be LA 605.

Each LIFEPAC is divided into 3 to 5 sections and begins with an introduction or overview of the booklet as well as a series of specific learning objectives to give a purpose to the study of the LIFEPAC. The introduction and objectives are followed by a vocabulary section which may be found at the beginning of each section at the lower levels, at the beginning of the LIFEPAC in the middle grades, or in the glossary at the high school level. Vocabulary words are used to develop word recognition and should not be confused with the spelling words introduced later in the LIFEPAC. The student should learn all vocabulary words before working the LIFEPAC sections to improve comprehension, retention, and reading skills.

Each activity or written assignment has a number for easy identification, such as 1.1. The first number corresponds to the LIFEPAC section and the number to the right of the decimal is the number of the activity.

Teacher checkpoints, which are essential to maintain quality learning, are found at various locations throughout the LIFEPAC. The teacher should check 1) neatness of work and penmanship, 2) quality of understanding (tested with a short oral quiz), 3) thoroughness of answers (complete sentences and paragraphs, correct spelling, etc.), 4) completion of activities (no blank spaces), and 5) accuracy of answers as compared to the answer key (all answers correct).

The self test questions are also number coded for easy reference. For example, 2.015 means that this is the 15th question in the self test of Section II. The first number corresponds to the LIFEPAC section, the zero indicates that it is a self test question, and the number to the right of the zero the question number.

The LIFEPAC test is packaged at the centerfold of each LIFEPAC. It should be removed and put aside before giving the booklet to the student for study.

Answer and test keys have the same numbering system as the LIFEPACs and appear at the back of this handbook. The student may be given access to the answer keys (not the test keys) under teacher supervision so that he can score his own work.

A thorough study of the Curriculum Overview by the teacher before instruction begins is essential to the success of the student. The teacher should become familiar with expected skill mastery and understand how these grade level skills fit into the overall skill development of the curriculum. The teacher should also preview the objectives that appear at the beginning of each LIFEPAC for additional preparation and planning.

TEST SCORING and GRADING

Answer keys and test keys give examples of correct answers. They convey the idea, but the student may use many ways to express a correct answer. The teacher should check for the essence of the answer, not for the exact wording. Many questions are high level and require thinking and creativity on the part of the student. Each answer should be scored based on whether or not the main idea written by the student matches the model example. "Any Order" or "Either Order" in a key indicates that no particular order is necessary to be correct.

Most self tests and LIFEPAC tests at the lower elementary levels are scored at 1 point per answer; however, the upper levels may have a point system awarding 2 to 5 points for various answers or questions. Further, the total test points will vary; they may not always equal 100 points. They may be 78, 85, 100, 105, etc.

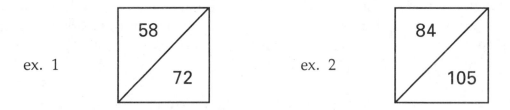

ex. 1   58 / 72   ex. 2   84 / 105

A score box similar to ex.1 above is located at the end of each self test and on the front of the LIFEPAC test. The bottom score, 72, represents the total number of points possible on the test. The upper score, 58, represents the number of points your student will need to receive an 80% or passing grade. If you wish to establish the exact percentage that your student has achieved, find the total points of his correct answers and divide it by the bottom number (in this case 72.) For example, if your student has a point total of 65, divide 65 by 72 for a grade of 90%. Referring to ex. 2, on a test with a total of 105 possible points, the student would have to receive a minimum of 84 correct points for an 80% or passing grade. If your student has received 93 points, simply divide the 93 by 105 for a percentage grade of 89%. Students who receive a score below 80% should review the LIFEPAC and retest using the appropriate Alternate Test found in the Teacher's Guide.

The following is a guideline to assign letter grades for completed LIFEPACs based on a maximum total score of 100 points.

LIFEPAC Test   =   60% of the Total Score (or percent grade)
Self Test   =   25% of the Total Score (average percent of self tests)
Reports   =   10% or 10* points per LIFEPAC
Oral Work   =   5% or 5* points per LIFEPAC
*Determined by the teacher's subjective evaluation of the student's daily work.

Example:

LIFEPAC Test Score	=	92%	92	x	.60	=	55 points
Self Test Average	=	90%	90	x	.25	=	23 points
Reports						=	8 points
Oral Work						=	4 points

TOTAL POINTS	=	90 points

Grade Scale based on point system:

100	–	94	=	A
93	–	86	=	B
85	–	77	=	C
76	–	70	=	D
Below		70	=	F

TEACHER HINTS and STUDYING TECHNIQUES

LIFEPAC Activities are written to check the level of understanding of the preceding text. The student may look back to the text as necessary to complete these activities; however, a student should never attempt to do the activities without reading (studying) the text first. Self tests and LIFEPAC tests are never open book tests.

Language arts activities (skill integration) often appear within other subject curriculum. The purpose is to give the student an opportunity to test his skill mastery outside of the context in which it was presented.

Writing complete answers (paragraphs) to some questions is an integral part of the LIFEPAC Curriculum in all subjects. This builds communication and organization skills, increases understanding and retention of ideas, and helps enforce good penmanship. Complete sentences should be encouraged for this type of activity. Obviously, single words or phrases do not meet the intent of the activity, since multiple lines are given for the response.

Review is essential to student success. Time invested in review where review is suggested will be time saved in correcting errors later. Self tests, unlike the section activities, are closed book. This procedure helps to identify weaknesses before they become too great to overcome. Certain objectives from self tests are cumulative and test previous sections; therefore, good preparation for a self test must include all material studied up to that testing point.

The following procedure checklist has been found to be successful in developing good study habits in the LIFEPAC curriculum.

1. Read the introduction and Table of Contents.
2. Read the objectives.
3. Recite and study the entire vocabulary (glossary) list.
4. Study each section as follows:
    a. Read the introduction and study the section objectives.
    b. Read all the text for the entire section, but answer none of the activities.
    c. Return to the beginning of the section and memorize each vocabulary word and definition.
    d. Reread the section, complete the activities, check the answers with the answer key, correct all errors, and have the teacher check.
    e. Read the self test but do not answer the questions.
    f. Go to the beginning of the first section and reread the text and answers to the activities up to the self test you have not yet done.
    g. Answer the questions to the self test without looking back.
    h. Have the self test checked by the teacher.
    i. Correct the self test and have the teacher check the corrections.
    j. Repeat steps a–i for each section.

5. Use the SQ3R* method to prepare for the LIFEPAC test.
6. Take the LIFEPAC test as a closed book test.
7. LIFEPAC tests are administered and scored under direct teacher supervision. Students who receive scores below 80% should review the LIFEPAC using the SQ3R* study method and take the Alternate Test located in the Teacher Handbook. The final test grade may be the grade on the Alternate Test or an average of the grades from the original LIFEPAC test and the Alternate Test.

    *SQ3R: Scan the whole LIFEPAC.
         Question yourself on the objectives.
         Read the whole LIFEPAC again.
         Recite through an oral examination.
         Review weak areas.

GOAL SETTING and SCHEDULES

Each school must develop its own schedule, because no single set of procedures will fit every situation. The following is an example of a daily schedule that includes the five LIFEPAC subjects as well as time slotted for special activities.

Possible Daily Schedule

8:15	–	8:25	Pledges, prayer, songs, devotions, etc.
8:25	–	9:10	Bible
9:10	–	9:55	Language Arts
9:55	–	10:15	Recess (juice break)
10:15	–	11:00	Mathematics
11:00	–	11:45	Social Studies
11:45	–	12:30	Lunch, recess, quiet time
12:30	–	1:15	Science
1:15	–		Drill, remedial work, enrichment*

*Enrichment: Computer time, physical education, field trips, fun reading, games and puzzles, family business, hobbies, resource persons, guests, crafts, creative work, electives, music appreciation, projects.

Basically, two factors need to be considered when assigning work to a student in the LIFEPAC curriculum.

The first is time. An average of 45 minutes should be devoted to each subject, each day. Remember, this is only an average. Because of extenuating circumstances a student may spend only 15 minutes on a subject one day and the next day spend 90 minutes on the same subject.

The second factor is the number of pages to be worked in each subject. A single LIFEPAC is designed to take 3 to 4 weeks to complete. Allowing about 3-4 days for LIFEPAC introduction, review, and tests, the student has approximately 15 days to complete the LIFEPAC pages. Simply take the number of pages in the LIFEPAC, divide it by 15 and you will have the number of pages that must be completed on a daily basis to keep the student on schedule. For example, a LIFEPAC containing 45 pages will require 3 completed pages per day. Again, this is only an average. While working a 45 page LIFEPAC, the student may complete only 1 page the first day if the text has a lot of activities or reports, but go on to complete 5 pages the next day.

Long range planning requires some organization. Because the traditional school year originates in the early fall of one year and continues to late spring of the following year, a calendar should be devised that covers this period of time. Approximate beginning and completion dates can be noted on the calendar as well as special occasions such as holidays, vacations and birthdays. Since each LIFEPAC takes 3-4 weeks or eighteen days to complete, it should take about 180 school days to finish a set of ten LIFEPACs. Starting at the beginning school date, mark off eighteen school days on the calendar and that will become the targeted completion date for the first LIFEPAC. Continue marking the calendar until you have established dates for the remaining nine LIFEPACs making adjustments for previously noted holidays and vacations. If all five subjects are being used, the ten established target dates should be the same for the LIFEPACs in each subject.

FORMS

The sample weekly lesson plan and student grading sheet forms are included in this section as teacher support materials and may be duplicated at the convenience of the teacher.

The student grading sheet is provided for those who desire to follow the suggested guidelines for assignment of letter grades found on page 3 of this section. The student's self test scores should be posted as percentage grades. When the LIFEPAC is completed the teacher should average the self test grades, multiply the average by .25 and post the points in the box marked self test points. The LIFEPAC percentage grade should be multiplied by .60 and posted. Next, the teacher should award and post points for written reports and oral work. A report may be any type of written work assigned to the student whether it is a LIFEPAC or additional learning activity. Oral work includes the student's ability to respond orally to questions which may or may not be related to LIFEPAC activities or any type of oral report assigned by the teacher. The points may then be totaled and a final grade entered along with the date that the LIFEPAC was completed.

The Student Record Book which was specifically designed for use with the Alpha Omega curriculum provides space to record weekly progress for one student over a nine week period as well as a place to post self test and LIFEPAC scores. The Student Record Books are available through the current Alpha Omega catalog; however, unlike the enclosed forms these books are not for duplication and should be purchased in sets of four to cover a full academic year.

WEEKLY LESSON PLANNER

Week of:

	Subject	Subject	Subject	Subject
Monday				
	Subject	Subject	Subject	Subject
Tuesday				
	Subject	Subject	Subject	Subject
Wednesday				
	Subject	Subject	Subject	Subject
Thursday				
	Subject	Subject	Subject	Subject
Friday				

WEEKLY LESSON PLANNER

Week of:

	Subject	Subject	Subject	Subject
Monday				
	Subject	Subject	Subject	Subject
Tuesday				
	Subject	Subject	Subject	Subject
Wednesday				
	Subject	Subject	Subject	Subject
Thursday				
	Subject	Subject	Subject	Subject
Friday				

Student Name _____   Year _____

Bible

LP #	Self Test Scores by Sections 1	2	3	4	5	Self Test Points	LIFEPAC Test	Oral Points	Report Points	Final Grade	Date
01											
02											
03											
04											
05											
06											
07											
08											
09											
10											

History & Geography

LP #	Self Test Scores by Sections 1	2	3	4	5	Self Test Points	LIFEPAC Test	Oral Points	Report Points	Final Grade	Date
01											
02											
03											
04											
05											
06											
07											
08											
09											
10											

Language Arts

LP #	Self Test Scores by Sections 1	2	3	4	5	Self Test Points	LIFEPAC Test	Oral Points	Report Points	Final Grade	Date
01											
02											
03											
04											
05											
06											
07											
08											
09											
10											

Student Name _____   Year _____

Mathematics

LP #	Self Test Scores by Sections 1	2	3	4	5	Self Test Points	LIFEPAC Test	Oral Points	Report Points	Final Grade	Date
01											
02											
03											
04											
05											
06											
07											
08											
09											
10											

Science

LP #	Self Test Scores by Sections 1	2	3	4	5	Self Test Points	LIFEPAC Test	Oral Points	Report Points	Final Grade	Date
01											
02											
03											
04											
05											
06											
07											
08											
09											
10											

Spelling/Electives

LP #	Self Test Scores by Sections 1	2	3	4	5	Self Test Points	LIFEPAC Test	Oral Points	Report Points	Final Grade	Date
01											
02											
03											
04											
05											
06											
07											
08											
09											
10											

TEACHER

NOTES

INSTRUCTIONS FOR SEVENTH GRADE MATHEMATICS

The LIFEPAC curriculum from grades two through twelve is structured so that the daily instructional material is written directly into the LIFEPACs. The student is encouraged to read and follow this instructional material in order to develop independent study habits. The teacher should introduce the LIFEPAC to the student, set a required completion schedule, complete teacher checks, be available for questions regarding both content and procedures, administer and grade tests, and develop additional learning activities as desired. Teachers working with several students may schedule their time so that students are assigned to a quiet work activity when it is necessary to spend instructional time with one particular student.

Mathematics is a subject that requires skill mastery, but skill mastery needs to be applied toward active student involvement. Measurements require measuring cups, rulers, empty containers. Boxes and other similar items help the study of solid shapes. Construction paper, beads, buttons, beans are readily available and can be used for counting, base ten, fractions, sets, grouping, and sequencing. Students should be presented with problem situtations and be given the opportunity to find their solutions.

Any workbook assignment that can be supported by a real world experience will enhance the student's ability for problem solving. There is an infinite challenge for the teacher to provide a meaningful environment for the study of mathematics. It is a subject that requires constant assessment of student progress. Do not leave the study of mathematics in the classroom.

The Teacher Notes section of the Teacher's Guide lists the required or suggested materials for the LIFEPACs and provides additional learning activities for the students. Additional learning activities provide opportunities for problem solving, encourage the student's interest in learning and may be used as a reward for good study habits.

I. MATERIALS NEEDED

Required:
none

Suggested:
inexpensive calculators for general classroom use
a display number line for comparison of numbers, rounding of numbers, and visual interpretation of addition and subtraction

II. ADDITIONAL LEARNING ACTIVITIES

Section I Number Concepts
1. What are names of place values beyond the hundred billions?
2. What is the largest number?
3. How can any number be rounded to zero?
4. A race can be held in which two numbers (from 0 to infinity) are said and the students write the correct symbol (< or >) between the numbers. The students could do this activity at their desks or several students at a time could do it at a chalkboard. Example: Say "10 and 15." The students would write " 10 < 15."
5. A student can research scientific notation for large (and small) numbers and explain the advantages of using scientific notation as compared to other methods.

Section II Addition
1. Discuss with your class what makes a magic square "magic."
2. An arithmetic bee can be held in which addition problems are given orally and solved mentally. The rules would be the same as in a spelling bee.
3. A student can research the use of Chisenbop, a method of adding large numbers by using their fingers.

Section III Subtraction
1. Discuss with your class how estimating differences is different than estimating sums.
2. An arithmetic bee can be held in which subtraction problems are given orally and solved mentally. The rules would be the same as in spelling bee.
3. A student can acquire an abacus, demonstrate its use, and relate a brief history of the abacus.
Note: This project is applicable for Section II also.

Section IV Applications
1. Discuss with your class how a simple calculator works.
2. Two or more students can create number patterns for the other students to solve.
3. Using a stop watch, a student can compare the time he takes to solve problems long hand versus the time he takes to solve the same problems with a calculator.

I. MATERIALS NEEDED

Required: Suggested:
none   multiplication and division flash
                                             cards
                                             calculators for classroom use

II. ADDITIONAL LEARNING ACTIVITIES

Section I Multiplication

1. An arithmetic bee can be held with multiplication of one-digit numbers. More difficult verbal problems, such as 304 x 12, may be used in the latter stages of the contest.
2. A further study of the commutative, associative, and distributive principles and their validity or nonvalidity with respect to the four basic operations can be done. Students should understand that mathematical laws have no exceptions to their rules.
   Note: This activity is applicable for Section II also.
3. Two or more students can demonstrate multiplication using base two numerals.
4. A student can research four temperature scales, Absolute (Kelvin), Celsius (Centigrade), Fahrenheit, and Reaumur, and derive formulas to convert from one scale to another scale. Note: This project is applicable for Section II also.

Section II Division

1. An arithmetic bee can be held with division of one-digit numbers. More difficult verbal problems, such as 5,463 ÷ 9. may be used in the latter stages of the contest.
2. A small group can determine the average age, height, weight, and so on of the class. Also, the average high and low temperatures, number of pages in the daily newspapers, and so on can be determined over a short period of time.
3. Two or more students can demonstrate division using base two numerals.
4. A student can determine the average area of floor space and volume of space in his classroom with relation to the number of people present.

I. MATERIALS NEEDED

Required: Suggested:
rulers having both metric   none
and English scales
protractor
compass

II. ADDITIONAL LEARNING ACTIVITIES

Section I Segments, Lines, and Angles
1. Discuss these questions with your class.
   a. What are the differences between a segment, a ray, and a line?
   b. Which of the three preceding terms contains the most points? Why? (Be careful! Each one has the same number of points, but the concepts of "infinitely many" and "one-to-one correspondence" may be too sophisticated at this level of study.)
   c. What is the measure of the largest possible angle? The smallest?

Section II Triangles
1. Discuss these questions with your class.
   a. What is the measure of each angle of a triangle with three equal sides (equilateral)?
   b. What is true about the angles of a triangle with two equal sides (isosceles)?
   c. What is the name given to a figure with five sides? seven sides? eight sides? nine sides? ten sides? twenty sides? an unknown number of sides?
2. Given the following measures, in which cases would a unique triangle (one and only one triangle) be possible to construct?
   a. three sides
   b. three angles
   c. two sides and the angle they form
   d. two sides and the angle opposite one of the sides
   e. two angles and the distance between their vertices
   f. two angles and the side opposite one of the angles
   g. any sides of a right triangle
   h. any two angles of a right triangle
3. Construct general figures with the given number of sides, and use a protractor to measure the angles. Then complete the following table.

Number of Sides	3	4	5	6	7	8	N
The Sum of the Measures of All Angles							

Study your results for Cases 3 through 8, and then try to form a rule for a figure with N sides.
Note: This project is applicable to Sections III and IV also.

Section III Quadrilaterals
1. Discuss these questions with your class.
   a. Is a square always a rectangle? Is a rectangle always a square?
   b. If one side of a rectangle is doubled, what is the relationship of the area of the new rectangle to that of the original rectangle?

Section IV Circles and Hexagons
1. Discuss these questions with your class.
   a. If the radius of a circle is doubled, what is the relationship of the circumference and area of the new circle to those of the original circle?
   b. If an equilateral triangle, square, and regular hexagon are all inscribed in the same circle, which figure will have the greatest area? the greatest perimeter?
2. Given a circle, square, rectangle (not a square), triangle (not equilateral), and an equilateral triangle, each of which has the same perimeter, which figure has the greatest area? second greatest area? third greatest area? and so on until the least area.
3. Using only a compass, construct a circle with a given diameter.

I. MATERIALS NEEDED

Required: Suggested:
none   fraction flash cards

II. ADDITIONAL LEARNING ACTIVITIES

Section I Common Fractions

1. Discuss these questions with your class.
   a. How does $\frac{1}{2}$ compare to $\frac{1}{4}$? to $\frac{1}{8}$? to $\frac{1}{16}$? to $\frac{1}{32}$?

   b. How does $\frac{1}{9}$ compare to $\frac{1}{18}$? $\frac{1}{18}$ to $\frac{1}{36}$? $\frac{1}{36}$ to $\frac{1}{72}$?

2. A fraction bee can be held in which fractions are given orally and students respond with equivalent fractions, reduce them to lowest terms, or raise them to higher terms. Improper fractions can be given in which students find the quotients, and mixed numbers can be given for the students to change to improper fractions. The rules would be the same as in a spelling bee.
3. Two or more students can use fraction flash cards to quiz each other on equivalent fractions.
4. A student can keep a log of what he does each day and how much time he spends doing each activity (figure each amount of time as fractional part of a day).

Section II Decimal Fractions

1. Ask the class, in cents, what is 10% of one dollar? 20% of one dollar? 25% of one dollar? 37% of one dollar? 50.8% of one dollar? 75% of one dollar? 88% of one dollar? 98.2% of one dollar? 100% of one dollar?
2. Conduct an arithmetic bee in which fractions are given orally and their respective decimal equivalents are given from memory. Use the DECIMAL EQUIVALENTS OF FRACTIONS table on page 17 in the LIFEPAC.
3. Research the use of place value in numeral systems such as the Roman, Greek, and Egyptian numeral systems as compared to the use of place value in our decimal system.

Section III Applications

1. Discuss these questions with your class.
   a. How many terminating decimals exist?
   b. How many repeating decimals exist?
2. Have groups of no more than four students measure their heights, lengths of arms, legs, and feet, and the lengths around their heads, arms, wrists, and ankles. Either use metric measurements or English measurements and convert them to metric measurements.

   To convert English measurements to metric measurements use 1 inch = 2.54 centimeters and 1 yard = 0.9144 meter.

3. Measure the dimensions of the classroom, door, windows, tables, desks, walls, and so on with meter sticks if possible; or measure with yardsticks and convert the measurements to metric measurements.

4. Use the scale of a state map to determine the number of miles from one city to another.
5. A student can measure the classroom or a room of his house and make a scale drawing of it. A more involved project can be for a student to make a scale drawing of his house.
6. Convert measurements in recipes to their metric equivalents. Use 1 cup = 0.2366 liter, 1 fluid ounce = 0.0295 liter, 1 ounce = 28.3495 grams, 1 tablespoon = 1.4786 centiliters, and 1 teaspoon = 0.4928 centiliter.
7. Use a ruler labeled with centimeters and millimeters to measure various parts of an automobile.

I. MATERIALS NEEDED

Required:
none

Suggested:
none

II. ADDITIONAL LEARNING ACTIVITIES

Section I Sets
1. Make a Venn diagram representing the intersection of football, basketball, and baseball players in your school. Show the number of each kind of athlete.

Section II Numbers
1. Conduct a powers of ten bee. Give each student a power of ten from 10^0 through 10^9. Have each student respond with the correct name for the number. Example: 10^4 equals 10,000. Or have each student tell what a number such as 1,000,000,000 equals as a power of ten. Numbers may be written beforehand on cards and one card given to each student or written on the chalkboard.
2. Make a chart of the place values of another base such as base 8. Write the first twenty-five numerals in base 8.

Section III Factors and Multiples
1. Conduct a race in which students are given fractions to reduce to lowest terms by finding the GCF. The same procedure may be used with two fractions to find the LCD.
2. Find as many different numbers as possible mentioned in the Bible. Separate the numbers into prime numbers or composite numbers. Then write the prime factors of the composite numbers.

I. MATERIALS NEEDED

Required: Suggested:
none   eight-digit or six-digit calculators

II. ADDITIONAL LEARNING ACTIVITIES

Section I Common Fractions
1. Take a trip to a local construction company to see how common fractions are used in construction.
2. Have students make a list of occupations that require a knowledge of the use of common fractions.
3. Conduct chalkboard speed drills for adding and subtracting fractions with like and unlike denominators.
4. Select a cookie recipe that uses fractional measurements. Have the students determine what the measurements are when the recipe is doubled. Use the new measurements to make the cookies. Then let the class enjoy eating them.

Section II Decimal Fractions
1. Take a field trip to a pharmacy to see how decimal fractions are used.
2. Read a list of decimal fractions, and have the students write them in numerical form.
3. Conduct chalkboard speed drills for adding and subtracting decimal fractions.
4. Using local newspaper advertisements, let each student make a collage of the items he would buy if he had $100.00 (or $1,000.00 and so on) to spend. The total should not be above his given amount of money. Prices should be included on clippings.
5. Memorize the most commonly used equivalents.

$$\frac{1}{2} = 0.5 \qquad\qquad \frac{2}{3} = 0.\overline{6} \qquad\qquad \frac{4}{5} = 0.8$$

$$\frac{1}{3} = 0.\overline{3} \qquad\qquad \frac{2}{5} = 0.4 \qquad\qquad \frac{5}{8} = 0.625$$

$$\frac{1}{4} = 0.25 \qquad\qquad \frac{3}{4} = 0.75 \qquad\qquad \frac{7}{8} = 0.875$$

$$\frac{1}{5} = 0.2 \qquad\qquad \frac{3}{5} = 0.6 \qquad\qquad \frac{1}{1}, \frac{2}{2}, \frac{3}{3}, \frac{4}{4}, \ldots = 1$$

$$\frac{1}{8} = 0.125 \qquad\qquad \frac{3}{8} = 0.375$$

I. MATERIALS NEEDED

Required: none	Suggested: calculators

II. ADDITIONAL LEARNING ACTIVITIES

Section I Common Fractions

1. Obtain seven sheets of construction paper, all of the same size, but each of a different color. Cut one piece into halves, one piece into thirds, one piece into fourths, and so on (you may leave out sevenths). Leave one sheet uncut to be a whole. Label the pieces appropriately. Show the students how $\frac{2}{2} = 1$, $\frac{3}{3} = 1$, and so on. Let them experiment with these cutouts until they are comfortable using them.

   If you cut several sheets of one kind (such as several sheets of fifths, making sure they are all the same color), they can be used to demonstrate improper fractions.

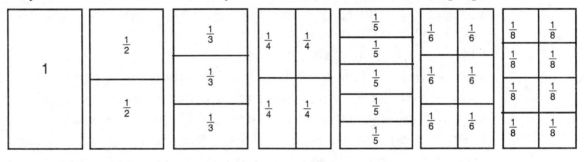

2. Pupils often have trouble working a problem such as $\frac{1}{3}$ divided by 6 or 4 divided by $\frac{1}{5}$ because they do not understand how to "invert" a whole number or a fraction. To illustrate inversion, cut from blank paper several strips about $\frac{1}{2}$ inch wide and 5 inches long. Cut squares of paper 3 by 3 inches square. Write a 1 on four of the squares, a 2 on three of the squares, a 3 on three of the squares, a 4 on three of the squares, and so on through nine.

   Put a fraction such as $\frac{2}{5}$ on a table using a blank strip as the fraction bar. Let the pupils invert the fraction by switching the numerator and the denominator. Drill the students several times with different fractions. Then put a whole number alone on the table and ask the pupils to invert it. Show them how to use the fraction bar with a 1 under it and then invert the fraction.

3. Two students at a time may play the following fraction game. The first player names a fraction, which may be proper, improper, or mixed. The second player names an operation (multiplication or division) and a second fraction. The first player must then answer the problem. If the answer is correct, he receives one point. The players alternate turns. Pencils and paper may be used. If neither player is sure of the correct answer of a problem, one of the players should write out the problem and should ask the teacher after the game. No points should be given for that round. A game consists of 25 points.

4. Three students at a time may play the following fraction game. The first player names a fraction. The second player names an operation (multiplication or division). The third player names a second fraction. The first player must then answer the problem. If the answer is correct, he receives one point. The players alternate turns. Pencils and paper may be used. If none of the players are sure of the correct answer of a problem, one of the players should write out the problem and should ask the teacher after the game. No points should be given for that round. A game consists of 25 points.

5. Four students at a time may play the following fraction game. The first player names a fraction. The second player names an operation (multiplication or division). The third player names a second fraction. The fourth player must then answer the problem. If the answer is correct, he receives one point. The players alternate turns. Pencils and paper may be used. If none of the players are sure of the correct answer of a problem, one of the players should write out the problem and should ask the teacher after the game. No points should be given for that round. A game consists of 35 points.

6. Obtain seven sheets of construction paper. Let the pupils cut six construction paper circles into two, three, four, five, six, and eight equal wedges. Leave one circle whole. Let the pupils experiment with the wedges to see how $\frac{2}{6} = \frac{1}{3}$ (two of the pieces cut into sixths are the same size as one of the pieces cut into thirds). The pupils can also compare the sizes of two fractions such as $\frac{3}{4}$ and $\frac{2}{3}$.

7. Instruct the pupils to scan a newspaper for instances of common fractions.

Section II Decimal Fractions
1. Use the construction paper fractions from Teacher-Directed Activity 1 in Section I. Demonstrate how to find the decimal equivalent of a fraction. Example: Pick 5 of the pieces cut into eights. Show how to change the fraction $\frac{5}{8}$ to 0.625 by dividing the denominator into the numerator.

2. Let four pupils go to the chalkboard, each writing a digit that he chooses. Then let a fifth pupil place a decimal point anywhere in the four digits to make a decimal number. Repeat this procedure with five more pupils to obtain a second decimal number. Then let the class multiply these two numbers on calculators. The numbers can also be divided.

3. Two students at a time may play the following decimal fractions game. The first player names a decimal fraction. The second player names an operation (multiplication or division) and a second decimal fraction. The first player must then answer the problem. If the answer is correct, he receives one point. The players alternate turns. Pencils and paper may be used. If neither player is sure of the correct answer to a problem, one of the players should write out the problem and should ask the teacher after the game. No points should be given for that round. A game consists of 25 points.

4. Three students at a time may play the following decimal fractions game. The first player names a decimal fraction. The second player names an operation (multiplication or division). The third player names a second decimal fraction. The first player must then answer the problem. If the answer is correct, he receives one point. The players alternate turns. Pencils and paper may be used. If none of the players are sure of the correct answer to a problem, one of the players should write out the problem and should ask the teacher after the game. No points should be given for that round. A game consists of 25 points.

5. Four students at a time may play the following decimal fractions game. The first player names a decimal fraction. The second player names an operation (multiplication or division). The third player names a second decimal fraction. The fourth player must then answer the problem. If the answer is correct, he receives one point. Pencils and paper may be used. If none of the players are sure of the correct answer to a problem, one of the players should write out the problem and should ask the teacher after the game. No points should be given for that round. A game consists of 35 points

6. Let the pupils search the newspapers for instances of uses of decimal fractions.

7. Let the pupils prepare a grocery list from the food advertisements in the newspaper and then calculate the cost of the groceries he chose.

Section III Percent

1. Using the actual numbers from your classroom for the current day, calculate with the pupils the following percentages:
   a. What percent of the class is boys?
   b. What percent of the class is girls?
   c. What percent of the class is absent?
   d. What percent of the class is present?
   e. What percent of the boys is present?
   f. What percent of the girls is present?

2. Show the pupils the formula $\frac{P}{B \times R}$. Explain that the formula is a shorthand way to remember the rules for finding the base, the rate, and the percentage. To find the formula for any one of the three terms, cover the symbol for the term you want to find with your finger. The remaining symbols are the formula you want. Example: to find the base, cover the B and the x. The P is left over the R, which means to divide the percentage by the rate.

3. Two students at a time may play the following percentage game. The first player names a number as the base, the rate, or the percentage. The second player names a number as one of the other two terms. The first player must then find the missing term. Pencils and paper or calculators may be used.

   Example: First player: The percentage is 230.
   Second player: The rate is 13%.
   First player: The base is the percentage divided by the rate, or 230 divided by
             0.13, which equals 1,769.2308.

4. Three students at a time may play the following percentage game. The first player names a number as the base, the rate, or the percentage. The second player names a number as one of the other terms. The third player must then calculate the missing term. Pencils and paper or calculators may be used.

5. Let the pupils search the newspaper for instances of uses of percentages.
6. Make one hundred small squares of paper, and mark each square with a different number from 0 through 99. Also make three other squares; one labeled base, one labeled rate, and the other labeled percentage. The pupil draws a number and one of the three squares to tell him whether that number is the base, the rate, or the percentage. Next he draws a second number and a second label. Then he must find the third term of the base-rate-percentage problem. Note that when the number the student draws is to be the rate, the pupil must add in his mind a % sign and change the percent to a decimal. A pencil and paper or a calculator may be used.

III. ADDITIONAL ACTIVITY

This activity is designed to give the students practice in dividing common fractions to arrive at percentages. The books of the Bible are divided into the Old Testament and the New Testament and are arranged by categories.

Books of the Bible
(arranged by categories)

OLD TESTAMENT

Pentateuch:	Genesis
	Exodus
	Leviticus
	Numbers
	Deuteronomy
History:	Joshua
	Judges
	Ruth
	1 Samuel
	2 Samuel
	1 Kings
	2 Kings
	1 Chronicles
	2 Chronicles
	Ezra
	Nehemiah
	Esther
Poetry:	Job
	Psalms
	Proverbs
	Ecclesiastes
	Song of Solomon
Major Prophets:	Isaiah
	Jeremiah
	Lamentations
	Ezekiel
	Daniel

Minor Prophets:	Hosea	Nahum
	Joel	Habakkuk
	Amos	Zephaniah
	Obadiah	Haggai
	Jonah	Zechariah
	Micah	Malachi

NEW TESTAMENT

Gospels:	Matthew
	Mark
	Luke
	John
History:	Acts
Paul's Letters:	Romans
	1 Corinthians
	2 Corinthians
	Galatians
	Ephesians
	Philippians
	Colossians
	1 Thessalonians
	2 Thessalonians
	1 Timothy
	2 Timothy
	Titus
	Philemon
General Letters:	Hebrews
	James
	1 Peter
	2 Peter
	1 John
	2 John
	3 John
	Jude
Prophecy:	Revelation

Answer the following questions.

1. What percent of the Old Testament is history?
2. What percent of the New Testament is Paul's letters?
3. What percent of the Old Testament is minor prophets?
4. What percent of the whole Bible is prophecy?
5. What percent of the New Testament is the Gospels?
6. What percent of the Old Testament is poetry?
7. What percent of the New Testament was written by John?
8. What percent of the Old Testament is the Pentateuch?
9. What percent of the Bible is history?
10. What percent of the Bible is the New Testament?

WORKSHEET, Solution Key

1. $\frac{12}{39} = $ 0.30769=30.77%

$$39\overline{)12.00000}$$
$$\underline{11\,7}$$
$$300$$
$$\underline{273}$$
$$270$$
$$\underline{234}$$
$$360$$
$$\underline{351}$$
$$9$$

2. $\frac{13}{27} = $ 0.48148=48.15%

$$27\overline{)13.00000}$$
$$\underline{10\,8}$$
$$2\,20$$
$$\underline{2\,16}$$
$$40$$
$$\underline{27}$$
$$130$$
$$\underline{108}$$
$$220$$
$$\underline{216}$$
$$4$$

3. $\frac{12}{39} = $ 0.30769=30.77%

$$39\overline{)12.00000}$$
$$\underline{11\,7}$$
$$300$$
$$\underline{273}$$
$$270$$
$$\underline{234}$$
$$360$$
$$\underline{351}$$
$$9$$

4. $\frac{18}{66} = $ 0.2727=27.27%

$$66\overline{)18.0000}$$
$$\underline{132}$$
$$480$$
$$\underline{462}$$
$$180$$
$$\underline{132}$$
$$480$$
$$\underline{462}$$
$$18$$

5. $\frac{4}{27} = $ 0.1481=14.81%

$$27\overline{)4.0000}$$
$$\underline{27}$$
$$130$$
$$\underline{108}$$
$$220$$
$$\underline{216}$$
$$40$$
$$\underline{27}$$
$$13$$

6. $\frac{5}{39} = $ 0.1282=12.82%

$$39\overline{)5.0000}$$
$$\underline{3\,9}$$
$$1\,10$$
$$\underline{78}$$
$$320$$
$$\underline{312}$$
$$80$$
$$\underline{78}$$
$$2$$

7. John wrote the gospel of John, 1 John, 2 John, 3 John, and Revelation.

$\frac{5}{27} = $ 0.18518=18.52%

$$27\overline{)5.00000}$$
$$\underline{2\,7}$$
$$2\,30$$
$$\underline{2\,16}$$
$$140$$
$$\underline{135}$$
$$50$$
$$\underline{27}$$
$$230$$
$$\underline{216}$$
$$14$$

8. $\frac{5}{39} = $ 0.1282=12.82%

$$39\overline{)5.0000}$$
$$\underline{3\,9}$$
$$1\,10$$
$$\underline{78}$$
$$320$$
$$\underline{312}$$
$$80$$
$$\underline{78}$$
$$2$$

9. $\frac{13}{66}$ =   0.196969=19.697% or 19.70%

$$66\overline{)13.000000}$$

     6 6
    6 40
    5 94
     460
     396
     640
     594
     460
     396
     640
     594
      46

10. $\frac{27}{66}$ =   0.40909=40.91%

$$66\overline{)27.00000}$$

    26 4
     600
     594
     600
     594
      6

I. MATERIALS NEEDED

Required: Suggested:
none   none

II. ADDITIONAL LEARNING ACTIVITIES

Section I Formulas
1. Use this activity to illustrate and practice substitution into a formula. Prepare
    ahead of time several circles, squares, and triangles cut from construction paper.
    Write a formula on the chalkboard (use one of the eight formulas from the
    LIFEPAC). Draw a circle around one letter, a square around another, and a
    triangle around the third. Decide on values for two of the unknowns (letters).
    Write the values on appropriately-shaped pieces of construction paper. Use
    masking tape to place them on the chalkboard over the appropriate unknowns.
    Then solve the problem.
    Example: Write on the chalkboard: $I = R \times P$

    Then draw: $(I) = \boxed{R} \times \triangle P$

    Now substitute: $(I) = \boxed{8\%} \times \triangle \$350$
    Then solve.
2. Conduct a memory drill on the eight formulas used in the LIFEPAC. Ask a pupil
    to say the first part of the formula and a second pupil to respond with the last
    part of the formula.
    Example: First student says: "The net price is equal to. . . ."
                 Second student says: "The list price minus the discount."
3. Ask the students to bring to class things that can be used as function machines.
    Let them use their imaginations. Any object can be a function machine. Let the
    students decide on a rule for each function machine (as simple or as elaborate as
    they like). Let them write the rule on paper and tape it to the function machine.
    Then let the students work problems using the function machine.
4. Let the students test each other on the function machines. One student names an
    input (or inputs) and another student answers with the output. The rules may be
    changed as often as the students wish.
5. Ask the student to find mathematical rules expressed in words in newspapers,
    books (such as elementary mathematics books), and magazines. Let him then
    make up formulas to express these rules.
6. Give the student the following rules. Let the student make up a formula to
    express each rule. Include other rules that you wish.
    a. The circumference of a circle is equal to 2 times the radius of the circle.
    b. The tangent of an angle is equal to the opposite side of a triangle divided by
        the adjacent side of the triangle.
    c. The volume of a cube is equal to the cube of a side.
    d. The volume of a box is equal to the length times the width times the height.
    e. The area of a triangle is equal to one-half the base times the altitude (height of
        the triangle).
    f. A person's weight on Saturn is equal to 1.17 times his weight on Earth.
    g. The velocity of an object is equal to its acceleration times the time.

Section II Equations

1. Write a variety of true, false, and open equations on the chalkboard. Ask the class to identify them as true, false, or open.
2. Illustrate "undoing" by using several examples of equations in pairs such as the following ones.

   $3 + 7 = 10; \ 3 = 10 - 7$
   $14 = 2 + 12; \ 14 - 2 = 12$
   $21 \div 3 = 7; \ 21 = 3 \times 7$
   $13 \times 2 = 26; \ 26 \div 13 = 2$

3. Provide a calculator for this game called GOTCHA! To play: The first player enters 50 on the calculator. Then he secretly presses one of the operation keys ($+$, $-$, $\times$, or $\div$). The second player enters any number except 1 and presses the equals key. If the display shows a negative number, a number greater than 200, or an error message, Player 2 is out of the game. Otherwise, Player 2 now secretly presses an operation key and passes the calculator to the next person, who enters a number (again not 1) and presses the equals key. Players take turns. Any players who get negatives, numbers greater than 200, or error messages are eliminated from the game until only one player is left; that player is the winner.
4. Make (or let the pupils make) a set of small paper squares with numbers, letters, operation signs, and an equal sign. Separate the numbers and letters from the operation signs in two piles, placing them face down. Leave the equals sign separate from the two piles. The first player chooses a number or a letter from the pile and turns it face up on the table. The same player chooses an operation sign and places it face up on the table to the right of the number or letter he drew. The second player chooses a number or a letter, turns it face up, and places it to the right of the operation sign. Then he places the equals sign after the number or letter he drew. The third player chooses a number or letter, turns it face up, places it to the right of the equals sign, and then must identify the equation as open, true, or false. If the equation is open, he may try to solve it. If the equation is false, he may try to correct it.
5. Ask the student to find equations in newspapers, magazines, books, and other sources.
6. Give the student the following formula: Degrees Centigrade equals degrees Fahrenheit minus 32, multiplied by 5, divided by 9. Each day, have the student read the temperature of the room or of the outdoors on a Fahrenheit thermometer and change it to degrees Centigrade. He may also use the newspaper weather report to find temperatures to convert to degrees centigrade.

Section III Ratios and Proportions

1. To illustrate that the product of the means always equals the product of the extremes in a proportion, write a variety of true proportions on the chalkboard. Let the pupils multiply the means times the extremes and see that they are equal.

2. Do some "word" or "logic" proportions. Introduce the idea of word proportions by referring to a simple number proportion such as 3 is to 6 as 7 is to 14. Present the following word proportions, and let the pupils complete the sentences.

     Ear is to hear as eye is to _____ .

     Finger is to hand as toe is to _____ .

     Up is to high as down is to _____ .

     A.M. is to morning as P.M. is to _____ .

     Dog is to bark as cat is to _____ .

     Carrot is to orange as spinach is to _____ .

     Puppy is to dog as kitten is to _____ .

You can make up many more of these word proportions.

3. Assign to a small group the task of finding out the eye color of every student in the class and of determining the ratio of each color.

4. Examine the classroom for ratios. Examples are to find the ratio of windows to doors; of desks to chalkboards; of wastebaskets to students; and of pencils to books.

5. Search newspapers, magazines, and other sources for ratios and proportions.

6. Provide a set of solid-color marbles (Chinese checker marbles are good). Put the marbles in a paper sack. Provide a shallow box such as a shoe box. Give the student the following instructions. Reach into the sack, get a handful of marbles, and put them in the box so that you can count them. Now find all possible ratios between the marbles. For example, suppose that you draw out nine marbles and that three of them are green, one is black, and five are red. You can write all of the following proportions:

        green to black -- 3:1

        green to red -- 3:5

        green to total -- 3:9 or 1:3

        black to green -- 1:3

        black to red -- 1:5

        black to total -- 1:9

        red to green -- 5:3

        red to black -- 5:1

        red to total -- 5:9

Next, put the marbles back into the sack, get another handful of marbles, and figure the new ratios; or reach into the sack, get another handful of marbles, add it to the marbles already in the box, and figure these new ratios.

I. MATERIALS NEEDED

Required: Suggested:
none   none

II. ADDITIONAL LEARNING ACTIVITIES

Section I Data

1. Write on the chalkboard the definition of random sample: a random sample is a sample in which each member of the population has an equal chance of being selected. Discuss several ways in which a random sample of the students in your classroom can be taken. Then take a random sample in two ways: (1) Put each pupil's name on a slip of paper, and put all the papers in a large bowl or box. Mix the names, and have someone draw names from the box, one name at a time. Be sure to put each name back in the bowl after recording it. After drawing a sample of six to ten students, record their heights. (2) Number the students. Then use the table of Random Digits on page 5 of the LIFEPAC, and pick six to ten numbers (representing students) at random. Record these students' heights. Compare the heights taken from sample (1) and sample (2).

2. To practice tallying, choose any column from the Random Digits table on page 5 of the LIFEPAC. On the chalkboard tally how many zeros, ones, twos, and so on through nines are in the column. Then make a frequency distribution, and calculate the relative frequency of each digit.

3. Take a survey in your classroom of how many hours each student studies per day for a week. Keep careful records surveying them each day. Then tally the totals.

4. Gather the following data about each student in the classroom: number of people living in each home; shoe size; height; weight. Save this data for later use (Section II, Group Activity 1).

5. Select a book, preferably a rather simple one, such as for grade two or three. Use the Random Digits table on page 5 of the LIFEPAC to select ten different pages. Count the words on each of these pages. Do you think the number of words on each of these pages is a representative sample of the number of words on all of the pages of the book?

6. Use the classified advertisements section of your newspaper for this activity. Choose a particular section, such as *Miscellaneous for Sale* or *Antique Automobiles*. List the prices for each of the items listed. Then arrange the items in order according to the cost of each item.

Section II Statistics

1. Using the following sets of data find the mean, the median, and the mode of each set. Notice how the measures relate to each other.
   - (1) 10, 10, 10, 10, 10, 10, 10, 10, 10, 10
   - (2) 1, 3, 5, 7, 10, 10, 13, 15, 17, 19
   - (3) 1, 1, 1, 1, 1, 1, 1, 1, 1, 91
   - (4) 1, 1, 1, 40, 41, 1, 1, 1, 13

2. Using the sets of data in the previous activity, show how the average deviation illustrates how spread out the data are.

    Example: The standard deviation of the data in Set (1) is $\frac{0}{10} = 0$. The standard deviation of the data in Set (3) is $\frac{91-1}{10} = \frac{90}{10} = 9$. The data in Set (3) have a larger average deviation than the data in Set (1) and therefore have a larger spread than the data in Set (1).

3. From the data gathered in Section I, Group Activity 2, find the mean, median, and mode of the number of people living in each student's home and each student's shoe size, height, and weight.

4. Take the pulse of every student in your classroom. To take a person's pulse, put your index and middle fingers lightly against the person's neck on the left side of the voice box. When you find the pulse count the number of times his heart beats in one minute. This number is his pulse rate. Make a tally and a frequency distribution of the students' pulses. Then find the mean, the median, and the mode of the students' pulses.

5. Devise four or five sets of data. Each set is to have the same mean but the average deviation is to be different. Draw some intelligent conclusions.

6. Survey the people in your class and find out how old they are. Ask them to tell you how many years and how many months old they are. If someone is 13 years and 7 months old, for example, write their age as $13 \frac{7}{12}$ (put the number of months over 12 to form a fractional part of a year). Change each age number to a decimal fraction. Then find the mean age and round it to the nearest hundredth. This activity is a good exercise to do with a handheld calculator.

I. MATERIALS NEEDED

Required: Suggested:
none   calculators

II. ADDITIONAL LEARNING ACTIVITIES

Section I Whole Numbers
1. Ask each pupil to write a four-digit number on a piece of paper. Then pair off the class and ask each pair
    a. to add their two numbers,
    b. to subtract the smaller number from the larger number, and
    c. to multiply the two numbers.
    If you have calculators available, let each pair check their work on calculators. Otherwise, let the pairs exchange problems and check each other's work.
2. Divide the pupils in this "trick." A 9 is hidden in every famous person's birthday. For example, Abraham Lincoln's birthday is February 12, 1809. To find the 9, write the birthday as one number: 2121809. The first digit(s) represents the month, the next digit(s) represents the day, and the last digits represent the year. Next, reverse these digits: 9081212. Now take these two numbers and subtract the smaller number from the larger number.

$$\begin{array}{r} 9081212 \\ -\ 2121809 \\ \hline 6959403 \end{array}$$

Add the digits in the answer.
$$6 + 9 + 5 + 9 + 4 + 0 + 3 = 36$$

Add the digits of that answer.
$$3 + 6 = 9$$

The author's birthday is December 10, 1935: 12101935. Reverse the digits and subtract the smaller number from the larger number.
$$\begin{array}{r} 53910121 \\ -\ 12101935 \\ \hline 41808186 \end{array}$$

Add the digits in the answer.
$$4 + 1 + 8 + 0 + 8 + 1 + 8 + 6 = 36$$

Add the digits of that answer.

$$3 + 6 = 9$$

The answer will always be 9 for a famous person. Perform the operations on your birthday. The answer will be 9. Then ask the pupils to perform the operations on their birthdays.

3. Provide each group with a calculator. One person in the group is to enter a five-digit number on the calculator. Then another person is to multiply that number by a two-digit number. Each person in the group is to use the result to:
   a. write the number in expanded form; and
   b. round the number to tens, hundreds, and thousands.
   Have the group compare their answers to see if they agree. Repeat as often as desired.
4. Take the class on a field trip to a bowling alley. Let the class bowl in groups of 3 or 4, and let each student keep the score of each member of his group.
5. Search newspapers for large numbers. Then write them in expanded form.
6. Go to a basketball game. Keep track of how many points each player on one team makes. At the end of the game, total the points.

Section II Geometry, Sets, and Number Systems
1. Provide each of the pupils with a ruler and a protractor. Then conduct a measurement scavenger hunt. Ask the students to find an object:
   a. exactly 3 feet 6 inches wide,
   b. exactly 8 inches long,
   c. about 14 $\frac{1}{2}$ inches high,
   d. with an angle of about 30°,
   e. with an angle of about 150°, and
   f. with a right angle and some part that measures 12 inches long.
   Add other items to the list as desired.
2. Cut squares of various colors of construction paper. Cut many squares of some colors and a few squares of other colors; cut enough squares so that each pupil can have three squares. Put the squares in a small container, and let each pupil draw three squares. Write the name of each color across the top of the chalkboard. Now, according to the color each pupil drew, let the students list their names on the chalkboard under the appropriate color name. If a student has more than one square of one color, have him list his name once under the appropriate color name(s).
   Several sets have now been defined. Lead the pupils to find the union and the intersection of several of these sets, two sets at a time. For example, say: "Those students in the intersection of the yellow set (Y) and the black set (B), please stand." (That is, those students that belong to both the yellow set and the black set.) If no one stands because you have defined the null set (no students are in both the yellow set and the black set), write $Y \cap B = \emptyset$ on the chalkboard.
3. Let one student in the group choose an object in the classroom. Each student in turn guesses what its measurements are. Then let a student measure the object with a ruler so the group can see who came the closest to the correct measurements.
4. Make a height chart for your classroom. Measure each student's height, and indicate it on the chart.
5. Measure two objects. List the objects and their respective heights. Then add these lengths, and list the combined length. Write this measurement in feet and inches.
6. Make a list of all the "sets" to which you belong. These sets may include groups you belong to or feel a part of at home, school, church, and so on.

Section III Fractions
1. Ask every pupil to write a common fraction at the top of a piece of paper with his name on it. Then have each pupil pass the paper to the pupil in front of him, who writes another common fraction on the paper. Next, have each pupil again pass the paper to the pupil in front of him, who adds the two fractions. Then have each pupil pass the paper to the pupil in front of him, who multiplies the two original fractions. Next, have each pupil pass the paper to the pupil in front of him, who divides the two original fractions. Then have each pupil pass the paper to the pupil in front of him, who changes both fractions to decimal fractions. Finally, have each pupil return the paper to the original student and have him check the work.
2. Make a fraction wheel using common fractions or decimal fractions such as the one shown. Put eight numbers on the small wheel and eight numbers on the large wheel. Add, multiply, and divide all the numbers. You may use a calculator for decimals.

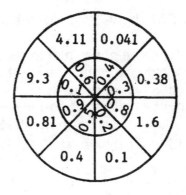

3. Make a large fraction wheel with numbers large enough for the class to see (you may use either common fractions or decimal fractions). Choose the fractions carefully to make good problems. You can have students lead a review of fractions using the fraction wheel.

Section IV Formulas, Ratios, Statistics, and Graphs
1. Involve the pupils in a "ratio hunt." Call out the following ratios, and have the students arrange themselves or the objects mentioned to make each ratio true.
   a. Ratio of boys to girls is 6:1.
      Example: Six boys and one girl should stand.
   b. Ratio of students wearing yellow clothes to students wearing blue clothes is 1:1.
   c. Ratio of pupils to books is 3:7.
   d. Ratio of pieces of chalk to pencils is 2:9.
   Include other ratios that you wish.
2. Use tinker toys to create some formula "function machines." Place small signs on the machines to tell what their functions are. Use different inputs for each formula.
   Example: Formula: $A = L \times H$.
              Input: 2 and 6.

Output: 12.

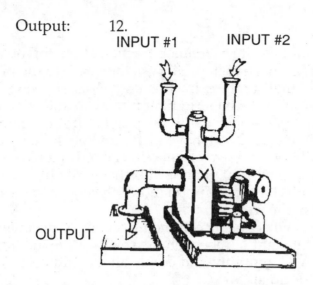

INPUT #1   INPUT #2

OUTPUT

3. Make a bar graph that shows how many books of each color are in your classroom.
4. Measure each student's height. Then measure each student's cubit (page 22 in the LIFEPAC). Make a bar graph illustrating these measurements. Then figure the ratio of each student's height to his cubit.
5. Measure each student's height. Then measure each student's wide-reach (page 22 in the LIFEPAC). Make a bar graph illustrating these measurements. Then figure the ratio of each student's height to his wide-reach.

TESTS

Reproducible Tests
for use with the Mathematics
700 Teacher's Guide

Name_____

Find the sum or difference (each answer, 3 points).

1.	76	2.	582	3.	5,694
	94		763		− 309
	38		+ 948		
	+ 22				

4. 6 + 23 + 342 + 5,106 = _____

5. 692 − 317 = _____

6. 70,005 − 6,947 = _____

Round the number to the places indicated (each answer, 3 points).

7. 6,457   a. ten _____   b. hundred _____

        c. thousand _____

Complete the table for the given sentences (each answer, 3 points).

8. $a + b = 20$

a	2	17	13	a. ____	b. ____
b	c. ____	d. ____	e. ____	9	6

9. $a - 5 = b$

a	7	16	5	a. ____	b. ____
b	c. ____	d. ____	e. ____	5	8

Write a sentence for the following sets of number pairs (each answer, 3 points).

10. {(17, 8), (9, 0), (29, 20)}   _____

11. {(8, 12), (3, 17), (15, 5)}   _____

Write the following numbers in expanded form (each answer, 3 points).

12. 7,010   _____

13. 23,698   _____

14. 540,006   _____

Construct the following numbers (each answer, 3 points).

15. 8(10,000) + 2(1,000) + 4(10)   _____

16. 3(1,000,000) + 7(1,000) + 6(100) + 9   _____

17. 1(100,000) + 5(10,000) + 1(100)   _____

Answer *true* or *false* (each answer, 3 points).

18. $8 + 15 > 22 - 8$ _____

19. $13 + 10 - 7 < 42 - 23$ _____

20. $48 + 75 = 162 - 49$ _____

Estimate to the nearest hundred (each answer, 3 points).

21. $652 + 1{,}320 + 75 =$ _____

22. $6{,}872 - 2{,}927 =$ _____

23. $350 + 1{,}649 + 800 =$ _____

Solve these problems on your calculator. If you do not have a calculator, solve the problems manually (each answer, 3 points).

24. $622 + 508 - 193 - 356 =$ _____

25. $37{,}212 - 4{,}321 + 22 - 999 =$ _____

26. $8{,}214 + 8{,}213 - 8{,}215 - 8{,}212 =$ _____

Solve the word problems. Show your work in the space below each problem, and circle your answer (each answer, 4 points).

27. If you purchased three items costing 22¢, 17¢, and 13¢ and gave the clerk $1.00, what will your change be?

28. If Mr. and Mrs. Smith were married in 1952 and their son Bill was born four years later, how old was Bill when his parents celebrated their twentieth wedding anniversary?

29. During a car wash held to raise money for summer church camp, a group earned $42 one day and $69 the next day. If their expenses for the two days were $26, how much profit did the group make?

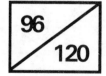

Date _____

Score _____

Name _____

Write in the blanks the correct numbers (each answer, 3 points).

1. In the division at the right, a. _____ is the
    quotient, b. _____ is the dividend, and c. _____
    is the divisor.

    $$9$$
    $$7\overline{)63}$$

2. In the multiplication at the right, a. _____ is
    the multiplier, b. _____ is the multiplicand,
    and c. _____ is the product.

    $$75$$
    $$\underline{\times\ 5}$$
    $$375$$

Use the function rule to write the second number for each ordered pair (each answer, 3 points).

3. Function rule: Multiply
    by 6, then subtract 3

Number	Function
1	a. _____
7	b. _____
11	c. _____

4. Function rule: Divide by
    7, then add 4

Number	Function
7	a. _____
21	b. _____
49	c. _____

Multiply. Express each answer as a single number (each answer, 3 points).

5. 2^4 _____

6. 5^3 _____

7. 7^3 _____

8. 3^4 _____

Write the next three numbers in each of these patterns (each answer, 3 points).

9. 0, 5, 10, 15, a. _____ , b. _____ , c. _____

10. 243, 81, 27, a. _____ , b. _____ , c. _____

11. 1, 5, 25, a. _____ , b. _____ , c. _____

12. 256, 128, 64, 32, a. _____ , b. _____ , c. _____

Work these word problems. Translate the words into a mathematical sentence. Be sure to label the answer with the proper units (dollars, degrees, feet, and so on). Show your work in the space below each problem, and circle your answer (each answer, 4 points).

13. If a car averages 7 kilometers per liter of gasoline, how far can it go on 45 liters?

14. A student made these scores on his LIFEPAC tests: 95, 85, 88, 86, 92, 94, 90. What is his average score?

15. If a worker receives a gross pay of $240 for working a 40-hour week, what is his pay per hour?

Multiply (each answer, 3 points).

16. 327
      75

17. 681
      86

18. 598
      91

19. 275
     902

20. 608
     752

Divide (each answer, 3 points).

21. $47\overline{)24,352}$

22. $76\overline{)31,825}$

23. $92\overline{)78,201}$

24.

$58 \overline{)475{,}013}$

25.

$89 \overline{)551{,}622}$

26.

$67 \overline{)541{,}045}$

27.

$948 \overline{)2{,}235{,}400}$

28.

$576 \overline{)4{,}075{,}200}$

29.

$9 \overline{)173{,}714}$

30.

$8 \overline{)427{,}507}$

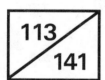

Date _____

Score _____

Name _____

Write true or false (each answer, 1 point).

1. _____ $\overleftrightarrow{RS}$ is the symbol for the line that passes through points *R* and *S*.
2. _____ We can measure the length of $\overleftrightarrow{RS}$ by using a ruler.
3. _____ A right angle has a measure of 90°.
4. _____ A hexagon has five sides.
5. _____ The perimeter of a figure is the distance around the figure.
6. _____ The hypotenuse is the longest side of a right triangle.
7. _____ $\angle ABC$ is the name of an angle whose vertex is point *C*.
8. _____ A protractor is used to measure the area of a rectangle.
9. _____ A parallelogram has opposite sides equal and parallel.
10. _____ The radius of a circle is the distance across the circle through its center.

Write on the blanks the letter for the correct figure (each answer, 2 points).

11. _____ line
12. _____ circle
13. _____ triangle
14. _____ parallelogram
15. _____ hexagon
16. _____ rectangle
17. _____ square
18. _____ segment
19. _____ angle
20. _____ trapezoid

a.
b.
c.
d.
e.
f.
g.
h.
i.
j.

Construct the following figures using a compass, ruler, and protractor (each figure, 4 points).

21. A parallelogram with sides of 1″ and $1\frac{1}{2}$″ and an angle of 40°

22. A right triangle with sides of 44 mm and 27 mm

23. A regular hexagon with sides of $1\frac{1}{2}$″

Find the following measures (each answer, 3 points).

24. The circumference of a circle with radius of 2″ (use π = 3.14).

C = _____

25. The area of a circle when the radius is 7 cm (use $\pi = 3\frac{1}{7}$).

A = _____

26. The sum of the angles of the parallelogram in Problem 21.

Sum of angles = _____

Match these items (each answer, 2 points).

27. _____ segment

28. _____ angle

29. _____ triangle ·

30. _____ hypotenuse

31. _____ quadrilateral

32. _____ radius

33. _____ diameter

34. _____ circumference

35. _____ hexagon

36. _____ regular hexagon

a. a three-sided figure formed by the union of three segments

b. the distance from the center to the circle itself

c. a quadrilateral with one set of opposite sides parallel

d. the distance around a circle

e. a portion of a line made up of two end-points and all the points in between

f. a hexagon with all sides equal and all angles equal

g. the side of a right triangle that is opposite the right angle

h. the distance across a circle through the center

i. the figure formed by joining two line segments or two rays at a common endpoint

j. a geometric figure with six sides and six angles

k. a geometric figure with four sides and four angles.

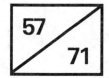

Date _____

Score _____

Name _____

Complete these items (each answer, 3 points).

1. Write the fraction thirty-four forty-thirds in numerals. _____

2. Reduce $\frac{39}{65}$ to lowest terms. _____

3. Raise $\frac{3}{7}$ to higher terms with a denominator of 56. _____

4. Find the quotient of $\frac{51}{9}$. _____

5. Change $4\frac{3}{8}$ to an improper fraction. _____

6. Which is the larger fraction, $\frac{5}{6}$ or $\frac{5}{7}$? _____

7. Change $4\frac{3}{5}$ to a decimal. _____

8. Write 0.57 as a fraction reduced to lowest terms. _____

9. Change $\frac{65}{4}$ to a mixed number. _____

10. How do 4.7 and 0.47 compare? _____

11. Write 56.39 in words. _____

12. Write 0.79 as a percent. _____

13. Show the ratio of 34 pennies and 9 nickels. _____

14. Write 76% as a fraction reduced to lowest terms. _____

15. Write 0.00082 as a percent. _____

16. Write 4.65 as a percent. _____

17. Change $\frac{33}{4}$ to a terminating decimal. _____

18. Write 0.31% as a decimal. _____

Write the letter for the correct answer on each line (each answer, 2 points).

19. The improper fraction $\frac{33}{4}$ written as a mixed number is _____ .
    a. $8\frac{1}{4}$ c. $8\frac{3}{4}$
    b. $7\frac{3}{4}$ d. $8\frac{1}{2}$

20. When two triangles are said to be similar, then they have the same _____ .
    a. measure   c. shape
    b. units   d. equivalence

21. The correct location of the decimal point in the fraction $\frac{9}{10,000}$ is _____ .
    a. 0.9   c. 0.009
    b. 0.09   d. 0.0009

22. An equivalent form of the fraction $\frac{5}{7}$ is _____ .

    a. $\frac{10}{7}$ c. $\frac{5}{14}$

    b. $\frac{10}{14}$ d. $\frac{7}{5}$

23. If 1,000 ml equals 1 liter, then 3,400 ml equals _____ .

    a. 3.4 liters   c. 0.34 liters

    b. 34 liters   d. 340 liters

24. The fraction $\frac{9}{8}$ written as a decimal is _____ .

    a. 1.25   c. 0.875

    b. 1.125   d. 1.2

25. If a certain tax takes 9¢ out of every dollar, the percent of tax is _____ .

    a. 90%   c. 9%

    b. 9¢   d. 0.9%

Match these items (each answer, 2 points).

26. _____ proportion

27. _____ similar

28. _____ mixed number

29. _____ improper fraction

30. _____ ratio

31. _____ denominator

32. _____ terminating decimal

33. _____ proper fraction

34. _____ percent

35. _____ repeating decimal

a. the answer in multiplication

b. 34.6666666

c. out of one hundred

d. 34.60

e. numerator equal to or greater than the denominator

f. two equal ratios

g. comparison using division

h. numerator smaller than the denominator

i. has the same shape

j. the number under the bar in a fraction

k. a whole number and a fraction

Complete the following activity (each answer, 1 point).

36. Raise the fraction $\frac{7}{12}$ to higher terms five different ways.

    a. _____   b. _____   c. _____   d. _____   e. _____

75 / 93

Date _____

Score _____

Name _____

Match these items (each answer, 2 points).

1. _____ 10^{-3}
2. _____ factoring
3. _____ intersection
4. _____ XXV
5. _____ improper subset
6. _____ 10^3
7. _____ composite number
8. _____ finite set
9. _____ union
10. _____ null set
11. _____ LCM
12. _____ base 5

a. finding factors
b. empty set
c. combining sets
d. countable elements
e. least common multiple
f. 0.001
g. common to both sets
h. 1,000
i. {0, 1, 2, 3, 4}
j. same number of elements
k. not prime
l. Roman numeral
m. lowest common denominator

Complete these items (each answer, 3 points).

13. The symbol for *is an element of* is _____.

14. Is the set of hairs on your head finite or infinite? _____

15. Write the set {a, e, i, o, u} as a rule. _____

16. Write 5×10^{-4} as a decimal. _____

17. Write 14 as a Roman numeral. _____

18. The number 100_2 is what in base ten? _____

19. The set of all possible factors of 32 is_____.

20. Explain why 7 is a prime number. _____

21. List the prime factors of 48. _____

22. Find the GCF of 15 and 25. _____

23. Name the elements in the set {+, -, x, ÷}. _____

24. Find the solution set of $7x = 35$. _____

25. Compare the size of 2 and 0.002. _____

26. The intersection of $A \cap B$ if $B \subset A$ is _____.

27. Write 6,000 as a power of ten. _____

28. Find the LCM for 3, 6, 12, and 18. _____

29. Write 30 as a number in base five. _____

30. Find $A \cap B$ if A = {1, 2, 3, 4} and B = {2, 4, 6, 8}. _____

Write the letter for the correct answer on each line (each answer, 2 points).

31. The number 0.001 x 8 can be written _____ .
    a. 8
    b. 0.8
    c. 0.08
    d. 0.008

32. A Venn diagram that shows an intersection of two sets is _____ .

    a.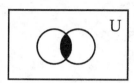
    b.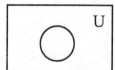
    c.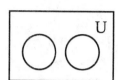
    d.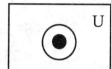

33. The number 1,461,515 is divisible by _____ .
    a. 2
    b. 3
    c. 4
    d. 5

34. The fraction $\frac{46}{69}$ reduced to lowest term is _____ .

    a. $\frac{3}{4}$
    b. $\frac{23}{34}$
    c. $\frac{2}{3}$
    d. $\frac{1}{2}$

35. The number 1011_2 in base ten is _____ .
    a. 9
    b. 11
    c. 13
    d. 15

36. The Roman numeral LXIV is _____ .
    a. 54
    b. 64
    c. 44
    d. 66

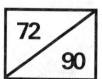

Date _____
Score _____

Name _____

Draw a line from the problem in Column I to its corresponding answer in Column II (each answer, 3 points).

Column I

Column II

1. $\frac{5}{8} + \frac{7}{8}$ $2\frac{3}{4}$

2. $4\frac{1}{3} - 1\frac{7}{12}$ 18.323

3. 3.25 $1\frac{1}{2}$

4. 11.632 + 6.691   18.317

5. 36.18 − 17.863 $3\frac{1}{4}$

Write a common fraction equivalent and a decimal fraction equivalent (each answer, 3 points).

6. $\frac{2}{5} =$ a. _____   b. _____

7. $3\frac{10}{16} =$ a. _____   b. _____

Write in numeral form (each answer, 3 points).

8. thirty-five and one hundred six ten thousandths   _____

9. seventy and one tenth   _____

Write the answer in simplified form (each answer, 4 points).

10. $\frac{1}{2}$ 11. $\frac{2}{3}$ 12. $\frac{1}{9}$ 13. $\frac{2}{3}$

        $\frac{2}{5}$ $\frac{5}{9}$ $3\frac{8}{9}$ $-\frac{1}{15}$

      $+\frac{1}{10}$ $+\frac{5}{6}$ $+19\frac{5}{9}$

14. $25\frac{2}{3}$ 15. $7\frac{2}{9}$ 16.   85.720   17.   6.0200

      $-13\frac{5}{8}$ $-4\frac{7}{9}$ 631.056 -0.3487

                                                        9.470

                                                   $+$ 52.000

Write in vertical form and add or subtract (each answer, 4 points).

18. 25.782 + 4.59 − 0.85 =

19. eight dollars and five cents plus sixteen dollars and forty cents plus three hundred dollars

Add and subtract using equivalent values (each answer, 5 points).

20. A dress designer charged $25.00 for a piece of cloth, $1.65 for a zipper, $3\frac{1}{2}$ dollars for buttons, and $\frac{4}{5}$ of a dollar for thread. The charge for labor for making the garment was $40.00.

   a. What was the total cost for the materials?

   b. How much more was charged for labor than for materials?

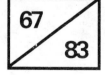

Date _____

Score _____

Name _____

Match these items (each answer, 2 points).

1. _____ percent

2. _____ proper fraction

3. _____ decimal

4. _____ numerator

5. _____ lowest terms

a. based on 10

b. a fraction with a value less than 1

c. a fraction such that the only common factor is 1

d. hundredths

e. the number above the line in a fraction

f. to write the numerator below the line and the denominator above it

Complete the following base-rate-percentage chart (each answer, 2 points).

base (3 decimal places)	rate (1 decimal place)	percentage (3 decimal places)
6. 741	0.8%	_____
7. 52	_____	29.6
8. _____	103%	258
9. _____	19%	48
10. 0.78	23%	_____
11. 0.369	_____	0.153

In the following problems, first multiply or divide using the common fractions. Then, change the fractions to decimals and multiply or divide. Finally, change the common fraction answer to a decimal and compare it with the decimal answer. You may use a calculator (three answers per problem; each answer, 2 points).

12. $16 \times 8\frac{1}{3}$

13. $1\frac{3}{4} \div \frac{2}{3}$

14. $8\frac{1}{3} \times 5\frac{4}{5}$

15. $\frac{7}{8} \div \frac{5}{3}$

Solve the following word problems (each answer, 4 points).

16. If 21% of the student body are juniors, and the juniors number 235, how many total students are in the student body? (Round to the nearest whole number.)

17. Alice paid $12 interest on a loan of $150. What percent interest was this amount?

18. Mary's insurance paid $270 of her $300 doctor bill. What percent of the bill did they pay?

19. Find the cost of 21 $\frac{1}{2}$ dozen eggs at $0.43 per dozen. (Round to the nearest cent.)

20. What is Phil's batting average if he got 47 hits in 145 times at bat? (Round to three decimal places.)

21. If you divide 47.8 grams of sulphur equally among 23 students, how much does each student get?

22. What is Marsha's tithe if her earnings are $97.30?

23. What is the cost of 7.3 gallons of gasoline at 63.9 cents per gallon? (Round to the nearest cent.)

24. A pair of shoes is reduced to sell for $18. This price is 75% of the original price. Find the original price.

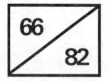

Date _____

Score _____

Name _____

Define the following words (each answer, 3 points).

1. formula _____

2. interest _____

3. solution _____

   _____

4. proportion _____

5. means _____

Use the given formula to complete the chart (each answer, 3 points).

$S = C + P$

	SELLING PRICE	COST	PROFIT	SHOW YOUR WORK
6.	$49.95	$30.00	_____	
7.	$1.50	_____	$0.60	
8.	_____	$23.50	$14.95	

Write *true, false,* or *open* for each of the following equations (each answer, 2 points).

9. _____ $3 = 3$

10. _____ $3 = 19$

11. _____ $3 = x \div 6$

Write the ratios (each answer, 3 points).

In a parking lot are twelve compact cars and twenty-one full-sized cars.

12. What is the ratio of compact cars? _____

13. What is the ratio of full-sized cars? _____

14. What is the ratio of compact cars to full-sized cars? _____

15. What is the ratio of full-sized cars to compact cars? _____

Check by multiplication to see if the following proportions are true proportions (each answer, 3 points).

16. 4:9 = 16:36 _____

17. 6:1 = 18:4 _____

18. 5:10 = 19:37 _____

Solve the following word problems (each answer, 4 points).

19. A certain appliance store pays $89.90 for a stereo set. When the store sells the set, the store earns a profit of $55.00. What is the selling price?

20. The ratio of boys to girls in a certain Sunday school class is 3:7. If the class has nine boys, how many girls does the class have?

Determine the outputs for each function machine shown (each answer, 3 points).

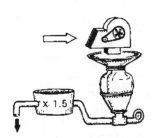

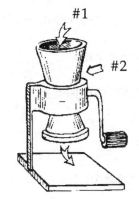

	input	output
21.	1.5	_____
22.	20	_____
23.	4	_____

	input #1	input #2	output
24.	18	11	_____
25.	1.5	0.8	_____
26.	125	30	_____

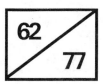

Date _____
Score _____

Name _____

Define the following terms (each answer, 3 points).

1. biased sample _____

2. random sample _____

    _____

On each coordinate axis, graph the points as stated (each graph, 3 points).

3. The set of all points in which the first coordinate is positive

4. The set of all points 2 units away from the origin.

5. The set of all points equally as far from the *x*-axis as from the *y*-axis

Complete the following activities based on the given sets of data (each answer, 4 points).

Set *A:* 13, 20, 16, 16, 18, 13, 15, 12
Set *B:* 20, 15, 12, 17, 15, 13, 15, 20

6. Make a tally for Set *A*.

7. Make a frequency distribution for Set *A*.

8. Calculate the relative frequencies for Set *A*.

9. Find the mode of Set *A*.

   _____

10. Find the median of Set *A*.

    _____

11. Find the mean of Set *A*.

    _____

12. Find the range of Set *A*.

    _____

13. Find the average deviation of Set *A*.

    _____

14. Draw a line graph for Set *A*.

15. Draw a bar graph for Set *A*.

16. Draw a comparison graph for Set *A* and Set *B*.

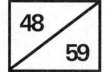

Date _____

Score _____

Name _____

Play "fraction baseball." Perform all four operations using the fractions indicated (each answer, 3 points).

	Single (+)	Double (−)	Triple (x)	Home Run (÷)

Fraction: $\frac{2}{5}$

1. $\frac{5}{6}$

        a. _____   b. _____   c. _____   d. _____

2. $1\frac{1}{3}$

        a. _____   b. _____   c. _____   d. _____

Decimal Fraction: 0.47 (Round to four decimal places.)

3. 1.62

        a. _____   b. _____   c. _____   d. _____

4. 0.23

        a. _____   b. _____   c. _____   d. _____

Consider the following sets.

$W = \{9, 10, 11, 12\}$ $Y = \{10, 12, 14\}$
$X = \{13, 14, 15, 16\}$ $Z = \{9, 11, 13\}$

Use the symbols $\in$ and $\notin$ to make the following sentences true (each answer, 2 points).

5. 13 _____ W   7. 13 _____ Y

6. 13 _____ X   8. 13 _____ Z

Write the elements of set W in base 2 (each answer, 2 points).

9. 9 = _____   11. 11 = _____

10. 10 = _____   12. 12 = _____

Write in set notation the sentence that is shown by each Venn diagram (each answer, 3 points).

13. _____

14. _____

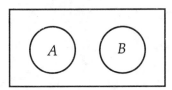

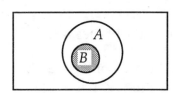

Match the following geometric figures with their definition (each answer, 2 points).

15. _____ rectangle

16. _____ trapezoid

17. _____ triangle

18. _____ parallelogram

19. _____ circle

20. _____ square

a. a quadrilateral with both sets of opposite sides equal and parallel, and opposite angles equal

b. a quadrilateral with four equal sides and four equal angles

c. a figure with all points equally distant from the center

d. a figure with three sides and three angles

e. a quadrilateral with one set of opposite sides parallel

f. a figure with seven sides and seven angles

g. a quadrilateral with both sets of opposite sides equal and parallel, and four right angles

Find the missing term in each of the following proportions (each answer, 3 points).

21. 2:14 = _____ :35

22. 25:_____ = 1:20

23. _____:32 = 6:2

24. 5:30 = 20:_____

25. 80:160 = _____:42

Match the following rules to the formulas that describe them (each answer, 2 points).

26. _____ The selling price is equal to the cost plus the profit.

a. $I = R \times P$

27. _____ The perimeter of a square equals 4 times a side.

b. $MPG = M \div G$

28. _____ The area of a rectangle is equal to the length times the width.

c. $A = L \times W$

d. $S = C + P$

29. _____ The interest is equal to the rate times the principle.

e. $P = 4 \times S$

30. _____ Weekly pay is equal to the hourly wage times the number of hours worked.

f. $P = W \times H$

31. _____ To find the distance an object travels, multiply the rate of travel by the time.

g. $D = R \times T$

Identify each equation as *true*, *false*, or *open* (each answer, 2 points).

32. _____ $9 \div 3 = 3$

33. _____ $9 - n = 3$

34. _____ $10 - 3 = 3$

35. _____ $? - 3 = 3$

36. _____ $18 - 4 = 3$

Consider the following set of data.

$$1 \ 2 \ 3 \ 4 \ 5 \ 6 \ 7 \ 8 \ 9 \ 10$$
$$1 \ 2 \ 3 \ 4 \ 5 \ 6 \ 7 \ 8 \ 9 \ 1$$
$$2 \ 3 \ 4 \ 5 \ 6 \ 7 \ 8 \ 1 \ 2 \ 3$$
$$4 \ 3 \ 4 \ 5 \ 4 \ 5 \ 4 \ 5 \ 5 \ 5$$

Complete these items based on the set of data (each answer, 4 points).

37. Make a tally for the data.

Score   Tally

38. Make a frequency distribution, including relative frequency, for the data.

Score	Frequency	Relative Frequency

39. Find the median. _____

40. Find the mode. _____

41. Find the mean. _____

42. Find the range. _____

43. Find the average deviation _____

44. Draw a bar graph for the data.

$\dfrac{121}{151}$

Date _____
Score _____

ANSWER KEYS

NUMBER SYMBOLS	NUMBER-OPERATION SYMBOLS			PUNCTUATION SYMBOLS		
Digits	Symbol	Meaning	Example	Symbol	Meaning	Example
0	+	addition	$1 + 2 = 3$	.	decimal point	$\pi = 3.1416$
1						
2	–	subtrac-tion	$3 - 2 = 1$	,	comma	$A = \{3, 4, 5\}$
3						
4	x	multipli-cation	$2 \times 3 = 6$	( )	paren-thesis	$2 + (3 + 1) = 6$
5	·		$2 \cdot 3 = 6$			
6	()		$(2)(3) = 6$	[ ]	brackets	$2 + [1 + (3 + 1)] = 7$
7	÷	division	$6 \div 3 = 2$			
8				{ }	braces	$\{1, 2\} = \{2, 1\}$
9	**SET-OPERATIONS SYMBOLS**					
	∪	union	$A \cup B = M$			
	∩	inter-section	$A \cap B = K$			

NUMBER-RELATION SYMBOLS			SET-RELATION SYMBOLS		
Symbol	Meaning	Example	Symbol	Meaning	Example
$=$	is equal to	$2 + 3 = 5$	$\in$	is an element of	$1 \in \{1, 2\}$
$\neq$	is not equal to	$8 + 1 \neq 7$	$\notin$	is not an element of	$3 \notin \phi$
$<$	is less than	$3 < 4$	$\subset$	is a subset of	$\phi \subset A$
$\not<$	is not less than	$3 \not< 2$	$\not\subset$	is not a subset of	$\{1, 2\} \not\subset \{2, 3\}$
$>$	is greater than	$9 > 7$	$\leftrightarrow$	is not equivalent to	$\{1, 2\} \leftrightarrow \{a, b\}$
$\not>$	is not greater than	$7 \not> 9$	$\not\leftrightarrow$	is not equivalent to	$A \not\leftrightarrow 3, 4$
$\leq$	is less than or equal to	$4 \leq 4$	$=$	is equal to	$A = A$
$\not\leq$	is not less than or equal to	$4 \not\leq 3$	$\neq$	is not equal to	$\{1, 2\} \neq \phi$
$\geq$	is greater than or equal to	$6 \geq 5$			
$\not\geq$	is not greater than or equal to	$5 \not\geq 6$			

I.   SECTION ONE

1.1 $7,321 = 7,000 + 300 + 20 + 1$

1.2 $5,692 = 5,000 + 600 + 90 + 2$

1.3 $741 = 700 + 40 + 1$

1.4 $72,655 = 70,000 + 2,000 + 600 + 50 + 5$

1.5 $33 = 30 + 3$

1.6 $921,733 = 900,000 + 20,000 + 1,000 + 700 + 30 + 3$

1.7 $1,380,010 = 1,000,000 + 300,000 + 80,000 + 10$

1.8 $2,001 = 2,000 + 1$

1.9 $602,057 = 600,000 + 2,000 + 50 + 7$

1.10 $430,006 = 400,000 + 30,000 + 6$

1.11 $5,000 + 200 + 30 + 4 = 5,234$

1.12 $10,000 + 400 + 30 + 6 = 10,436$

1.13 $4,000,000 + 6,000 + 500 = 4,006,500$

1.14 $2,000,000,000 + 80,000,000 + 300 = 2,080,000,300$

1.15 $1,000,000,000 + 1,000,000 + 1,000 + 100 = 1,001,001,100$

1.16 $80,000 + 3,000 + 40 = 83,040$

1.17 $4,000 + 300 + 2 = 4,302$

1.18 $500,000 + 300 + 2 = 500,302$

1.19 $7,000 + 400 + 30 + 5 = 7,435$

1.20 $90,000 + 800 + 1 = 90,801$

1.21   1 hundred + 2 tens = 100 + 20 = 120

1.22   15 tens and 12 ones = 10 tens + 5 tens + 10 ones + 2 ones = 1 hundred + 5 tens + 1 ten + 2 ones = 1 hundred + 6 tens + 2 ones = 100 + 60 + 2 = 162

1.23   13 hundreds and 40 ones = 10 hundreds + 3 hundreds + 4 tens = 1 thousand + 3 hundreds + 4 tens = 1,000 + 300 + 40 = 1,340

1.24   17 tens and 17 ones = 10 tens + 7 tens + 10 ones + 7 ones = 1 hundred + 7 tens + 1 ten + 7 ones = 1 hundred + 8 tens + 7 ones = 100 + 80 + 7 = 187

1.25   15 hundreds and 8 tens = 10 hundreds + 5 hundreds + 8 tens = 1 thousand + 5 hundreds + 8 tens = 1,000 + 500 + 80 = 1,580

1.26   66, 67, 68

1.27   139,

1.28   1,027; 1,028; 1,029

1.29   16,254

1.30   106,522

1.31   false

1.32   false

1.33   false

1.34   true

1.35   true

1.36   <

1.37   <, <

1.38   <

1.39   >

1.40	>, <	1.66	a. 6,300
1.41	4 < 6 < 8		b. 6,000
1.42	22 > 21	1.67	a. 6,300
1.43	66 < 70		b. 6,000
1.44	4 < 5 < 6	1.68	a. 9,700
1.45	18 > 16		b. 10,000
1.46	40	1.69	a. 7,300
1.47	270		b. 7,000
1.48	5,280	1.70	a. 7,600
1.49	62,150		b. 8,000
1.50	50	1.71	a. 12,900
1.51	600		b. 13,000
1.52	6,320	1.72	a. 24,300
1.53	23,470		b. 24,000
1.54	120	1.73	a. 55,600
1.55	420		b. 56,000
1.56	4,830	1.74	a. 73,500
1.57	82,550		b. 73,000
1.58	260	1.75	a. 52,000
1.59	770		b. 52,000
1.60	3,170	1.76	a. 623,500
1.61	75,660		b. 623,000
1.62	200	1.77	a. 764,500
1.63	380		b. 765,000
1.64	4,400	1.78	a. 836,800
1.65	80,960		b. 837,000
		1.79	a. 492,000
			b. 492,000
		1.80	a. 331,400
			b. 331,000

II.   SECTION TWO

2.1

+	5	2	1	7	6	3	4	8	9
7	12	9	8	14	13	10	11	15	16
3	8	5	4	10	9	6	7	11	12
8	13	10	9	15	14	11	12	16	17
2	7	4	3	9	8	5	6	10	11
9	14	11	10	16	15	12	13	17	18
1	6	3	2	8	7	4	5	9	10
4	9	6	5	11	10	7	8	12	13
5	10	7	6	12	11	8	9	13	14
6	11	8	7	13	12	9	10	14	15

2.2 $12 + 5 = 17$

2.3 $44 + 3 = 47$

2.4 $9 + 15 = 24$

2.5 $18 + 6 = 24$

2.6 $48 + 5 = 53$

2.7 $8 + 16 = 24$

2.8 $20 + 7 = 27$

2.9 $62 + 9 = 71$

2.10 $7 + 10 = 17$

2.11 $22 + 9 = 31$

2.12 $23 + 8 = 31$

2.13 $6 + 13 = 19$

2.14 $35 + 6 = 41$

2.15 $35 + 4 = 39$

2.16 $88 + 8 = 96$

2.17 $73 + 6 = 79$

2.18 $63 + 5 = 68$

2.19 $86 + 9 = 95$

2.20 $69 + 5 = 74$

2.21 $44 + 9 = 53$

2.22 $92 + 5 = 97$

2.23 $84 + 8 = 92$

2.24 $38 + 6 = 44$

2.25 $91 + 6 = 97$

2.26 $86 + 4 = 90$

2.27 $45 + 9 = 54$

2.28 $95 + 9 = 104$

2.29 $95 + 7 = 102$

2.30 $39 + 5 = 44$

2.31 $6 + 8 + 3 = 17$

2.32 $6 + 5 + 4 + 8 = 23$

2.33 $5 + 9 + 4 = 18$

2.34 $9 + 2 + 3 + 1 = 15$

2.35 $6 + 5 + 10 = 21$

2.36 $5 + 8 + 6 + 5 = 24$

2.37 $5 + 8 + 2 = 15$

2.38 $4 + 7 + 9 + 1 = 21$

2.39 $9 + 7 + 3 = 19$

2.40 $2 + 3 + 5 + 9 = 19$

2.41	$6 + 4 + 5 + 1 + 7 = 23$
2.42	$3 + 2 + 8 + 9 + 4 = 26$
2.43	$9 + 8 + 1 + 2 + 6 = 26$
2.44	$10 + 2 + 3 + 7 + 6 = 28$
2.45	$12 + 3 + 5 + 1 + 2 = 23$
2.46	$6 + 2 + 5 + 4 + 7 = 24$
2.47	$5 + 7 + 1 + 10 + 8 = 31$
2.48	$8 + 3 + 6 + 5 + 4 = 26$
2.49	$9 + 6 + 6 + 5 + 2 + 7 = 35$
2.50	$3 + 10 + 4 + 9 + 8 + 6 = 40$
2.51	$12 + 4 + 3 + 9 + 8 + 6 + 10 + 2 = 54$
2.52	$8 + 5 + 11 + 3 + 12 + 4 + 5 = 48$
2.53	$10 + 2 + 12 + 13 + 7 + 10 + 1 = 55$
2.54	$14 + 9 + 8 + 3 + 6 + 4 + 3 = 47$
2.55	$8 + 2 + 5 + 6 + 4 + 5 + 11 = 41$
2.56	$10 + 6 + 12 + 15 + 7 = 50$
2.57	$15 + 6 + 10 + 4 + 12 = 47$
2.58	$16 + 8 + 5 + 11 + 12 = 52$
2.59	$20 + 9 + 9 + 9 + 9 + = 56$
2.60	$8 + 9 + 8 + 9 + 8 = 42$
2.61	$5 + 8 + 10 + 3 + 12 + 5 + 10 + 2 + 8 + 7 = 70$
2.62	$9 + 2 + 3 + 1 + 5 + 8 + 12 + 3 + 2 + 9 = 54$
2.63	$6 + 2 + 5 + 12 + 10 + 11 + 3 + 8 + 4 + 7 = 68$

2.64	$7 + 3 + 4 + 10 + 5 + 11 + 6 + 15 + 4 + 13 = 78$
2.65	$4 + 13 + 5 + 12 + 7 + 18 + 3 + 4 + 7 + 10 = 83$
2.66	$\begin{array}{r} 63 \\ \underline{14} \\ 77 \end{array}$
2.67	$\begin{array}{r} 93 \\ \underline{25} \\ 118 \end{array}$
2.68	$\begin{array}{r} {}^1 67 \\ \underline{34} \\ 101 \end{array}$
2.69	$\begin{array}{r} 88 \\ \underline{11} \\ 99 \end{array}$
2.70	$\begin{array}{r} 83 \\ \underline{41} \\ 124 \end{array}$
2.71	$\begin{array}{r} 93 \\ \underline{21} \\ 114 \end{array}$
2.72	$\begin{array}{r} 24 \\ \underline{35} \\ 59 \end{array}$
2.73	$\begin{array}{r} 73 \\ \underline{45} \\ 118 \end{array}$
2.74	$\begin{array}{r} {}^1 66 \\ \underline{25} \\ 91 \end{array}$
2.75	$\begin{array}{r} {}^1 82 \\ \underline{38} \\ 120 \end{array}$
2.76	$\begin{array}{r} 45 \\ \underline{63} \\ 108 \end{array}$
2.77	$\begin{array}{r} 74 \\ \underline{91} \\ 165 \end{array}$

2.78	^{188} $\underline{62}$ 150		2.91	43 $\underline{62}$ 105
2.79	73 $\underline{52}$ 125		2.92	73 $\underline{55}$ 128
2.80	^{148} $\underline{45}$ 93		2.93	^{177} $\underline{78}$ 155
2.81	93 $\underline{61}$ 154		2.94	93 $\underline{61}$ 154
2.82	75 $\underline{22}$ 97		2.95	82 $\underline{54}$ 136
2.83	73 $\underline{53}$ 126		2.96	95 $\underline{91}$ 186
2.84	^{146} $\underline{25}$ 71		2.97	55 $\underline{62}$ 117
2.85	82 $\underline{71}$ 153		2.98	73 $\underline{74}$ 147
2.86	27 $\underline{61}$ 88		2.99	^{154} $\underline{58}$ 112
2.87	^{128} $\underline{33}$ 61		2.100	62 $\underline{72}$ 134
2.88	92 $\underline{31}$ 123		2.101	93 $\underline{81}$ 174
2.89	^{197} $\underline{63}$ 160		2.102	42 $\underline{75}$ 117
2.90	^{155} $\underline{56}$ 111		2.103	43 $\underline{52}$ 95

2.104	45
	71
	116

2.105	48
	51
	99

2.106	13
	25
	38

2.107	72
	15
	87

2.108	^{1}65
	18
	83

2.109	22
	12
	34

2.110	83
	51
	134

2.111	^{1}75
	25
	100

2.112	92
	36
	128

2.113	^{1}35
	37
	72

2.114	621
	55
	676

2.115	433
	62
	495

2.116	^{1}591
	75
	666

2.117	821
	23
	844

2.118	^{1}453
	62
	515

2.119	^{1}546
	28
	574

2.120	^{1}963
	62
	1,025

2.121	^{1}546
	28
	574

2.122	123
	65
	188

2.123	^{1}153
	82
	235

2.124	^{11}175
	55
	230

2.125	^{1}196
	82
	278

2.126	365
	22
	387

2.127	^{1}485
	62
	547

2.128	721 55 776		2.140	1 624 93 717
2.129	836 43 879		2.141	1 528 62 590
2.130	1 591 73 664		2.142	1 543 84 627
2.131	1 728 36 764		2.143	762 33 795
2.132	1 928 39 967		2.144	1 62 33 45 140
2.133	1 483 65 548		2.145	2 98 76 58 232
2.134	1 521 92 613		2.146	1 43 97 58 198
2.135	1 623 84 707		2.147	2 79 69 72 220
2.136	526 73 599		2.148	1 51 23 17 91
2.137	822 73 895		2.149	1 93 58 32 183
2.138	1 792 54 846		2.150	1 52 33 65 94 244
2.139	1 963 82 1,045			

2.151
$$\begin{array}{r} 1 \\ 21 \\ 58 \\ 41 \\ \underline{33} \\ 153 \end{array}$$

2.152
$$\begin{array}{r} 1 \\ 75 \\ 82 \\ 61 \\ \underline{48} \\ 266 \end{array}$$

2.153
$$\begin{array}{r} 1 \\ 22 \\ 33 \\ 52 \\ \underline{96} \\ 203 \end{array}$$

2.154
$$\begin{array}{r} 1 \\ 93 \\ 72 \\ 88 \\ \underline{45} \\ 298 \end{array}$$

2.155
$$\begin{array}{r} 2 \\ 37 \\ 65 \\ 47 \\ \underline{81} \\ 230 \end{array}$$

2.156
$$\begin{array}{r} 6 \\ \underline{4} \\ 10 \end{array}$$

2.157
$$\begin{array}{r} 1\,0 \\ \underline{2\,0} \\ 3\,0 \end{array}$$

2.158
$$\begin{array}{r} 2\underline{6} \\ \underline{93} \\ 119 \end{array}$$

2.159
$$\begin{array}{r} \underline{37} \\ \underline{42} \\ 79 \end{array}$$

2.160
$$\begin{array}{r} 98 \\ \underline{76} \\ 1\underline{74} \end{array}$$

2.161
$$\begin{array}{r} 7 \\ \underline{18} \\ \underline{8} \\ 33 \end{array}$$

2.162
$$\begin{array}{r} 10 \\ 9 \\ \underline{6} \\ 25 \end{array}$$

2.163
$$\begin{array}{r} 1\,2 \\ \underline{1\,7} \\ 7 \\ 36 \end{array}$$

2.164
$$\begin{array}{r} 22 \\ \underline{1\,9} \\ 3\,8 \\ 7\,9 \end{array}$$

2.165
$$\begin{array}{r} 5\,3 \\ 4\,4 \\ \underline{7\,1} \\ 168 \end{array}$$

2.166
$$\begin{array}{r} 2\underline{6} \\ \underline{33} \\ \underline{5}9 \end{array}$$

2.167
$$\begin{array}{r} \underline{41} \\ \underline{75} \\ 1\underline{16} \end{array}$$

2.168
$$\begin{array}{r} \underline{35} \\ \underline{48} \\ 8\underline{3} \end{array}$$

2.169   6<u>5</u>
         41
         <u>33</u>
         139

2.170   <u>88</u>
         35
         <u>21</u>
         144

2.171   <u>465</u>
         <u>263</u>
         728

2.172   6<u>9</u>2
         584
         1,<u>2</u>76

2.173   59<u>8</u>
         <u>8</u>21
         <u>635</u>
         2,054

2.174   <u>621</u>
         322
         <u>596</u>
         1,5<u>3</u>9

2.175   6,213
         5,<u>4</u>18
         693
         <u>415</u>
         12,7<u>3</u>9

2.176

Q	D	N	P
1	1	0	5
1	0	1	10
1	1	1	0
1	0	2	5
1	0	3	0
1	0	0	15
0	4	0	0
0	3	2	0
0	3	1	5
0	3	0	10
0	2	4	0
0	2	3	5
0	2	2	10
0	2	1	15
0	2	0	20
0	1	6	0
0	1	5	5
0	1	4	10
0	1	3	15
0	1	2	20
0	1	1	25
0	1	0	30
0	0	8	0
0	0	7	5
0	0	6	10
0	0	5	15
0	0	4	20
0	0	3	25
0	0	2	30
0	0	1	35
0	0	0	40

2.177   11
         651
         236
         <u>489</u>
         1,376

2.178   1
         216
         305
         <u>112</u>
         633

2.179   11
         768
         432
         <u>175</u>
         1,375

2.180   11
         526
         831
         <u>786</u>
         2,143

2.181
$$
\begin{array}{r}
\overset{1}{7}23 \\
327 \\
415 \\
\hline
1,465
\end{array}
$$

2.182
$$
\begin{array}{r}
\overset{1\,1}{7},865 \\
2,763 \\
\hline
10,628
\end{array}
$$

2.183
$$
\begin{array}{r}
\overset{1}{4},316 \\
5,870 \\
\hline
10,186
\end{array}
$$

2.184
$$
\begin{array}{r}
9,273 \\
625 \\
\hline
9,898
\end{array}
$$

2.185
$$
\begin{array}{r}
\overset{1}{4}15 \\
7,368 \\
\hline
7,783
\end{array}
$$

2.186
$$
\begin{array}{r}
\overset{1\,1}{8},992 \\
9,432 \\
\hline
18,424
\end{array}
$$

2.187
$$
\begin{array}{r}
\overset{2\,21}{6},970 \\
4,983 \\
3,321 \\
5,468 \\
\hline
20,742
\end{array}
$$

2.188
$$
\begin{array}{r}
\overset{2\,21}{9},683 \\
4,952 \\
7,369 \\
1,763 \\
\hline
23,767
\end{array}
$$

2.189
$$
\begin{array}{r}
\overset{1\,22}{7},385 \\
496 \\
33 \\
5,469 \\
\hline
13,383
\end{array}
$$

2.190
$$
\begin{array}{r}
\overset{1}{}721 \\
5,436 \\
310 \\
402 \\
\hline
6,869
\end{array}
$$

2.191
$$
\begin{array}{r}
\overset{1}{7},001 \\
415 \\
9,623 \\
960 \\
\hline
17,999
\end{array}
$$

2.192
$$
\begin{array}{r}
\overset{32}{6}97 \\
498 \\
123 \\
654 \\
782 \\
\hline
2,754
\end{array}
$$

2.193
$$
\begin{array}{r}
\overset{1}{5}46 \\
103 \\
36 \\
500 \\
201 \\
\hline
1,386
\end{array}
$$

2.194
$$
\begin{array}{r}
\overset{2\,11}{7},254 \\
312 \\
415 \\
768 \\
8,840 \\
\hline
17,589
\end{array}
$$

2.195
$$
\begin{array}{r}
\overset{1\,12}{8},526 \\
3,279 \\
543 \\
4,012 \\
5,002 \\
\hline
21,362
\end{array}
$$

2.196
$$
\begin{array}{r}
\overset{21}{1}25 \\
326 \\
\overset{2}{}971 \\
4,596 \\
8,271 \\
\hline
14,289
\end{array}
$$

2.197
$$
\begin{array}{r}
\overset{1\,11}{6},572 \\
8,231 \\
5,468 \\
\hline
20,271
\end{array}
$$

2.198
  1 1
  5,846
  9,610
  6,005
  21,461

2.199
  1 11
  9,432
    675
  8,214
  18,321

2.200
   11
  4,263
  6,143
    564
  10,970

2.201
  1 11
  1,692
  5,438
  2,005
  9,135

2.202
  1 11
  6,082
    973
  5,406
  6,215
  18,676

2.203
   11
  7,038
  6,251
    456
     22
  13,767

2.204
  2 12
  9,638
  4,871
  2,305
  7,306
  24,120

2.205
   11
    721
  1 56
  8,205
    326
  9,308

2.206
  1 1
  1,506
    307
  8,020
     93
  9,926

2.207
   11
  62,154
  96,915
  82,630
  241,699

2.208
  21 11
  68,416
  98,325
   7,368
  174,109

2.209
  21 21
  48,621
  54,986
  38,294
  141,901

2.210
  1 1 21
  78,543
  21,698
  51,488
  151,729

2.211
  1   1
  68,214
   8,216
  30,514
  106,944

2.212
  11 2
  62,150
   5,362
      33
   5,480
     720
  73,745

2.213
  21 21
  69,496
   8,256
     431
      75
   8,510
  86,768

2.214
$$\begin{array}{r} {}^{31} \\ {}_{12}\,621 \\ 52{,}763 \\ 4{,}593 \\ 86 \\ \underline{4{,}852} \\ 62{,}915 \end{array}$$

2.215
$$\begin{array}{r} {}^{21\ 22} \\ 43{,}651 \\ 73{,}082 \\ 85{,}069 \\ 88{,}213 \\ \underline{77{,}856} \\ 367{,}871 \end{array}$$

2.216
$$\begin{array}{r} {}^{2\ 12} \\ 10{,}506 \\ 20{,}708 \\ 5{,}055 \\ 730 \\ \underline{60{,}251} \\ 97{,}250 \end{array}$$

2.217   Example:

36 a.	31 b.	38
37	35	33 c.
32	39 d.	34

2.218   Example:

59	52 a.	57 b.
54 c.	56	58
55	60 d.	53

2.219   Example:

17	24	1	8 a.	15
23	5	7	14 b.	16 c.
4 d.	6	13	20	22
10	12 e.	19 f.	21	3 g.
11	18	25 h.	2	9

2.220   62 + 34
    a. estimate = 60 + 30 = 90
    b. sum = 96

2.221   78 + 52
    a. estimate = 80 + 50 = 130
    b. sum = 130

2.222   43 + 51
    a. estimate = 40 + 50 = 90
    b. sum = 94

2.223   74 + 87
    a. estimate = 70 + 90 = 160
    b. sum = 161

2.224   27 + 45
    a. estimate = 30 + 50 = 80
    b. sum = 72

2.225   91 + 76
    a. estimate = 90 + 80 = 170
    b. sum = 167

2.226   521 + 76
    a. estimate = 500 + 100 = 600
    b. sum = 597

2.227   325 + 94
    a. estimate = 300 + 100 = 400
    b. sum = 419

2.228   722 + 123
    a. estimate = 700 + 100 = 800
    b. sum = 845

2.229   695 + 522
    a. estimate = 700 + 500 = 1,200
    b. sum = 1,217

2.230   821 + 496
    a. estimate = 800 + 500 = 1,300
    b. sum = 1,317

2.231   7,265 + 372
      a.   estimate = 7,300 + 400
           = 7,700
      b.   sum = 7,637

2.232   521 + 636 + 498
      a.   estimate = 500 + 600 + 500
           = 1,600
      b.   sum = 1,655

2.233   695 + 455 + 322
      a.   estimate = 700 + 500 + 300
           = 1,500
      b.   sum = 1,472

2.234   759 + 622 + 963
      a.   estimate = 800 + 600
           + 1,000 = 2,400
      b.   sum = 2,344

2.235   5,280 + 961
      a.   estimate = 5,000 + 1,000
           = 6,000
      b.   sum = 6,241

2.236   7,622 + 4,365
      a.   estimate = 8,000 + 4,000
           = 12,000
      b.   sum = 11,987

2.237   9,925 + 4,863
      a.   estimate = 10,000 + 5,000
           = 15,000
      b.   sum = 14,788

2.238   10,265 + 1,650 + 5,690
      a.   estimate = 10,000 + 2,000
           + 6,000 = 18,000
      b.   sum = 17,605

2.239   15,500 + 8,622 + 3,960
      a.   estimate = 16,000 + 9,000
           + 4,000 = 29,000
      b.   sum = 28,082

2.240
$$10 + N = 28$$
$$10 + N - 10 = 28 - 10$$
$$N = 28 - 10$$
$$N = 18$$

2.241
$$15 + 7 = N$$
$$22 = N$$

2.242
$$N + 12 = 32$$
$$N + 12 - 12 = 32 - 12$$
$$N = 32 - 12$$
$$N = 20$$

2.243
$$8 + 6 = 12 + N$$
$$14 = 12 + N$$
$$14 - 12 = 12 + N - 12$$
$$14 - 12 =$$
$$2 = N$$

2.244
$$33 + 5 = 22 + N$$
$$38 = 22 + N$$
$$38 - 22 = 22 + N - 22$$
$$38 - 22 =$$
$$16 = N$$

2.245
$$N + 5 = N + 5$$
Since both expressions are the same, any number will make the sentence true.

2.246
$$12 + N = 13 + N$$
$$12 + N - 12 = 13 + N - 12$$
$$N = 1 + N$$
N cannot equal $1 + N$; therefore, no number will make the sentence true.

2.247
$$50 + 32 = N + 60$$
$$82 = N + 60$$
$$82 - 60 = N + 60 - 60$$
$$82 - 60 = N$$
$$22 = N$$

2.248
$$60 + N < 70$$
$$60 + N - 60 < 70 - 60$$
$$N < 70 - 60$$
$$N < 10$$
Any number less than 10 will make the sentence true.

2.249
$$N + 4 > 22$$
$$N + 4 - 4 > 22 - 4$$
$$N > 22 - 4$$
$$N > 18$$
Any number greater than 18 will make the sentence true.

2.250
$$12 + 10 > N + 8$$
$$22 > N + 8$$
$$22 - 8 > N + 8 - 8$$
$$22 - 8 > N$$
$$14 > N$$
Since 14 is greater than N, N must be less than 14 ($N < 14$). Therefore, any number less than 14 will make the sentence true.

2.251
$$15 + N < 62$$
$$15 + N - 15 < 62 - 15$$
$$N \quad < 62 - 15$$
$$N < 47$$

Any number less than 47 will make the sentence true.

2.252
$$5 \_ 6$$
$$5 < 6$$

2.253
$$22 \_ 30$$
$$22 < 30$$

2.254
$$15 + 10 \_ 25$$
$$25 \_ 25$$
$$25 = 25$$

2.255
$$12 + 38 \_ 15 + 30$$
$$50 \_ 45$$
$$50 > 45$$

2.256
$$102 + 33 \_ 105 + 10$$
$$135 \_ 115$$
$$135 > 115$$

2.257
$$50 \_ 7 + 8 + 22$$
$$50 \_ 37$$
$$50 > 37$$

2.258
$$16 + 50 \_ 60 + 6$$
$$66 \_ 66$$
$$66 = 66$$

2.259
$$19 + 79 \_ 98$$
$$98 \_ 98$$
$$98 = 98$$

III.   SECTION THREE

3.1 $10 - 3 = 7$

3.2 $12 - 6 = 6$

3.3 $7 - 2 = 5$

3.4 $54 - 6 = 48$

3.5 $8 - 6 = 2$

3.6 $11 - 5 = 6$

3.7 $27 - 4 = 23$

3.8 $67 - 6 = 61$

3.9 $12 - 4 = 8$

3.10 $37 - 2 = 35$

3.11 $29 - 9 = 20$

3.12 $69 - 6 = 63$

3.13 $13 - 5 = 8$

3.14 $45 - 5 = 40$

3.15 $35 - 6 = 29$

3.16 $76 - 5 = 71$

3.17 $14 - 7 = 7$

3.18 $40 - 6 = 34$

3.19 $41 - 6 = 35$

3.20 $72 - 4 = 68$

3.21 $76 - 6 = 70$

3.22 $77 - 6 = 71$

3.23 $83 - 5 = 78$

3.24 $94 - 5 = 89$

3.25 $76 - 7 = 69$

3.26 $77 - 8 = 69$

3.27 $83 - 6 = 77$

3.28 $94 - 6 = 88$

3.29 $76 - 8 = 68$

3.30 $77 - 9 = 68$

3.31	$83 - 7 = 76$		3.49	96
3.32	$94 - 7 = 87$			$\underline{65}$
3.33	$76 - 9 = 67$			31
3.34	$83 - 3 = 80$		3.50	27
3.35	$83 - 8 = 75$			$\underline{17}$
3.36	$94 - 8 = 86$			10
3.37	$77 - 7 = 70$		3.51	38
3.38	$83 - 4 = 79$			$\underline{25}$
3.39	$83 - 9 = 74$			13
3.40	$94 - 9 = 85$		3.52	52

3.41
  a. $13 - 5 = 8$
  b. $13 - 8 = 5$
  c. $8 + 5 = 13$

3.42
  a. $42 - 9 = 33$
  b. $42 - 33 = 9$
  c. $33 + 9 = 42$

3.43
  a. $21 - 7 = 14$
  b. $21 - 14 = 7$
  c. $7 + 14 = 21$

3.44
  a. $55 - 9 = 46$
  b. $55 - 46 = 9$
  c. $9 + 46 = 55$

3.45
  a. $72 - 4 = 68$
  b. $72 - 68 = 4$
  c. $68 + 4 = 72$

3.46 59
       $\underline{26}$
       33

3.47 77
       $\underline{36}$
       41

3.48 84
       $\underline{53}$
       31

3.52 52
       $\underline{22}$
       30

3.53 $4\,^{1}5\!\!\!/5$
       $\underline{38}$
       17

3.54 $6\,^{1}7\!\!\!/3$
       $\underline{29}$
       44

3.55 $3\,^{1}4\!\!\!/5$
       $\underline{36}$
       9

3.56 $6\,^{1}7\!\!\!/7$
       $\underline{68}$
       9

3.57 $4\,^{1}5\!\!\!/5$
       $\underline{18}$
       37

3.58 $4\,^{1}5\!\!\!/3$
       $\underline{14}$
       39

3.59 $6\,^{1}7\!\!\!/6$
       $\underline{29}$
       47

3.60 86
       $\underline{63}$
       23

3.61	8_1 $\frac{}{95}$ $\frac{48}{47}$		3.73	28 $\frac{18}{10}$
3.62	8_1 $\frac{}{91}$ $\frac{56}{35}$		3.74	49 $\frac{27}{22}$
3.63	6_1 $\frac{}{73}$ $\frac{64}{9}$		3.75	49 $\frac{28}{21}$
3.64	4_1 $\frac{}{53}$ $\frac{28}{25}$		3.76	49 $\frac{29}{20}$
3.65	4_1 $\frac{}{51}$ $\frac{15}{36}$		3.77	49 $\frac{30}{19}$
3.66	4_1 $\frac{}{50}$ $\frac{27}{23}$		3.78	4_1 $\frac{}{57}$ $\frac{29}{28}$
3.67	6_1 $\frac{}{73}$ $\frac{49}{24}$		3.79	4_1 $\frac{}{57}$ $\frac{28}{29}$
3.68	82 $\frac{61}{21}$		3.80	57 $\frac{27}{30}$
3.69	7_1 $\frac{}{80}$ $\frac{56}{24}$		3.81	4_1 $\frac{}{55}$ $\frac{28}{27}$
3.70	2_1 $\frac{}{33}$ $\frac{17}{16}$		3.82	5_1 $\frac{}{68}$ $\frac{29}{39}$
3.71	28 $\frac{13}{15}$		3.83	73 $\frac{62}{11}$
3.72	1_1 $\frac{}{28}$ $\frac{19}{9}$		3.84	7_1 $\frac{}{84}$ $\frac{77}{7}$
			3.85	8_1 $\frac{}{93}$ $\frac{89}{4}$

3.86
$$\begin{array}{r} 3\,1 \\ \cancel{4}57 \\ \underline{283} \\ 174 \end{array}$$

3.87
$$\begin{array}{r} 3\,1 \\ \cancel{5}42 \\ \underline{123} \\ 419 \end{array}$$

3.88
$$\begin{array}{r} 8\,1 \\ 6\cancel{9}4 \\ \underline{356} \\ 338 \end{array}$$

3.89
$$\begin{array}{r} 485 \\ \underline{371} \\ 114 \end{array}$$

3.90
$$\begin{array}{r} 8\,1 \\ 5\cancel{9}8 \\ \underline{489} \\ 109 \end{array}$$

3.91
$$\begin{array}{r} 6\,1 \\ \cancel{7}23 \\ \underline{432} \\ 291 \end{array}$$

3.92
$$\begin{array}{r} 815\,1 \\ \cancel{9}\cancel{6}3 \\ \underline{485} \\ 478 \end{array}$$

3.93
$$\begin{array}{r} 1\,1 \\ 8\cancel{2}2 \\ \underline{617} \\ 205 \end{array}$$

3.94
$$\begin{array}{r} 7\,1 \\ 78\cancel{3} \\ \underline{529} \\ 254 \end{array}$$

3.95
$$\begin{array}{r} 7\,1 \\ 585 \\ \underline{437} \\ 148 \end{array}$$

9.96
$$\begin{array}{r} 8\,1 \\ 69\cancel{3} \\ \underline{287} \\ 406 \end{array}$$

3.97
$$\begin{array}{r} 8\,1 \\ 4\cancel{9}6 \\ \underline{369} \\ 127 \end{array}$$

3.98
$$\begin{array}{r} 4\,1 \\ 5\cancel{5}5 \\ \underline{238} \\ 317 \end{array}$$

3.99
$$\begin{array}{r} 4\,1 \\ \cancel{5}69 \\ \underline{384} \\ 185 \end{array}$$

3.100
$$\begin{array}{r} 212\,1 \\ \cancel{3}\cancel{3}2 \\ \underline{163} \\ 169 \end{array}$$

3.101
$$\begin{array}{r} 999 \\ \underline{383} \\ 616 \end{array}$$

3.102
$$\begin{array}{r} 5\,1 \\ 465 \\ \underline{328} \\ 137 \end{array}$$

3.103
$$\begin{array}{r} 8\,1 \\ 3\cancel{9}8 \\ \underline{229} \\ 169 \end{array}$$

3.104
$$\begin{array}{r} 511\,1 \\ \cancel{6}{,}234 \\ \underline{584} \\ 5{,}650 \end{array}$$

3.105
$$\begin{array}{r} 6\;12\;14\,1 \\ 7{,}\cancel{3}\cancel{5}4 \\ \underline{486} \\ 6{,}868 \end{array}$$

3.106
$$\begin{array}{r} 8\;11\;18\,1 \\ \cancel{9}{,}2\cancel{3}6 \\ \underline{549} \\ 8{,}687 \end{array}$$

3.107
$$\begin{array}{r} 7\;11\;18\,1 \\ \cancel{8}{,}2\cancel{9}6 \\ \underline{499} \\ 7{,}797 \end{array}$$

3.108
$$\begin{array}{r} 2\;1\;8\,1 \\ \cancel{3}{,}6\cancel{9}2 \\ \underline{825} \\ 2{,}867 \end{array}$$

3.109
$$\begin{array}{r} 8\;11\,1 \\ \cancel{9}{,}2\cancel{6}5 \\ \underline{493} \\ 8{,}772 \end{array}$$

3.110
211_1
3,254
1,682
1,572

3.111
4,856
2,731
2,125

3.112
413_1
5,468
1,692
3,776

3.113
6,626
5,416
1,210

3.114
410_1
5,151
2,361
2,790

3.115
$7\,15_1$
8,265
4,318
3,947

3.116
511_1
62,563
4,831
57,732

3.117
3_1
59,438
7,251
52,187

3.118
7_1
69,483
7,325
62,158

3.119
$3\,1\,5_1$
94,863
3,948
90,915

3.120
412_1
82,531
2,345
80,186

3.121
2_1
49,632
9,123
40,509

3.122
6111215_1
72,365
43,596
28,769

3.123
4111512_1
52,634
16,798
35,836

3.124
$511_1\,2_1$
62,436
23,918
38,518

3.125
8_1
92,548
76,514
16,034

3.126 Rounding up:
a. 70
− 50
20

Rounding down:
a. 60
− 40
20

68
− 45
b. 23

3.127 Rounding up:
a. 80
− 50
30

Rounding down:
a. 70
− 40
30

$6\,1$
76
− 48
b. 28

3.128 Rounding up:
a. 60
− 30
30

Rounding down:
a. 50
− 20
30

58
− 22
b. 36

3.129 Rounding up:
a. 100
− 30
70

Rounding down:
a. 90
− 20
70

$8\,1$
95
− 27
b. 68

3.130   Rounding up: Rounding down:
    a. 100   a. 90
$$\begin{array}{r} 100 \\ -\ 50 \\ \hline 50 \end{array} \qquad \begin{array}{r} 90 \\ -\ 40 \\ \hline 50 \end{array}$$

$$\begin{array}{r} ^{8}\!\!\!\!\!\not{9}\,^{1}\!3 \\ -\ 45 \\ \hline \end{array}$$
    b. 48

3.131   Rounding up: Rounding down:
    a. 80   a. 70
$$\begin{array}{r} 80 \\ -\ 70 \\ \hline 10 \end{array} \qquad \begin{array}{r} 70 \\ -\ 60 \\ \hline 10 \end{array}$$

$$\begin{array}{r} 78 \\ -\ 62 \\ \hline \end{array}$$
    b. 16

3.132   Rounding up: Rounding down:
    a. 90   a. 80
$$\begin{array}{r} 90 \\ -\ 50 \\ \hline 40 \end{array} \qquad \begin{array}{r} 80 \\ -\ 40 \\ \hline 40 \end{array}$$

$$\begin{array}{r} ^{7}\!\!\!\!\not{8}\,^{1}\!5 \\ -\ 48 \\ \hline \end{array}$$
    b. 37

3.133   Rounding up: Rounding down:
    a. 60   a. 50
$$\begin{array}{r} 60 \\ -\ 30 \\ \hline 30 \end{array} \qquad \begin{array}{r} 50 \\ -\ 20 \\ \hline 30 \end{array}$$

$$\begin{array}{r} ^{4}\!\!\!\!\not{5}\,^{1}\!6 \\ -\ 29 \\ \hline \end{array}$$
    b. 27

3.134   Rounding up: Rounding down:
    a. 400   a. 300
$$\begin{array}{r} 400 \\ -\ 200 \\ \hline 200 \end{array} \qquad \begin{array}{r} 300 \\ -\ 100 \\ \hline 200 \end{array}$$

$$\begin{array}{r} 362 \\ -\ 152 \\ \hline \end{array}$$
    b. 210

3.135   Rounding up: Rounding down:
    a. 600   a. 500
$$\begin{array}{r} 600 \\ -\ 300 \\ \hline 300 \end{array} \qquad \begin{array}{r} 500 \\ -\ 200 \\ \hline 300 \end{array}$$

$$\begin{array}{r} 596 \\ -\ 283 \\ \hline \end{array}$$
    b. 313

3.136   Rounding up: Rounding down:
    a. 900   a. 800
$$\begin{array}{r} 900 \\ -\ 500 \\ \hline 400 \end{array} \qquad \begin{array}{r} 800 \\ -\ 400 \\ \hline 400 \end{array}$$

$$\begin{array}{r} ^{7}\!\!\!\!\not{8}\,^{18}\!\!\!\!\not{9}\,^{1}\!5 \\ -\ 496 \\ \hline \end{array}$$
    b. 399

3.137   Rounding up: Rounding down:
    a. 900   a. 800
$$\begin{array}{r} 900 \\ -\ 700 \\ \hline 200 \end{array} \qquad \begin{array}{r} 800 \\ -\ 600 \\ \hline 200 \end{array}$$

$$\begin{array}{r} 823 \\ -\ 601 \\ \hline \end{array}$$
    b. 222

IV.   SECTION FOUR

4.1   6, 8, 10: The pattern is to add 2 each time.
    a. $10 + 2 = 12$
    b. $12 + 2 = 14$
    c. $14 + 2 = 16$
    d. $16 + 2 = 18$
    e. $18 + 2 = 20$

4.2   9, 12, 15: The pattern is to add 3 each time.
    a. $15 + 3 = 18$
    b. $18 + 3 = 21$
    c. $21 + 3 = 24$
    d. $24 + 3 = 27$
    e. $27 + 3 = 30$

4.3   10, 8, 6: The pattern is to subtract 2 each time:
    a. $6 - 2 = 4$
    b. $4 - 2 = 2$
    c. $2 - 2 = 0$

4.4   20, 19, 17, 14: 1, 2, and 3 have been subtracted; subtract increasing consecutive numbers 4 and 5.
    a. $14 - 4 = 10$
    b. $10 - 5 = 5$

4.5 1, 3, 6, 10: The pattern is to add 2, add 3, and add 4; therefore, add increasing consecutive numbers 5, 6, 7, and 8.
 a. $10 + 5 = 15$
 b. $15 + 6 = 21$
 c. $21 + 7 = 28$
 d. $28 + 8 = 36$

4.6 1, 2, 4, 7: The pattern is to add 1, add 2, and add 3; therefore, add increasing consecutive numbers 4, 5, 6, and 7.
 a. $7 + 4 = 11$
 b. $11 + 5 = 16$
 c. $16 + 6 = 22$
 d. $22 + 7 = 29$

4.7 49, 1, 48, 2: Two patterns are used. The first one is to subtract 1 ($49 - 1 = 48$). The second one is to add 1 ($1 + 1 = 2$).
 a. $48 - 1 = 47$
 b. $2 + 1 = 3$
 c. $47 - 1 = 46$
 d. $3 + 1 = 4$

4.8 99, 93, 88: The pattern is to subtract 6 and then subtract 5, or to subtract decreasing consecutive numbers 4, 3, 2, and 1.
 a. $88 - 4 = 84$
 b. $84 - 3 = 81$
 c. $81 - 2 = 79$
 d. $79 - 1 = 78$

4.9 $a + 22 = b$
 a. $2 + 22 = b$
 $24 = b$
 b. $3 + 22 = b$
 $25 = b$
 c. $5 + 22 = b$
 $27 = b$
 d. $8 + 22 = b$
 $30 = b$
 e. $10 + 22 = b$
 $32 = b$
 f. $12 + 22 = b$
 $34 = b$

4.9 cont.
The table for $a + 22 = b$ is

a	2	3	5	8	10	12
$b.$	a. 24	b. 25	c. 27	d. 30	e. 32	f. 34

4.10 $36 + a = b$
 a. $36 + 0 = b$
 $36 = b$
 b. $36 + 1 = b$
 $37 = b$
 c. $36 + 8 = b$
 $44 = b$
 d. $36 + 10 = b$
 $46 = b$
 e. $36 + 15 = b$
 $51 = b$
 f. $36 + 20 = b$
 $56 = b$

The table for $36 + a = b$ is

a	0	1	8	10	15	20
$b.$	a. 36	b. 37	c. 44	d. 46	e. 51	f. 56

4.11 $50 - a = b$
 a. $50 - 10 = b$
 $40 = b$
 b. $50 - 15 = b$
 $35 = b$
 c. $50 - a = 30$
 $a = 20$
 d. $50 - a = 50$
 $a = 0$
 e. $50 - 6 = b$
 $44 = b$
 f. $50 - a = 10$
 $a = 40$

The table for $50 - a = b$ is

a	10	15	c. 20	d. 0	6	f. 40
$b.$	a. 40	b. 35	30	50	e. 44	10

4.12 $a - b = 20$
 a. $50 - b = 20$
 $b = 30$
 b. $20 - b = 20$
 $b = 0$
 c. $70 - b = 20$
 $b = 50$
 d. $a - 60 = 20$
 $a = 80$
 e. $a - 100 = 20$
 $a = 120$

4.12 cont.

  f. $a - 10 = 20$

    $a = 30$

  The table for $a - b = 20$ is

a	50	20	70	d. 80	e. 120	f. 30
$b.$	a. 30	b. 0	c. 50	60	100	10

4.13 $a + b = 10$

  a. $2 + b = 10$

    $b = 8$

  The number pair is (2, 8).

  b. $3 + b = 10$

    $b = 7$

  The number pair is (3, 7).

  c. $a + 5 = 10$

    $a = 5$

  The number pair is (5, 5).

4.14 $a - b = 20$

  a. $25 - b = 20$

    $b = 5$

  The number pair is (25, 5).

  b. $35 - b = 20$

    $b = 15$

  The number pair is (35, 15).

  c. $52 - b = 20$

    $b = 32$

  The number pair is (52, 32).

4.15 $x + 15 = y$

  a. $7 + 15 = y$

    $22 = y$

  The number pair is (7, 22).

  b. $10 + 15 = y$

    $25 = y$

  The number pair is (10, 25).

  c. $x + 15 = 30$

    $x = 15$

  The number pair is (15, 30).

  d. $x + 15 = 35$

    $x = 20$

  The number pair is (20, 35).

4.16 $r + 5 = s$

  a. $2 + 5 = s$

    $7 = s$

  The number pair is (2, 7).

  b. $8 + 5 = s$

    $13 = s$

  The number pair is (8, 13).

  c. $r + 5 = 20$

    $r = 15$

  The number pair is (15, 20).

  d. $r + 5 = 15$

    $r = 10$

  The number pair is (10, 15).

4.17   {(0, 16), (1, 17), (2, 18),

  (3, b.___), (4, c.___)}

  a. Each number pair has a difference of 16. Since the first number is smaller than the second number in each number pair, the first number plus 16 equals the second number. Therefore, the sentence is $a + 16 = b$ (or $b - 16 = a$, or $b - a = 16$)

  b. $3 + 16 = b$

    $19 = b$

  The number pair is (3, 19).

  c. $4 + 16 = b$

    $20 = b$

  The number pair is (4, 20).

4.18   {(1, 3), (2, 4), (3, 5),

  (4, b. ___), (5, c.___)}

  a. Each number pair has a difference of 2. Since the first number is smaller than the second number in each number pair, the first number plus 2 equals the second number. Therefore, the sentence is $a + 2 = b$ (or $b - a = 2$, or $b - 2 = a$)

  b. $4 + 2 = b$

    $6 = b$

  The number pair is (4, 6).

  c. $5 + 2 = b$

    $7 = b$

  The number pair is (5, 7).

4.19   {(38, 4), (40, 6), (50, 16),
(60, b.___), (70, c.___)}
a. Each number pair has a difference of 34. Since the
first number is larger than
the second number in each
number pair, the first number
minus 34 equals the second
number or the first number
minus the second number
equals 34.
Therefore, the sentence is
$a - 34 = b$ (or $a - b = 34$, or $b + 34 = a$)
b. $60 - 34 = b$
$26 = b$
The number pair is (60, 26).
c. $70 - 34 = b$
$36 = b$
The number pair is (70, 36).

4.20   {(10, 5), (25, 20), (40, 35),
(5, b.___), (22, c. ___)}
a. Each number pair has a
difference of 5. Since the
first number is larger than
the second number in each
number pair, the first
number minus 5 equals the
second number or the first
number minus the second
number equals 5. Therefore,
the number sentence is
$a - 5 = b$ (or $a - b = 5$, or $b + 5 = a$)
b. $5 - 5 = b$
$0 = b$
The number pair is (5, 0).
c. $22 - 5 = b$
$17 = b$
The number pair is (22, 17).

4.21 16:   a. $15 + n = 16$
$n = 1$
b. $32 - n = 16$
$n = 16$
c. $22 - n = 16$
$n = 6$
d. $5 + n = 16$
$n = 11$

4.22 23:   a. $3 + n = 23$
$n = 20$
b. $8 + n = 23$
$n = 15$
c. $43 - n = 23$
$n = 20$
d. $30 - n = 23$
$n = 7$

4.23 100: a. $97 + n = 100$
$n = 3$
b. $103 - n = 100$
$n = 3$
c. $45 + n = 100$
$n = 55$
d. $55 + n = 100$
$n = 45$

4.24 66: a. $30 + n = 66$
$n = 36$
b. $70 - n = 66$
$n = 4$
c. $99 - n = 66$
$n = 33$
d. $22 + n = 66$
$n = 44$

4.25 37   a. $25 + n = 37$
$n = 12$
b. $47 - n = 37$
$n = 10$
c. $55 - n = 37$
$n = 18$
d. $17 + n = 37$
$n = 20$

4.26   a. 3

$$\begin{array}{cc} 25¢ & 80¢ \\ \times\ 3 & -75¢ \\ \hline 75¢ & 5¢ \end{array}$$

b. 0
c. 1
d. 0

4.27   a. 1

$$\begin{array}{r} 42¢ \\ -25¢ \\ \hline 17¢ \end{array}$$

b. 1

$$\begin{array}{r} 17¢ \\ -10¢ \\ \hline 7¢ \end{array}$$

c. 1

$$\begin{array}{r} 7¢ \\ -5¢ \\ \hline 2¢ \end{array}$$

d. 2

4.28   a. 1

$$\begin{array}{r} 27¢ \\ -\ 25¢ \\ \hline 2¢ \end{array}$$

b. 0
c. 0
d. 2

4.29   a. 0
b. 2

$$\begin{array}{r} 23¢ \\ -\ 20¢ \\ \hline 3¢ \end{array}$$

c. 0
d. 3

4.30   a. 2

$$\begin{array}{r} 65¢ \\ -\ 50¢ \\ \hline 15¢ \end{array}$$

b. 1

$$\begin{array}{r} 15¢ \\ -\ 10¢ \\ \hline 5¢ \end{array}$$

c. 1

$$\begin{array}{r} 5¢ \\ -\ 5¢ \\ \hline 0¢ \end{array}$$

d. 0

4.31   a. 5

$$\begin{array}{r} \$1.32 \\ -\ 1.25 \\ \hline \$0.07 \end{array}$$

b. 0
c. 1

$$\begin{array}{r} 7¢ \\ -\ 5¢ \\ \hline 2¢ \end{array}$$

d. 2

4.32   a. 10

$$\begin{array}{r} \$\,2.73 \\ -\ 2.50 \\ \hline \$\,0.23 \end{array}$$

b. 2

$$\begin{array}{r} 23¢ \\ -\ 20¢ \\ \hline 3¢ \end{array}$$

c. 0
d. 3

4.33   a. 17

$$\begin{array}{r} \$4.48 \\ -\ 4.25 \\ \hline \$0.23 \end{array}$$

b. 2

$$\begin{array}{r} 23¢ \\ -\ 20¢ \\ \hline 3¢ \end{array}$$

c. 0
d. 3

4.34   a. 7

$$\begin{array}{r} \$1.92 \\ -\ 1.75 \\ \hline \$0.17 \end{array}$$

b. 1

$$\begin{array}{r} 17¢ \\ -\ 10¢ \\ \hline 7¢ \end{array}$$

c. 1

$$\begin{array}{r} 7¢ \\ -\ 5¢ \\ \hline 2¢ \end{array}$$

d. 2

4.35   a. 15

$$\begin{array}{r} \$3.97 \\ -\ 3.75 \\ \hline \$0.22 \end{array}$$

b. 2

$$\begin{array}{r} 22¢ \\ -\ 20¢ \\ \hline 2¢ \end{array}$$

c. 0
d. 2

4.36   a. 41

$$\begin{array}{r} \$10.31 \\ -\ 10.25 \\ \hline \$0.06 \end{array}$$

b. 0
c. 1

$$\begin{array}{r} 6¢ \\ -\ 5¢ \\ \hline 1¢ \end{array}$$

d. 1

4.37   a. 4

$$\begin{array}{r} \$1.07 \\ -\ 1.00 \\ \hline \$0.07 \end{array}$$

        b. 0

        c. 1

$$\begin{array}{r} 7¢ \\ -\ 5¢ \\ \hline 2¢ \end{array}$$

        d. 2

4.38   a. 3
           67¢ + 3¢ = 70¢
        b. 1
           70¢ + 5¢ = 75¢
        c. 0
        d. 1
           75¢ + 25¢ = $1.00

4.39   a. 3
           72¢ + 3¢ = 75¢
        b. 0
        c. 0
        d. 1
           75¢ + 25¢ = $1.00

4.40   a. 0
        b. 1
           85¢ + 5¢ = 90¢
        c. 1
           90¢ + 10¢ = $1.00
        d. 0

4.41   a. 2
           23¢ + 2¢ = 25¢
        b. 0
        c. 0
        d. 3
           25¢ + 75¢ = $1.00

4.42   a. 0
        b. 1
           35¢ + 5¢ = 40¢
        c. 1
           40¢ + 10¢ = 50¢
        d. 2
           50¢ + 50¢ = $1.00

4.43   a. 3
           47¢ + 3¢ = 50
        b. 0
        c. 0
        d. 2
           50¢ + 50¢ = $1.00

4.44   a. 2
           58¢ + 2¢ = 60¢
        b. 1
           60¢ + 5¢ = 65¢
        c. 1
           65¢ + 10¢ = 75¢
        d. 1
           75¢ + 25¢ = $1.00

4.45   a. 4
           26¢ + 4¢ = 30¢
        b. 0
        c. 2
           30¢ + 20¢ = 50¢
        d. 2
           50¢ + 50¢ = $1.00

4.46   The word "total" means to add.

$$\begin{array}{r} \overset{1\ 2}{1}05\ \text{lbs.} \\ 135\ \text{lbs.} \\ 145\ \text{lbs.} \\ \underline{127\ \text{lbs.}} \\ 512\ \text{lbs.}\ \ \text{total} \end{array}$$

4.47   The word "left" is a signal to subtract.

$$\begin{array}{r} \overset{1\ 91}{2{,}\cancel{0}15} \\ -\ 1{,}025 \\ \hline 990\ \ \text{tourists stayed} \end{array}$$

4.48   The word "total" is a signal to add.

$$\begin{array}{r} \$625 \\ +\ \ 63 \\ \hline \$688\ \ \text{total cost} \end{array}$$

4.49   The word "altogether" is a clue to add.

$$\begin{array}{r} \$6 \\ 4 \\ +\ 5 \\ \hline \$15 \end{array}$$

4.50   The word "difference" means subtract.

$$\begin{array}{r} \overset{8\ 151}{\cancel{9}\cancel{6}3} \\ -\ 485 \\ \hline 478 \end{array}$$

4.51   The words "How much more" are a
       clue to subtract.

$$\overset{61}{\$7\!\!\!/5}$$
$$\underline{-37}$$
$$\$38 \text{ more needed to save}$$

4.52   The word "total" is a clue to
       add.

$$\$5,260$$
$$\underline{+\ 210}$$
$$\$5,470 \text{ total paid}$$

4.53
$$\overset{61}{19\!\!\!/75}$$
$$\underline{-1947}$$
$$28 \text{ yrs.}$$

4.54
$$\overset{0\ 91}{\$1.\!\!\!/00}$$
$$\underline{-0.67}$$
$$\$0.33 \text{ change}$$

4.55   The word "sum" means to add.

$$\overset{1\ 1}{625}$$
$$45$$
$$32$$
$$\underline{+\ 64}$$
$$766$$

4.56
$$\overset{1}{32}\ \cent$$
$$\underline{\times 5}$$
$$160\cent = \$1.60$$

4.57
$$\$4,587$$
$$\underline{-\ 4,332}$$
$$\$\ 255 \text{ saved}$$

4.58

4.58 cont.

HORIZONTAL

1.   The digit that has a
     numerical value of
     one is 1.
2.   Two thousand, seven
     hundred fifty-six
     = 2,000 + 700 + 56
     = 2,756.
6.   57 = 60 rounded to
     the nearest ten.
8. $\underline{26}$ tens are contained
     $10\overline{)260}$ in 260.
$$\underline{20}$$
$$60$$
$$\underline{60}$$
$$0$$

10. Seventy-four thousand,
     nine hundred four
     = 74,000 + 900 + 4
     = 74,904.
12. 33 rounded to the nearest
     ten is 30.
13. The smallest even digit
     is 2.
14. The tens' digit is the
     second digit to the left
     of a whole number and the
     units' digit is the digit
     on the far right of a
     whole number. Therefore,
     the number is 52.
15. Four thousand, eight
     hundred ninety-six
     = 4,000 + 800 + 96
     = 4,896.
17. The number contains
     three digits (from the
     puzzle), all of which
     are 3's; the number is
     333.
19. The smallest odd digit
     is 1.
20. Thirty thousand, fifty-
     two = 30,000 + 52
     = 30,052.
22. $\underline{32}$ hundreds are
     $100\overline{)3,200}$ contained in
     $\underline{3\ 00}$ 3,200.
$$200$$
$$\underline{200}$$
$$0$$

4.58 cont.

23. The tens' digit is 9 and the ones' digit is 2. The number is 92.

24. One hundred two thousand, one hundred five = 102,000 + 105 = 102,105.

27. The next consecutive digit larger than eight (8) is 9.

28. 349 = 350 rounded to the nearest ten.

30.

$$1,000 \overline{)1,000,000}$$

1,000 thousands are necessary to make a million.

```
      1,000
1,000)1,000,000
      1000
       0000
       0000
          0
```

31. The next consecutive odd number larger than seven (7) is 9.

32.

```
    40
10)400
   40
   00
   00
    0
```

40 tens are contained in 400.

33. The next consecutive odd integer larger than three (3) is 5.

34. The units' digit is the digit on the far right of a whole number and the tens' digit is one digit to the left of the units' digit. Therefore, the number is 83.

35. Thirty-seven thousand, four = 37,000 + 4 = 37,004.

38. 8,499 = 8,000 rounded to the nearest thousand. With the last two digits dropped, 8,000 becomes 80.

39. 3,792,904 = 4,000,000 rounded to the nearest million.

40. The digit having the largest numerical value is 9.

4.58 cont.

VERTICAL

1. Beginning with 1, four consecutive digits are 1, 2, 3, 4.

2. The numbers whose numerical values are less than 3 are 1 and 2. The only even number whose numerical value is less than 3 is 2.

3. The units' digit and the tens' digit are both 7; the number is 77.

4. Five hundred forty-five thousand, three hundred twenty = 545,000 + 320 = 545,320.

5.
```
 6,335
   588
 6,923
```

6.
```
    64
10)640
   60
   40
   40
    0
```

64 tens are contained in 640.

7. Zero = 0.

9. Sixty thousand, eight hundred thirty-two = 60,000 + 832 = 60,832.

11. The number symbol when written alone that has no numerical value is 0.

13. 256 = 260 rounded to the nearest ten.

16. Tens   Units   The number
     9   0   is 90.

18. The next odd number larger than one (1) is 3.

19. One million, two hundred fifty thousand, three hundred nine = 1,000,000 + 250,000 + 309 = 1,250,309.

21. 505,499 = 510,000 rounded to the nearest ten thousand.

22. Thirty thousand, eighty-eight = 30,000 + 88 = 30,088.

4.58 cont.

23. From Number 23 Horizontal in the puzzle, the first number is 9. Since all the digits are the same in this problem, the number is 999.

25.
$$100\overline{)21,500} = 215$$
hundreds are contained in 21,500.

$$
\begin{array}{r}
215 \\
100\overline{)21,500} \\
\underline{200} \\
150 \\
\underline{100} \\
500 \\
\underline{500} \\
0
\end{array}
$$

26. The numbers that are larger than one billion (1,000,000,000) but smaller than ten billion are (1,000,000,001 through 9,999,999,999. The number of digits in each of these numbers is 10.

28. The next consecutive odd number larger than one (1) is 3.

29. 5,449 = 5,400 rounded to the nearest hundred.

35.
$$
\begin{array}{r}
30 \\
100,000\overline{)3,000,000} \\
\underline{300000} \\
00 \\
\underline{00} \\
0
\end{array}
$$
hundred thousands are contained in 3,000,000.

36. 69,527 = 70,000 rounded to the nearest thousand. Dropping the last three places in 70,000 results in 70.

37. The largest even number that is less than forty-two (42) is 40.

39. The next consecutive even number larger than two (2) is 4.

Note: Puzzles 4.59 and 4.60 have been deleted and Problems 4.61 through 4.70 renumbered to 4.59 through 4.68.

Note: The answers to Problems 4.59 through 4.68 are worked out for students who do not have access to calculators.

4.59

Enter	Display	Comments
C	O	Clears calc.
592	592	First term
+	592	Add
483	483	Second term
–	1,075	Subtract
721	721	Third term
=	354	Answer

$$
\begin{array}{r}
\overset{1}{592} \\
+\ 483 \\
\hline
1,075
\end{array}
\qquad
\begin{array}{r}
\overset{0}{\cancel{1}}\overset{1}{,075} \\
-\ 721 \\
\hline
354
\end{array}
$$

4.60

Enter	Display	Comments
C	O	Clears calc.
622	622	First term
+	622	Add
983	983	Second term
–	1,605	Subtract
654	654	Third term
=	951	Answer

$$
\begin{array}{r}
\overset{1}{622} \\
+\ 983 \\
\hline
1,605
\end{array}
\qquad
\begin{array}{r}
\overset{015}{\cancel{1}}\overset{1}{,605} \\
-\ 654 \\
\hline
951
\end{array}
$$

4.61

Enter	Display	Comments
C	O	Clears calc.
521	521	First term
–	521	Subtract
263	263	Second term
+	258	Add
496	496	Third term
=	754	Answer

$$
\begin{array}{r}
\overset{4\ 11}{\cancel{5}\cancel{2}1} \\
-\ 263 \\
\hline
258
\end{array}
\qquad
\begin{array}{r}
\overset{11}{258} \\
+\ 496 \\
\hline
754
\end{array}
$$

4.62

Enter	Display	Comments
C	0	Clears calc.
826	826	First term
+	826	Add
753	753	Second term
–	1,579	Subtract
1,056	1,056	Third term
=	523	Answer

$$\begin{array}{r} 826 \\ +\ 753 \\ \hline 1,579 \end{array} \qquad \begin{array}{r} 1,579 \\ -\ 1,056 \\ \hline 523 \end{array}$$

4.63

Enter	Display	Comments
C	0	Clears calc.
762	762	First term
–	762	Subtract
521	521	Second term
–	241	Subtract
43	43	Third term
=	198	Answer

$$\begin{array}{r} 762 \\ -\ 521 \\ \hline 241 \end{array} \qquad \begin{array}{r} {}^{113}\llap{2}41 \\ -\ 43 \\ \hline 198 \end{array}$$

4.64

Enter	Display	Comments
C	0	Clears calc.
6,543	6,543	First term
+	6,543	Add
7,263	7,263	Second term
–	13,806	Subtract
8,514	8,514	Third term
–	5,292	Subtract
1,000	1,000	Fourth term
=	4,292	Answer

$$\begin{array}{r} 1 \\ 6,543 \\ +\ 7,263 \\ \hline 13,806 \end{array} \qquad \begin{array}{r} 71 \\ 13,806 \\ -\ 8,514 \\ \hline 5,292 \end{array} \qquad \begin{array}{r} 5,292 \\ -\ 1,000 \\ \hline 4,292 \end{array}$$

4.65

Enter	Display	Comments
C	0	Clears calc.
5,544	5,544	First term
+	5,544	Add
3,548	3,548	Second term
–	9,092	Subtract
1,265	1,265	Third term
–	7,827	Subtract
5,124	5,124	Fourth term
=	2,703	Answer

$$\begin{array}{r} 1\ \ 1 \\ 5,544 \\ +\ 3,548 \\ \hline 9,092 \end{array} \qquad \begin{array}{r} 8\ _1 8_1 \\ 9,092 \\ -1,265 \\ \hline 7,827 \end{array}$$

$$\begin{array}{r} 7,827 \\ -\ 5,124 \\ \hline 2,703 \end{array}$$

4.66

Enter	Display	Comments
C	0	Clears calc.
6,543	6,543	First term
–	6,543	Subtract
1,255	1,255	Second term
+	5,288	Add
3,150	3,150	Third term
–	8,438	Subtract
2,010	2,010	Fourth term
=	6,428	Answer

$$\begin{array}{r} {}^{413}\llap{6,5}43 \\ -\ 1,255 \\ \hline 5,288 \end{array} \qquad \begin{array}{r} 1 \\ 5,288 \\ +\ 3,150 \\ \hline 8,438 \end{array}$$

$$\begin{array}{r} 8,438 \\ -2,010 \\ \hline 6,428 \end{array}$$

4.67

Enter	Display	Comments
C	O	Clears calc.
25,623	25,623	First term
+	25,623	Add
10,543	10,543	Second term
–	36,166	Subtract
12,060	12,060	Third term
=	24,106	Subtract
10,112	10,112	Fourth term
=	13,994	Answer

$$\begin{array}{r} \overset{1}{}25{,}623 \\ +\ 10{,}543 \\ \hline 36{,}166 \end{array} \qquad \begin{array}{r} 36{,}166 \\ -\ 12{,}060 \\ \hline 24{,}106 \end{array}$$

$$\begin{array}{r} \overset{3\ 1\ 01}{24{,}106} \\ -\ 10{,}112 \\ \hline 13{,}994 \end{array}$$

4.68

Enter	Display	Comments
C	O	Clears calc.
253,541	253,541	First term
+	253,541	Add
55,698	55,698	Second term
–	309,239	Subtract
22,655	22,655	Third term
–	286,584	Subtract
10,500	10,500	Fourth term
=	276,084	Answer

$$\begin{array}{r} \overset{1\ \ 1\ \ 1}{253{,}541} \\ +\ 55{,}698 \\ \hline 309{,}239 \end{array} \qquad \begin{array}{r} \overset{2_1\ 811_1}{309{,}239} \\ -\ 22{,}655 \\ \hline 286{,}584 \end{array}$$

$$\begin{array}{r} 286{,}584 \\ -\ 10{,}500 \\ \hline 276{,}084 \end{array}$$

I. SECTION ONE

1.1

X	3	0	2	6	4	9	7	1	8	5
1	3	0	2	6	4	9	7	1	8	5
5	15	0	10	30	20	45	35	5	40	25
3	9	0	6	18	12	27	21	3	24	15
9	27	0	18	54	36	81	63	9	72	45
8	24	0	16	48	32	72	56	8	64	40
6	18	0	12	36	24	54	42	6	48	30
2	6	0	4	12	8	18	14	2	16	10
4	12	0	8	24	16	36	28	4	32	20
7	21	0	14	42	28	63	49	7	56	35
0	0	0	0	0	0	0	0	0	0	0

1.2 $6 \times 3 = 18$

1.3 $6 \times 4 = 24$

1.4 $6 \times 6 = 36$

1.5 $6 \times 7 = 42$

1.6 $6 \times 8 = 48$

1.7 $6 \times 9 = 54$

1.8 $7 \times 3 = 21$

1.9 $7 \times 4 = 28$

1.10 $7 \times 6 = 42$

1.11 $7 \times 7 = 49$

1.12 $7 \times 8 = 56$

1.13 $7 \times 9 = 63$

1.14 $8 \times 7 = 56$

1.15 $8 \times 8 = 64$

1.16 $8 \times 9 = 72$

1.17 $9 \times 7 = 63$

1.18 $9 \times 8 = 72$

1.19 $9 \times 9 = 81$

1.20
$$\begin{array}{r} 2 \\ 75 \\ \underline{5} \\ 375 \end{array}$$

1.21
$$\begin{array}{r} 42 \\ \underline{4} \\ 168 \end{array}$$

1.22
$$\begin{array}{r} 3 \\ 85 \\ \underline{7} \\ 595 \end{array}$$

1.23
$$\begin{array}{r} 2 \\ 23 \\ \underline{8} \\ 184 \end{array}$$

1.24
$$\begin{array}{r} 5 \\ 99 \\ \underline{6} \\ 594 \end{array}$$

1.25
$$\begin{array}{r} 3 \\ 64 \\ \underline{9} \\ 576 \end{array}$$

1.26
$$\begin{array}{r} 2 \\ 88 \\ \underline{3} \\ 264 \end{array}$$

1.27
$$\begin{array}{r} 8 \\ 79 \\ \underline{9} \\ 711 \end{array}$$

1.28
$$\begin{array}{r} 54 \\ \underline{21} \\ 54 \\ 1 \\ 108 \\ \underline{} \\ 1,134 \end{array}$$

1.29
$$
\begin{array}{r}
2 \\
1 \\
78 \\
\underline{32} \\
156 \\
\underline{234} \\
2{,}496
\end{array}
$$

1.30
$$
\begin{array}{r}
6 \\
5 \\
49 \\
\underline{76} \\
^{1}294 \\
\underline{343} \\
3{,}724
\end{array}
$$

1.31
$$
\begin{array}{r}
1 \\
0 \\
94 \\
\underline{42} \\
^{1}188 \\
\underline{376} \\
3{,}948
\end{array}
$$

1.32
$$
\begin{array}{r}
2 \\
2 \\
45 \\
\underline{54} \\
^{1}180 \\
\underline{225} \\
2{,}430
\end{array}
$$

1.33
$$
\begin{array}{r}
1 \\
72 \\
\underline{29} \\
^{1}648 \\
\underline{^{1}144} \\
2{,}088
\end{array}
$$

1.34
$$
\begin{array}{r}
7 \\
6 \\
38 \\
\underline{98} \\
304 \\
\underline{342} \\
3{,}724
\end{array}
$$

1.35
$$
\begin{array}{r}
5 \\
7 \\
79 \\
\underline{68} \\
^{1}632 \\
\underline{474} \\
5{,}372
\end{array}
$$

1.36
$$
\begin{array}{r}
2 \\
1 \\
65 \\
\underline{43} \\
195 \\
\underline{260} \\
2{,}795
\end{array}
$$

1.37
$$
\begin{array}{r}
3 \\
2 \\
89 \\
\underline{43} \\
^{1}267 \\
\underline{356} \\
3{,}827
\end{array}
$$

1.38
$$
\begin{array}{r}
60 \\
\underline{87} \\
420 \\
^{1}480 \\
\underline{} \\
5{,}220
\end{array}
$$

1.39
$$
\begin{array}{r}
2 \\
1 \\
95 \\
\underline{53} \\
^{1}285 \\
\underline{^{1}475} \\
5{,}035
\end{array}
$$

1.40
$$
\begin{array}{r}
3 \\
3 \\
56 \\
\underline{65} \\
^{1}280 \\
\underline{336} \\
3{,}640
\end{array}
$$

1.41
$$
\begin{array}{r}
1 \\
0 \\
83 \\
\underline{40} \\
00 \\
\underline{332} \\
3{,}320
\end{array}
$$

1.42
$$
\begin{array}{r}
8 \\
7 \\
49 \\
\underline{98} \\
^{1}392 \\
\underline{441} \\
4{,}802
\end{array}
$$

1.43
$$
\begin{array}{r}
90 \\
\underline{79} \\
^{1}810 \\
\underline{630} \\
7{,}110
\end{array}
$$

1.44
$$
\begin{array}{r}
2 \\
0 \\
47 \\
\underline{31} \\
47 \\
\underline{141} \\
1{,}457
\end{array}
$$

1.45
$$
\begin{array}{r}
\scriptstyle 3 \\
\scriptstyle 1 \\
79 \\
\underline{{}^1\,42} \\
158 \\
\underline{316} \\
3{,}318
\end{array}
$$

1.46
$$
\begin{array}{r}
80 \\
\underline{53} \\
240 \\
\underline{400} \\
4{,}240
\end{array}
$$

1.47
$$
\begin{array}{r}
91 \\
\underline{{}^1\,64} \\
364 \\
\underline{546} \\
5{,}824
\end{array}
$$

1.48
$$
\begin{array}{r}
444 \\
\underline{{}^1\,22} \\
{}^1888 \\
\underline{{}^1888} \\
9{,}768
\end{array}
$$

1.49
$$
\begin{array}{r}
\scriptstyle 31 \\
\scriptstyle 10 \\
452 \\
\underline{72} \\
904 \\
\underline{3^1164} \\
32{,}544
\end{array}
$$

1.50
$$
\begin{array}{r}
\scriptstyle 67 \\
\scriptstyle 78 \\
379 \\
\underline{89} \\
3411 \\
\underline{3032} \\
33{,}731
\end{array}
$$

1.51
$$
\begin{array}{r}
\scriptstyle 51 \\
\scriptstyle 41 \\
482 \\
\underline{76} \\
{}^12^1892 \\
\underline{3374} \\
36{,}632
\end{array}
$$

1.52
$$
\begin{array}{r}
\scriptstyle 76 \\
\scriptstyle 87 \\
698 \\
\underline{89} \\
{}^16^1282 \\
\underline{{}^15584} \\
62{,}122
\end{array}
$$

1.53
$$
\begin{array}{r}
\scriptstyle 64 \\
\scriptstyle 53 \\
896 \\
\underline{76} \\
{}^15376 \\
\underline{6272} \\
68{,}096
\end{array}
$$

1.54
$$
\begin{array}{r}
\scriptstyle 43 \\
\scriptstyle 32 \\
754 \\
\underline{87} \\
5278 \\
\underline{6032} \\
65{,}598
\end{array}
$$

1.55
$$
\begin{array}{r}
\scriptstyle 46 \\
\scriptstyle 24 \\
547 \\
\underline{96} \\
3^1282 \\
\underline{{}^14923} \\
52{,}512
\end{array}
$$

1.56
$$
\begin{array}{r}
\scriptstyle 24 \\
\scriptstyle 35 \\
859 \\
\underline{56} \\
{}^15^1154 \\
\underline{4295} \\
48{,}104
\end{array}
$$

1.57
$$
\begin{array}{r}
810 \\
\underline{90} \\
000 \\
\underline{7290} \\
72{,}900
\end{array}
$$

1.58
$$
\begin{array}{r}
\scriptstyle 66 \\
\scriptstyle 77 \\
789 \\
\underline{78} \\
6312 \\
\underline{{}^15523} \\
61{,}542
\end{array}
$$

1.59
$$
\begin{array}{r}
\scriptstyle 11 \\
\scriptstyle 11 \\
555 \\
\underline{33} \\
{}^11^1665 \\
\underline{1665} \\
18{,}315
\end{array}
$$

1.60
$$
\begin{array}{r}
\scriptstyle 52 \\
\scriptstyle 10 \\
563 \\
\underline{83} \\
1^1689 \\
\underline{4504} \\
46{,}729
\end{array}
$$

1.61
$$\begin{array}{r} 7 \\ 7 \\ 480 \\ 99 \\ \hline 4320 \\ 4320 \\ \hline 47{,}520 \end{array}$$

1.62
$$\begin{array}{r} 72 \\ 62 \\ 593 \\ 87 \\ \hline 4151 \\ ^14744 \\ \hline 51{,}591 \end{array}$$

1.63
$$\begin{array}{r} 8 \\ 709 \\ 90 \\ \hline 000 \\ 6381 \\ \hline 63{,}810 \end{array}$$

1.64
$$\begin{array}{r} 5 \\ 4 \\ 907 \\ 87 \\ \hline 6^1349 \\ 7256 \\ \hline 78{,}909 \end{array}$$

1.65
$$\begin{array}{r} 54 \\ 54 \\ 865 \\ 98 \\ \hline ^16920 \\ ^77785 \\ \hline 84{,}770 \end{array}$$

1.66
$$\begin{array}{r} 57 \\ 45 \\ 658 \\ 97 \\ \hline 4606 \\ ^15922 \\ \hline 63{,}826 \end{array}$$

1.67
$$\begin{array}{r} 14 \\ 415 \\ 19 \\ \hline 3735 \\ 415 \\ \hline 7{,}885 \end{array}$$

1.68
$$\begin{array}{r} 01 \\ 23 \\ 12 \\ 647 \\ 253 \\ \hline ^11941 \\ ^13235 \\ 1294 \\ \hline 163{,}691 \end{array}$$

1.69
$$\begin{array}{r} 12 \\ 14 \\ 01 \\ 839 \\ 352 \\ \hline ^21^1678 \\ 4195 \\ 2517 \\ \hline 295{,}328 \end{array}$$

1.70
$$\begin{array}{r} 42 \\ 63 \\ 32 \\ 874 \\ 695 \\ \hline ^14^1370 \\ ^77866 \\ ^15244 \\ \hline 607{,}430 \end{array}$$

1.71
$$\begin{array}{r} 55 \\ 33 \\ 66 \\ 478 \\ 748 \\ \hline ^13824 \\ ^11912 \\ 3346 \\ \hline 357{,}544 \end{array}$$

1.72
$$\begin{array}{r} 41 \\ 41 \\ 20 \\ 752 \\ 984 \\ \hline 3008 \\ 6016 \\ ^16768 \\ \hline 739{,}968 \end{array}$$

1.73
$$\begin{array}{r} 10 \\ 63 \\ 73 \\ 684 \\ 289 \\ \hline ^16156 \\ ^15472 \\ 1368 \\ \hline 197{,}676 \end{array}$$

1.74
$$\begin{array}{r} 43 \\ 43 \\ 43 \\ 486 \\ 555 \\ \hline 2430 \\ 2430 \\ 2430 \\ \hline 269{,}730 \end{array}$$

1.75
$$\begin{array}{r} 53 \\ 63 \\ 21 \\ 395 \\ 673 \\ \hline 1^1185 \\ ^12765 \\ 2370 \\ \hline 265{,}835 \end{array}$$

1.76
```
        23
        24
        12
       736
       684
  ²2¹944
  ¹5888
 ¹4416
 503,424
```

1.77
```
        21
        53
        32
       264
       486
  ¹1¹584
   2112
   1056
 128,304
```

1.78
```
        01
        14
        13
       426
       375
  ¹2130
  ¹2982
   1278
 159,750
```

1.79
```
       321
       301
       321
       000
       963
   96,621
```

1.80
```
       121
       011
       353
     2,475
     1,327
  ²1¹7325
    4950
  ¹7425
  ¹2475
 3,284,325
```

1.81
```
       541
       651
       000
       330
     4,982
     6,714
  ²1²9928
    4982
  ¹3²4874
  ¹29892
 33,449,148
```

1.82
```
        12
        34
        23
       758
       364
   3¹032
  ¹4548
   2274
 275,912
```

1.83
```
         1
         2
         1
       940
       463
  ¹2820
  ¹5640
  ¹3760
 435,220
```

1.84
```
        53
        00
        53
       985
       706
  ¹5910
   000
  6895
 695,410
```

1.85
```
        77
        44
        88
       589
       859
      5301
  ¹2945
  ¹4712
 505,951
```

1.86
```
        52
        52
        31
       863
       995
  ¹4315
  ¹7767
  ¹7767
 858,685
```

1.87
```
        21
        84
        84
       795
       399
  ¹7¹155
  ¹7155
  ¹2385
 317,205
```

1.88

$$54$$
$$54$$
$$54$$
$$597$$
$$666$$
$$^13^1582$$
^{13582}
$$3582$$
$$\overline{397,602}$$

1.89

$$64$$
$$74$$
$$32$$
$$496$$
$$784$$
$$^11^1984$$
^{13968}
$$3472$$
$$\overline{388,864}$$

1.90

$$34$$
$$46$$
$$23$$
$$847$$
$$795$$
^{14235}
^{17623}
^{15929}
$$\overline{673,365}$$

1.91

$$32$$
$$64$$
$$53$$
$$375$$
$$587$$
^{12625}
^{13000}
^{11875}
$$\overline{220,125}$$

1.92

$$12$$
$$25$$
$$24$$
$$537$$
$$486$$
^{13222}
^{14296}
$$2148$$
$$\overline{260,982}$$

1.93

$$1$$
$$0$$
$$0$$
$$432$$
$$412$$
$$864$$
^{1432}
$$1728$$
$$\overline{177,984}$$

1.94

$$111$$
$$232$$
$$121$$
$$464$$
$$3,586$$
$$2,438$$
$$^22^18^1688$$
^{110758}
$$14344$$
$$7172$$
$$\overline{8,742,668}$$

1.95

$$662$$
$$772$$
$$110$$
$$441$$
$$5,893$$
$$7,825$$
$$^22^19^1465$$
^{111786}
^{147144}
$$41251$$
$$\overline{46,112,725}$$

1.96 3, 6, 9, a. ___ , b. ___ ,
c. ___ , d. ___ , ...
Each member is a multiple
of 3.
3 x 1 = 3
3 x 2 = 6
3 x 3 = 9
The next four members are
a.   4 x 3 = 12
b.   5 x 3 = 15
c.   6 x 3 = 18
d.   7 x 3 = 21

1.97 1, 2, 4, a. ___ , b. ___ ,
c. ___ , d. ___ , ...
Multiplying each member by
2 results in the member
following it.
1 x 2 = 2
2 x 2 = 4
The next four members are
a.   4 x 2 = 8
b.   8 x 2 = 16
c.   16 x 2 = 32
d.   32 x 2 = 64

1.98   5, 10, 15, a. ___ , b. ___ ,
        c. ___ , d. ___ , ...
        Each member is a multiple of 5.
        5 x 1 = 5
        5 x 2 = 10
        5 x 3 = 15
        The next four members are
        a.   5 x 4 = 20
        b.   5 x 5 = 25
        c.   5 x 6 = 30
        d.   5 x 7 = 35

1.99   1, 4, 16, a. ___ , b. ___ ,
        c. ___ , d. ___ , ...
        Multiplying each member by 4
        results in the member
        following it.
        1 x 4 = 4
        4 x 4 = 16
        The next four members are
        a.   16 x 4 = 64
        b.   64 x 4 = 256
        c.   256 x 4 = 1,024
        d.   1,024 x 4 = 4,096

1.100   1, 7, 49, a. ___ , b. ___ ,
        c. ___ , d. ___ , ...
        Multiplying each member by 7
        results in the member
        following it.
        1 x 7 = 7
        7 x 7 = 49
        The next four members are
        a.   49 x 7 = 343
        b.   343 x 7 = 2,401
        c.   2,401 x 7 = 16,807
              16,807 x 7 = 117,649

1.101   24 x 38
        a.   Rounded: 20 x 40
        b.   Estimated Product:
              20 x 40 = 800

1.102   18 x $4.15
        a.   Rounded: 20 x $4
        b.   Estimated Product:
              20 x $4 = $80

1.103   250 x 442
        a.   Rounded: 300 x 400
        b.   Estimated Product:
              300 x 400 = 120,000

1.104   27 x 324
        a.   Rounded: 30 x 300
        b.   Estimated Product:
              30 x 300 = 9,000

1.105   44,352 x 8
        a.   Rounded: 40,000 x 10
        b.   Estimated Product:
              40,000 x 10 = 400,000

1.106 $27.75 x 24
        a.   Rounded: $30 x 20
        b.   Estimated Product:
              $30 x 20 = $600

1.107   4 x 204 = 816

1.108   5 x 307 = 1,535

1.109   7 x 908 = 6,356

1.110   6 x 702 = 4,212

1.111   8 x 901 = 7,208

1.112   9 x 608 = 5,472

1.113 $7 \times 197 = 7(200 - 3)$
        $= 7(200) - 7(3)$
        $= 1,400 - 21$
        $= 1,379$

1.114 $5 \times 396 = 5(400 - 4)$
        $= 5(400) - 5(4)$
        $= 2,000 - 20 = 1,980$

1.115 $8 \times 498 = 8(500 - 2)$
        $= 8(500) - 8(2)$
        $= 4,000 - 16 = 3,984$

1.116 $6 \times 793 = 6(800 - 7)$
        $= 6(800) - 6(7)$
        $= 4,800 - 42 = 4,758$

1.117 $8 \times \$3.95 = 8(\$4.00 - 0.05)$
        $= 8(\$4.00) - 8(0.05)$
        $= \$32.00 - 0.40$
        $= \$31.60$

1.118 $4 \times 896 = 4(900 - 4)$
        $= 4(900) - 4(4)$
        $= 3,600 - 16 = 3,584$

1.119 3^4 = a. 3 x 3 x 3 x 3
b. 81

1.120 6^2 = a. 6 x 6
b. 36

1.121 2^6 = a. 2 x 2 x 2 x 2 x 2 x 2
b. 64

1.122 5^4 = a. 5 x 5 x 5 x 5
b. 625

1.123 4^5 = a. 4 x 4 x 4 x 4 x 4
b. 1,024

1.124 10^5 = a. 10 x 10 x 10 x 10 x 10
b. 100,000

1.125 3 x 10^5 = a. 3 x 10 x 10 x 10 x 10 x 10
b. 300,000

1.126 42 x 10^8 = a. 42 x 10 x 10 x 10 x 10 x 10 x 10 x 10 x 10
b. 4,200,000,000

1.127 372 x 10^9 = a. 372 x 10 x 10 x 10 x 10 x 10 x 10 x 10 x 10 x 10
b. 372,000,000,000

1.128 7^4 = a. 7 x 7 x 7 x 7
b. 2,401

1.129 8^5 = a. 8 x 8 x 8 x 8 x 8
b. 32,768

1.130 9^4 = a. 9 x 9 x 9 x 9
b. 6,561

1.131 11^5 = a. 11 x 11 x 11 x 11 x 11
b. 161,051

1.132 Function rule: Multiply by 6

Number	Function
0	0 x 6 = 0
2	2 x 6 = 12
7	7 x 6 = 42
12	12 x 6 = 72
13	13 x 6 = 78
21	21 x 6 = 126

1.133 Function rule: Multiply by 4, then add 7

Number	Function
0	0 x 4 = 0
	0 + 7 = ⑦
1	1 x 4 = 4
	4 + 7 = ⑪
6	6 x 4 = 24
	24 + 7 = ㉛
12	12 x 4 = 48
	48 + 7 = (55)
14	14 x 4 = 56
	56 + 7 = (63)
22	22 x 4 = 88
	88 + 7 = (95)

1.134 Function rule: Multiply by 7, then add 2

Number	Function
0	0 x 7 = 0
	0 + 2 = ②
4	4 x 7 = 28
	28 + 2 = ㉚
7	7 x 7 = 49
	49 + 2 = (51)
9	9 x 7 = 63
	63 + 2 = (65)
12	12 x 7 = 84
	84 + 2 = (86)
20	20 x 7 = 140
	140 + 2 = (142)

1.135   Function rule: Multiply by
5, then subtract 3

Number	Function
1	1 x 5 = 5
	5 – 3 = ②
2	2 x 5 = 10
	10 – 3 = ⑦
5	5 x 5 = 25
	25 – 3 = ㉒
11	11 x 5 = 55
	55 – 3 = ㊾52
15	15 x 5 = 75
	75 – 3 = ㊲72
22	22 x 5 = 110
	110 – 3 = ⑩⑦107

1.136
$$\begin{array}{r} \overset{1}{} \\ 344 \\ \times\ 4 \\ \hline 1{,}376 \end{array}\ \text{oranges}$$

1.137   2 half-dollars
        = 2($0.50) or 2 x 50¢ = $1.00
     6 quarters
        = 6($0.25) or 6 x 25¢ =   1.50
   12 dimes
        = 12($0.10) or
        12 x 10¢   =   1.20
     8 nickels
        = 8($0.05) or 8 x 5¢   =   0.40
     Altogether Sally has $4.10

1.138

Student	Joe	Bill	Jane	Mary
Number missed	3	5	2	6
Number correct	a. 20 – 3 = 17	b. 20 – 5 = 15	c. 20 – 2 = 18	d. 20 – 6 = 14
Score in points	e. 3 x 5 = 15	f. 5 x 5 = 25	g. 2 x 5 = 10	h. 6 x 5 = 30
	100 – 15 = ㊗85	100 – 25 = ㊟75	100 – 10 = ⑨⓪90	100 – 30 = ⑦⓪70

1.139

$$
\begin{array}{lcll}
4 \times 45¢ &=& 180¢ &= \$1.80 \\
2 \times 89¢ &=& 178¢ &= 1.78 \\
2 \times 58¢ &=& 116¢ &= 1.16 \\
1 \times 98¢ &=& 98¢ &= \underline{\ 0.98} \\
& & & \ \ \$5.72
\end{array}
$$

1.140

1st day	= 1	= 1¢
2nd day	= 2^1	= 2¢
3rd day	= 2^2	= 4¢
4th day	= 2^3	= 8¢
5th day	= 2^4	= 16¢
6th day	= 2^5	= 32¢
7th day	= 2^6	= 64¢
8th day	= 2^7	= 128¢
		= $1.28
9th day	= 2^8	= 256¢
		= $2.56
10th day	= 2^9	= 512¢
		= $5.12
11th day	= 2^{10}	= 1,024¢
		= $10.24
12th day	= 2^{11}	= 2,048¢
		= $20.48
13th day	= 2^{12}	= 4,096¢
		= $40.96
14th day	= 2^{13}	= 8,192¢
		= $81.92
15th day	= 2^{14}	= 16,384¢
		= $163.84
16th day	= 2^{15}	= 32,768¢
		= $327.68
17th day	= 2^{16}	= 65,536¢
		= $655.36
18th day	= 2^{17}	= 131,072¢
		= $1,310.72
19th day	= 2^{18}	= 262,144¢
		= $2,621.44
20th day	= 2^{19}	= 524,288¢
		= $5,242.88
21st day	= 2^{20}	= 1,048,576¢
		= $10,485.76
22nd day	= 2^{21}	= 2,097,152¢
		= $20,971.52

1.141

$$
\begin{array}{r}
1 \\
24 \\
\underline{\times 3} \\
72 \text{ yds.}
\end{array}
$$

1.142

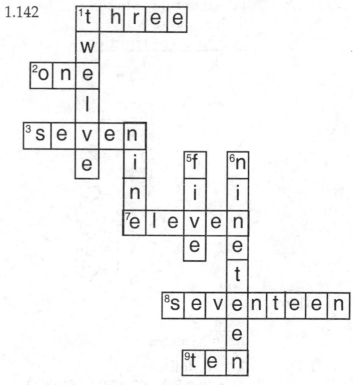

ACROSS

1. $(2 \times 2) - 1 = 4 - 1 = 3$
= three
2. $(2 \times 3) - 5 = 6 - 5 = 1$
= one
3. $(2 \times 3) + 1 = 6 + 1 = 7$
= seven
7. $(3 \times 3) + 2 = 9 + 2 = 11$
= eleven
8. $(4 \times 4) + 1 = 16 + 1 = 17$
= seventeen
9. $3^2 + 1 = 9 + 1 = 10$
= ten

DOWN

1. $3 \times 4 = 12 = $ twelve
4. $(3 \times 4) - 3 = 12 - 3 = 9$
= nine
5. $2^2 + 1 = 4 + 1 = 5 = $ five
6. $4^2 + 3 = 16 + 3 = 19$
= nineteen

II. SECTION TWO

2.1 $2 \over 2\overline{)4}$

2.2 $4 \over 2\overline{)8}$

2.3 $3 \over 2\overline{)6}$

126

2.4
$$5\overline{)10} = 2$$

2.5
$$6\overline{)12} = 2$$

2.6
$$7\overline{)21} = 3$$

2.7
$$7\overline{)28} = 4$$

2.8
$$8\overline{)64} = 8$$

2.9
$$9\overline{)81} = 9$$

2.10
$$8\overline{)72} = 9$$

2.11
$$9\overline{)63} = 7$$

2.12
$$6\overline{)54} = 9$$

2.13
$$8\overline{)48} = 6$$

2.14
$$5\overline{)40} = 8$$

2.15
$$8\overline{)32} = 4$$

2.16
$$7\overline{)56} = 8$$

2.17
$$7\overline{)49} = 7$$

2.18
$$6\overline{)42} = 7$$

2.19
$$7\overline{)35} = 5$$

2.20
$$6\overline{)30} = 5$$

2.21
$$6\overline{)36} = 6$$

2.22
$$6\overline{)24} = 4$$

2.23
$$3\overline{)18} = 6$$

2.24
$$4\overline{)16} = 4$$

2.25

```
      752
 2)1,504      Check:
    14            1
    10          752
    10          x 2
    04        1,504
     4
     0
```

2.26

```
    1,947
 3)5,841      Check:
    3           2 12
    28        1,947
    27          x 3
    14        5,841
    12
    21
    21
     0
```

2.27

```
    7,083
 4)28,332     Check:
    28           31
    033        7,083
    32           x 4
    12        28,332
    12
     0
```

2.28

```
     744
 5)3,720      Check:
    35           22
    22          744
    20          x 5
    20        3,720
    20
     0
```

2.29

```
   97,250
 6)583,500    Check:
    54          41 3
    43        97,250
    42          x 6
    15       583,500
    12
    30
    30
    00
    00
     0
```

2.30
```
      42,507
   7)297,549
      28
      17
      14
       35
       35
        049
         49
          0
```

Check:
```
   13  4
  42,507
    x 7
 297,549
```

2.31
```
      1,325
   8)10,600
      8
      26
      24
       20
       16
        40
        40
         0
```

2.32
```
      1,234
   9)11,106
      9
      21
      18
       30
       27
        36
        36
         0
```

2.33
```
       861
   4)3,444
      32
      24
      24
       04
        4
        0
```

2.34
```
      1,725
   2)3,450
      2
      14
      14
       05
        4
        10
        10
         0
```

2.35
```
      9,014
   3)27,042
      27
      004
        3
       12
       12
        0
```

2.36
```
      9,073
   8)72,584
      72
      058
       56
       24
       24
        0
```

2.37
```
      24,261
   5)121,305
      10
      21
      20
       13
       10
        30
        30
         05
          5
          0
```

2.38
```
      7,209
   7)50,463
      49
      14
      14
       063
        63
         0
```

2.39
```
      4,317
   6)25,902
      24
      19
      18
       10
        6
        42
        42
         0
```

2.40

```
        5,921
     9)53,289
       45
       82
       81
        18
        18
        09
         9
         0
```

2.41

```
        123,456,789
     9)1,111,111,101
       9
       21
       18
        31
        27
         41
         36
          51
          45
           61
           54
            71
            63
             80
             72
              81
              81
               0
```

2.42

```
        1,543
     37)57,091      Check:
        37            1 1
        200           3 3 2
        185           1,543
        159          x  37
        148          10801
        111          4629
        111          57,091
          0
```

2.43

```
        7,645
     42)321,090     Check:
        294           2 12
        270           1 01
        252           7,645
        189          x  42
        168          ¹1¹5290
        210          30580
        210          321,090
          0
```

2.44

```
        978
     59)57,702      Check:
        531           34
        460           77
        413           978
        472          x  59
        472          ¹8802
          0          ¹4890
                     57,702
```

2.45

```
        7,085
     96)680,160     Check:
        672           7 4
        816           5 3
        768           7,085
        480          x  96
        480          ¹4¹2510
          0          63765
                     680,160
```

2.46

```
        1,157
     74)85,618      Check:
        74            1 3 4
        116           0 2 2
        74            1,157
        421          x  74
        370          ¹4¹628
        518          8099
        518          85,618
          0
```

2.47

```
        924
     69)63,576      Check:
        621           12
        165           23
        138           924
        276          x  69
        276          8316
          0          ¹5544
                     63,756
```

2.48

```
        2,654
     26)69,004      Check:
        52            1 1
        170           3 3 2
        156           2,654
        140          x  26
        130          15924
        104          5308
        104          69,004
          0
```

2.49
```
        8,756
   31)271,436        Check:
      248              2 1 1
      234              0 0 0
      217              8,756
      173              x 31
      155             ¹8¹756
      186             2¹6268
      186             271,436
        0
```

2.50
```
        1,089
   60)65,340         Check:
      60               55
      534              00
      480              1,089
      540              x 60
      540              0000
        0             6534
                      65,340
```

2.51
```
        8,196
   85)696,660        Check:
      680              1 7 4
      166              0 4 3
       85              8,196
      816              x 85
      765             4¹ 0¹980
      510             65568
      510             696,660
        0
```

2.52
```
        2,268
   63)142,884        Check:
      126              1 4 4
      168              0 2 2
      126              2,268
      428              x 63
      378             ¹6804
      504             13608
      504             142,884
        0
```

2.53
```
        1,035
   58)60,030         Check:
      58               1 2
      203              2 4
      174              1,035
      290              x 58
      290             ¹8¹280
        0             ¹5175
                      60,030
```

2.54
```
        6,797
   71)482,587        Check:
      426              5 6 4
      565              0 0 0
      497              6,797
      688              x 71
      639             ¹6¹797
      497             4¹7579
      497             482,587
        0
```

2.55
```
        7,562
   38)287,356        Check:
      266              1 1
      213              4 4 1
      190              7,562
      235              x 38
      228             6¹0¹496
       76             22686
       76             287,356
        0
```

2.56
```
        947
   88)83,336         Check:
      792              35
      413              35
      352              947
      616              x 88
      616             ¹7¹576
        0             ¹7576
                      83,336
```

2.57
```
        9,751
   22)214,522        Check:
      198              1 1
      165              1 1
      154              9,751
      112              x 22
      110             ¹19502
       22             ¹19502
       22             214,522
        0
```

2.58
```
        925
   25)23,125         Check:
      225              1
       62              12
       50              925
      125              x 25
      125             ¹4625
        0             ¹1850
                      23,125
```

2.59

$$
\begin{array}{r}
7,095 \\
82\overline{)581,790} \\
\underline{574} \\
779 \\
\underline{738} \\
410 \\
\underline{410} \\
0
\end{array}
$$

Check:

$$
\begin{array}{r}
7\ 4 \\
1\ 1 \\
7,095 \\
\times\ 82 \\
\hline
^{1}14190 \\
56760 \\
\hline
581,790
\end{array}
$$

2.60

$$
\begin{array}{r}
11,452 \\
49\overline{)561,148} \\
\underline{49} \\
71 \\
\underline{49} \\
221 \\
\underline{196} \\
254 \\
\underline{245} \\
98 \\
\underline{98} \\
0
\end{array}
$$

Check:

$$
\begin{array}{r}
1\ 2 \\
14\ 4\ 1 \\
11,452 \\
\times\ 49 \\
\hline
1^{1}03^{1}068 \\
45808 \\
\hline
561,148
\end{array}
$$

2.61

$$
\begin{array}{r}
8,457 \\
67\overline{)566,619} \\
\underline{536} \\
306 \\
\underline{268} \\
381 \\
\underline{335} \\
469 \\
\underline{469} \\
0
\end{array}
$$

Check:

$$
\begin{array}{r}
2\ 3\ 4 \\
3\ 3\ 4 \\
8,457 \\
\times\ 67 \\
\hline
^{1}59^{1}199 \\
50742 \\
\hline
566,619
\end{array}
$$

2.62

$$
\begin{array}{r}
2,481 \\
99\overline{)245,619} \\
\underline{198} \\
476 \\
\underline{396} \\
801 \\
\underline{792} \\
99 \\
\underline{99} \\
0
\end{array}
$$

Check:

$$
\begin{array}{r}
4\ 7 \\
4\ 7 \\
2,481 \\
\times\ 99 \\
\hline
22^{1}329 \\
22329 \\
\hline
245,619
\end{array}
$$

2.63

$$
\begin{array}{r}
12,461 \\
475\overline{)5,918,975} \\
\underline{475} \\
1168 \\
\underline{950} \\
2189 \\
\underline{1900} \\
2897 \\
\underline{2850} \\
475 \\
\underline{475} \\
0
\end{array}
$$

Round the divisor, 475, to the nearest hundred, 500, to make the answer easier to estimate.

2.64

$$
\begin{array}{r}
77,952 \\
727\overline{)56,671,104} \\
\underline{5089} \\
5781 \\
\underline{5089} \\
6921 \\
\underline{6543} \\
3780 \\
\underline{3635} \\
1454 \\
\underline{1454} \\
0
\end{array}
$$

Round the divisor, 727, to the nearest hundred, 700, to make the answer easier to estimate.

2.65

$$
\begin{array}{r}
9\ \text{R7} \\
9\overline{)88} \\
\underline{81} \\
7
\end{array}
$$

Check:

$$
\begin{array}{r}
9 \\
\times\ 9 \\
\hline
81 \\
+\ 7 \\
\hline
88
\end{array}
$$

2.66

$$
\begin{array}{r}
27\ \text{R12} \\
19\overline{)525} \\
\underline{38} \\
145 \\
\underline{133} \\
12
\end{array}
$$

Check:

$$
\begin{array}{r}
6 \\
27 \\
\times\ 19 \\
\hline
^{1}243 \\
27 \\
\hline
513 \\
+12 \\
\hline
525
\end{array}
$$

2.67

$$
\begin{array}{r}
453\ \text{R0} \\
78\overline{)35,334} \\
\underline{312} \\
413 \\
\underline{390} \\
234 \\
\underline{234} \\
0
\end{array}
$$

Check:

$$
\begin{array}{r}
32 \\
42 \\
453 \\
\times\ 78 \\
\hline
^{1}3624 \\
3171 \\
\hline
35,334
\end{array}
$$

2.68

$$
\begin{array}{r}
1,942\ \text{R20} \\
653\overline{)1,268,146} \\
\underline{653} \\
6151 \\
\underline{5877} \\
2744 \\
\underline{2612} \\
1326 \\
\underline{1306} \\
20
\end{array}
$$

Check:

$$
\begin{array}{r}
5\ 2\ 1 \\
4\ 2\ 1 \\
2\ 1\ 0 \\
1,942 \\
\times\ 653 \\
\hline
^{1}5826 \\
^{1}9710 \\
^{1}11652 \\
\hline
1,268,126 \\
+\ 20 \\
\hline
1,268,146
\end{array}
$$

2.69

```
            3,411 R19    Check:
  571)1,947,700            2
      1713                 2
      2347                 0
      2284               3,411
       630              x  571
       571              ¹3411
       590              2¹3877
       571              17055
        19             1,947,¹6¹81
                           + 19
                       1,947,700
```

2.70

```
            4,072 R751
  915)3,726,631          Check:
      3660                 6 1
      6663                 0 0
      6405                 3 1
      2581               4,072
      1830              x  915
       751              2¹0360
                         4072
                       3¹6648
                       372¹5¹880
                          + 751
                       3,726,631
```

2.71

```
            607 R6
  78)47,352             Check:
     468                   4
     552                   5
     546                 607
       6              x  78
                      ¹4¹856
                       4249
                      47,3¹46
                        + 6
                      47,352
```

2.72

```
           700 R4
  47)32,904             Check:
     329                 700
     004              x  47
       0              4900
       4              ¹2800
                      32,900
                        + 4
                      32,904
```

2.73

```
          1,352 R10
  91)123,042            Check
     91                  3 4 1
     320                 0 0 0
     273               1,352
     474              x  91
     455              ¹1352
     192              12168
     182              123,032
      10                + 10
                      123,042
```

2.74

```
           821 R40
  67)55,047             Check:
     536                   1
     144                   1
     134                 821
     107              x  67
      67              ¹5¹747
      40              ¹4926
                      55,007
                       + 40
                      55,047
```

2.75

```
          1,162 R18
  58)67,414             Check:
     58                  31
     94                 141
     58               1,162
     361              x  58
     348              9296
     134              ¹5810
     116              67,¹3¹96
      18                + 18
                      67,414
```

2.76

```
          931 R70
  82)76,412             Check:
     738                 2
     261                 0
     246               931
     152              x  82
      82              ¹1¹862
      70              7448
                      76,¹342
                       + 70
                      76,412
```

2.77

```
          1,346 R25
  56)75,401             Check:
     56                 1 2 3
     194                2 2 3
     168              1,346
     260              x  56
     224              8076
     361              ¹6730
     336              75,¹3¹76
      25                + 25
                      75,401
```

2.78

```
          905 R6
  69)62,451             Check:
     621                 3
     351                 4
     345               905
       6              x  69
                      8145
                      ¹5430
                      62,4¹45
                        + 6
                      62,451
```

2.79

$$77)\overline{68,707} \quad 892 \text{ R23}$$

```
        892  R23
77)68,707          Check:
   616                 61
   710                 61
   693                892
   177              x  77
   154              6244
    23              6244
                   6¹8,684
                   +   23
                   68,707
```

2.80

```
       24,017  R0
45)1,080,765        Check:
   90                1   2
   180               2   3
   180              24,017
   076              x  45
    45             120¹085
   315              96068
   315            1,080,765
     0
```

2.81

```
       24,033  R65
70)1,682,375        Check:
   140               2 2 2
   282               0 0 0
   280              24,033
   237              x  70
   210              00000
   275             168231
   210            1,682,310
    65             +  65
                 1,682,375
```

2.82

```
       70,315  R25
88)6,187,745        Check:
   616               2 1 4
   277               2 1 4
   264              70,315
   134              x  88
    88             562520
   465            ¹562520
   440           6,187,720
    25             +  25
                 6,187,745
```

2.83

```
        423  R12
417)176,403         Check:
    1668              1
     960              0
     834             12
    1263            423
    1251            x 417
      12           ¹2961
                    423
                  1¹692
                  176,391
                  +   12
                  176,403
```

2.84

```
        921  R100
707)651,247         Check:
    6363              1
    1494              0
    1414              1
     807            921
     707            x 707
     100           ¹6447
                    000
                  6¹447
                  651,147
                  +  100
                  651,247
```

2.85

```
        798  R591
942)752,307         Check:
    6594             87
    9290             33
    8478             11
    8127            798
    7536            x 942
     591           ¹1¹596
                  ¹3192
                   7182
                 75¹1,¹716
                 +   591
                 752,307
```

2.86

```
        653  R230
528)345,014         Check:
    3168             21
    2821             10
    2640             42
    1814            653
    1584            x 528
     230            5224
                  ¹1306
                   3265
                 34¹4¹784
                 +   230
                 345,014
```

2.87

```
        913  R83
818)746,917         Check:
    736 2            12
    1071             00
     818             12
    2537            913
    2454            x 818
      83            7304
                    913
                  7¹304
                  746,¹834
                  +   83
                  746,917
```

2.88

$$831\overline{)901{,}243}\quad\begin{array}{r}1{,}084\ \text{R}439\end{array}$$

```
          1,084  R439
  831)901,243        Check:
      831               63
     7024               21
     6648               00
     3763            1,084
     3324           x  831
      439            1ꞌ084
                     1ꞌ3252
                     1ꞌ8672
                   90ꞌ0,8ꞌ04
                    +  439
                    901,243
```

2.89

```
         10,622  R358
 1,643)17,452,304     Check:
       1643            3 1 1
      10223            2 0 0
       9858            1 0 0
       3650           10,622
       3286          x 1,643
       3644          1ꞌ3ꞌ1ꞌ866
       3286          1ꞌ42488
        358          1ꞌ63732
                      10622
                   17,45ꞌ1,9ꞌ46
                    +   358
                   17,452,304
```

2.90

```
          4,962  R5,983
 7,204)35,752,231     Check:
       28816           6 4 1
       69362           1 1 0
       64836           0 0 0
       45263           3 2 0
       43224           4,962
       20391          x 7,204
       14408          1ꞌ1ꞌ9848
        5983           0000
                      1ꞌ9924
                      3ꞌ4734
                   35,7ꞌ4ꞌ6,ꞌ2ꞌ48
                    +  5,983
                   35,752,231
```

2.91

```
          9,587  R1,908
 5,137)49,250,327     Check:
       46233           2 4 3
       30173           0 0 0
       25685           1 2 2
       44882            464
       41096           9,587
       37867          x 5,137
       35959          ꞌ6ꞌ7109
       1908           2ꞌ28761
                       9587
                      4ꞌ7935
                   49,2ꞌ4ꞌ8,4ꞌ19
                    +  1,908
                   49,250,327
```

2.92 243; 81; 27; a. ___ ; b. ___ ; c. ___
To find the divisor of the pattern, divide any member by the one following it.
$243 \div 81 = 3$
The divisor is 3. The next three members are
a. $27 \div 3 = 9$
b. $9 \div 3 = 3$
c. $3 \div 3 = 1$

2.93 7,168; 1,792; 448; a. ___ ;
b. ___ ; c. ___
To find the divisor of the pattern, divide any member by the one following it.
$7{,}168 \div 1{,}792 = 4$
The divisor is 4. The next three members are
a. $448 \div 4 = 112$
b. $112 \div 4 = 28$
c. $28 \div 4 = 7$

2.94 161,051; 14,641; 1,331;
a. ___ ; b. ___ ; c ___ .
To find the divisor of the pattern, divide any member by the one following it.
$14{,}641 \div 1{,}331 = 11$
The divisor is 11. The next three members are
a. $1{,}331 \div 11 = 121$
b. $121 \div 11 = 11$
c. $11 \div 11 = 1$

2.95 16,384; 4,096; 1,024; a. ___ ;
b. ___ ; c ___ .
To find the divisor of the pattern, divide any member by the one following it.
$4{,}096 \div 1{,}024 = 4$
The divisor is 4. The next three members are
a. $1{,}024 \div 4 = 256$
b. $256 \div 4 = 64$
c. $64 \div 4 = 16$

2.96   78,125; 15,625; 3,125; a. ___ ;
b. ___ ; c ___ .
To find the divisor of the
pattern, divide any member
by the one following it.
$15,625 \div 3,125 = 5$
The divisor is 5. The next
three members are
a. $3,125 \div 5 = 625$
b. $625 \div 5 = 125$
c. $125 \div 5 = 25$

2.97   Function rule: Divide by
7, then add 4

Number	Function
7	$7 \div 7 = 1$
	$1 + 4 = ⑤$
28	$28 \div 7 = 4$
	$4 + 4 = ⑧$
49	$49 \div 7 = 7$
	$7 + 4 = ⑪$
77	$77 \div 7 = 11$
	$11 + 4 = ⑮$
84	$84 \div 7 = 12$
	$12 + 4 = ⑯$

2.98   Function rule: Subtract 4,
then divide by 3

Number	Function
4	$4 - 4 = 0$
	$0 \div 3 = ⓪$
10	$10 - 4 = 6$
	$6 \div 3 = ②$
19	$19 - 4 = 15$
	$15 \div 3 = ⑤$
31	$31 - 4 = 27$
	$27 \div 3 = ⑨$
70	$70 - 4 = 66$
	$66 \div 3 = ㉒$

2.99   Function rule: Divide by 5,
then add 11

Number	Function
0	$0 \div 5 = 0$
	$0 + 11 = ⑪$
10	$10 \div 5 = 2$
	$2 + 11 = ⑬$
25	$25 \div 5 = 5$
	$5 + 11 = ⑯$
75	$75 \div 5 = 15$
	$15 + 11 = ㉖$
95	$95 \div 5 = 19$
	$19 + 11 = ㉚$

2.100   Function rule: Add 7,
then divide by 6

Number	Function
5	$5 + 7 = 12$
	$12 \div 6 = ②$
17	$17 + 7 = 24$
	$24 \div 6 = ④$
35	$35 + 7 = 42$
	$42 \div 6 = ⑦$
71	$71 + 7 = 78$
	$78 \div 6 = ⑬$
83	$83 + 7 = 90$
	$90 \div 6 = ⑮$

2.101   The word "per" is a clue
to divide.

$$\begin{array}{r} 752 \\ 160\overline{)120,320} \\ \underline{1120} \\ 832 \\ \underline{800} \\ 320 \\ \underline{320} \\ 0 \end{array}$$

Mr. Jones paid
$752 per acre.

2.102   A year has 12 months.

$$\begin{array}{r} 975 \\ 12\overline{)11,700} \\ \underline{108} \\ 90 \\ \underline{84} \\ 60 \\ \underline{60} \\ 0 \end{array}$$

Mr. Smith gets
$975 per month.

2.103   A year has 52 weeks.

$$\begin{array}{r} 225 \\ 52\overline{)11,700} \\ \underline{104} \\ 130 \\ \underline{104} \\ 260 \\ \underline{260} \\ 0 \end{array}$$

Mr. Smith's
weekly salary
is $225.

2.104

$$\begin{array}{r} 7 \\ 50\overline{)350} \\ \underline{350} \\ 0 \end{array}$$

You will need 7
hours for the trip.

2.105   Attendance   38
          on Monday: – 1
                              37 pupils

          Attendance
          on Tuesday: 38
                             – 2
                              36 pupils

          Attendance
          on Wednesday: 38
                             – 4
                              34 pupils

          Attendance
          on Thursday: 38
                             – 0
                              38 pupils

          Attendance
          on Friday: 38
                             – 3
                              35 pupils

To find the average, add
the numbers and then
divide that total by the
number of members in the
set.

          3   36 pupils is the
          37   5)180 average daily
          36   15   attendance.
          34   30
          38   30
        + 35   0
          180

2.106   The total of the maximum
          temperatures is
          2
          30
          33
          35
          34
          32
          31
        + 29
          224.

          32   The average
        7)224 maximum temp-
          21   erature for
          14   the week was
          14   32° Celsius.
           0

2.107   From 10 a.m. to 12 noon is
          2 hours. From 12 noon to
          6 p.m. is 6 hours. The
          total time was 8 hours.

2.108   To find the average
          speed of the car in
          Problem 2.107, divide the
          number of miles by the
          number of hours driven.
             50
          8)400   The average speed
             40   was 50 mph.
             00
             00
              0

2.109   To find the number of miles
          Mr. Brown averaged per
          gallon (mpg), subtract the
          smaller mileage reading
          from the larger mileage
          reading.
            6 131
           65,740
         – 65,452
             288
          He traveled 288 miles. Next,
          divide 288 by the number of
          gallons required to fill the
          tank.
              18
          16)288   Mr. Brown
              16   averaged 18 mpg.
             128
             128
               0

2.110   The word "per" is a hint
          to divide (price divided by
          number of pounds).
             1.19
          6)7.14   The price per
             6   pound was $1.19.
            11
             6
            54
            54
             0

2.111 To find Joe's pay,
multiply 7 by $1.50.

   3
$1.50
x 7
$10.50

To find Bill's pay,
multiply 6 by $1.70.

   4
$1.70
 x 6
$10.20

To find out who received
the most pay, subtract the
smaller amount from the
larger amount.

Joe: $10.50
Bill: − 10.20
   $ 0.30

Joe received 30¢ more
than Bill.

2.112 The word "per" is a clue to
divide (cost divided by number
of ounces).

           $0.06 per oz.
Corn flakes: 10)0.60
            00
            60
            60
             0

Granola (16 oz. are in a
pound):

  $0.05 per oz.
16)0.80
  00
  80
  80
   0

2.113 In a year, Mr. Jones
receives

     2
     1
  $150
 x 52
  300
  750
$7,800.

In a month, Mr.
Jones receives

    $650.
12)7,800
  72
  60
  60
  00
  00
   0

2.114 To find the total cost,
find the costs of the
corn and the green beans
and add the costs
together. 2

Corn: $0.40
       x 6
      $2.40

        2 2
Green $0.33
beans: x 8
      $2.64

        1
Total cost: $2.40
       +2.64
       $5.04

Divide this total cost by
the number of cans of
vegetables.

  $0.36
14)5.04
  42
  84
  84
   0

The average cost per can
of vegetables was $0.36.

2.115
       4
17 lbs. 5 oz.
    x 7
119 lbs. 35 oz.

     2 lbs. 3 oz. Divide by
16)35 oz.   16 since
  32   16 ounces
   3 oz.   = 1 pound.
119 lbs.
+ 2 lbs. 3 oz.
121 lbs. 3 oz.

2.116
$$
\begin{array}{r}
^{4} \\
45 \text{ ft. } 11 \text{ in.} \\
\times\ 8 \\
\hline
360 \text{ ft. } 88 \text{ in.}
\end{array}
$$

$$
\begin{array}{r}
7 \text{ ft. } 4 \text{ in.} \\
12\overline{)88 \text{ in.}} \\
\underline{84} \\
4 \text{ in.}
\end{array}
$$ Divide by 12 since 12 inches = 1 foot.

360 ft
+ 7 ft 4 in
———
367 ft 4 in

2.117
$$
\begin{array}{r}
^{1} \quad\quad ^{1} \\
65 \text{ hrs. } 37 \text{ min.} \\
\times\ 12 \\
\hline
130 \quad _{1}74 \\
65 \quad\ 37 \\
\hline
780 \text{ hrs. } 444 \text{ min.}
\end{array}
$$

$$
\begin{array}{r}
7 \text{ hrs. } 24 \text{ min.} \\
60\overline{)444 \text{ min.}} \\
\underline{420} \\
24 \text{ min.}
\end{array}
$$ Divide by 60 since 60 minutes = 1 hour.

780 hrs.
+ 7 hrs. 24 min.
———
787 hrs. 24 min.

2.118
$$
\begin{array}{r}
^{2} \\
35 \text{ yds. } 2 \text{ ft.} \\
\times\ 14 \\
\hline
140 \quad 28 \\
35 \\
\hline
490 \text{ yds. } 28 \text{ ft.}
\end{array}
$$

$$
\begin{array}{r}
9 \text{ yds. } 1 \text{ ft.} \\
3\overline{)28 \text{ ft.}} \\
\underline{27} \\
1 \text{ ft.}
\end{array}
$$ Divide by 3 since 3 feet = 1 yard.

490 yds.
+ 9 yds. 1 ft.
———
499 yds. 1 ft.

2.119
$$
\begin{array}{r}
^{5} \quad\quad ^{4,\ 41} \\
16 \text{ tons } 1,452 \text{ lbs.} \\
\times\ 9 \\
\hline
144 \text{ tons } 13,068 \text{ lbs.}
\end{array}
$$

$$
\begin{array}{r}
6 \text{ tons } 1,068 \text{ lbs.} \\
2,000\overline{)13,068 \text{ lbs.}} \\
\underline{12000} \\
1,068 \text{ lbs.}
\end{array}
$$ Divide by 2,000 since 2,000 pounds = 1 ton.

144 tons
+ 6 tons 1,068 lbs.
———
150 tons 1,068 lbs.

2.120
$$
\begin{array}{r}
^{3} \quad\quad ^{1} \\
45 \text{ lbs. } 13 \text{ oz.} \\
\times\ 6 \\
\hline
270 \text{ lbs. } 78 \text{ oz.}
\end{array}
$$

$$
\begin{array}{r}
4 \text{ lbs. } 14 \text{ oz.} \\
16\overline{)78 \text{ oz.}} \\
\underline{64} \\
14 \text{ oz.}
\end{array}
$$ Divide by 16 since 16 ounces = 1 pound.

270 lbs.
+ 4 lbs. 14 oz.
———
274 lbs. 14 oz.

2.121
$$
\begin{array}{r}
72 \text{ hrs. } 42 \text{ min.} \\
\times\ 10 \\
\hline
00 \quad\quad 00 \\
72 \quad\quad 42 \\
\hline
720 \text{ hrs. } 420 \text{ min.}
\end{array}
$$

$$
\begin{array}{r}
7 \text{ hrs. } 0 \text{ min.} \\
60\overline{)420 \text{ min.}} \\
\underline{420} \\
0 \text{ min.}
\end{array}
$$ Divide by 60 since 60 minutes = 1 hour.

720 hrs.
+ 7 hrs.
———
727 hrs.

2.122

$$
\begin{array}{r}
\overset{1}{}\overset{2}{} \\
\text{65 ft.} \qquad \text{9 in.} \\
\times\ 24 \\
\hline
260) \quad 216 \\
130 \\
\hline
\text{1,560 ft. 216 in.}
\end{array}
$$

$$
\begin{array}{r}
18 \text{ ft. } 0 \text{ in.} \\
12\overline{)216} \text{ in.} \\
12 \\
\hline
96 \\
96 \\
\hline
0 \text{ in.}
\end{array}
$$
Divide by 12 since 12 inches = 1 foot.

1,560 ft.
+ 18 ft.
1,578 ft.

2.123

$$
\begin{array}{r}
\overset{1}{}\overset{1}{} \\
\text{75 yds.} \qquad \text{1 ft.} \\
\times\ 32 \\
\hline
^{1}150 \quad 32 \\
225 \\
\hline
\text{2,400 yds. 32 ft.}
\end{array}
$$

$$
\begin{array}{r}
10 \text{ yds. 2 ft.} \\
3\overline{)32} \text{ ft.} \\
3 \\
\hline
02 \\
0 \\
\hline
2 \text{ ft.}
\end{array}
$$
Divide by 3 since 3 feet = 1 yard.

2,400 yds.
+ 10 yds. 2 ft.
2,410 yds. 2 ft.

2.124

$$
\begin{array}{r}
\overset{3}{} \qquad \overset{3}{}\overset{1}{} \\
\text{35 tons} \qquad \text{752 lbs.} \\
\times\ 17 \\
\hline
245 \quad 5264 \\
35 \quad\ 752 \\
\hline
\text{595 tons 12,784 lbs.}
\end{array}
$$

$$
\begin{array}{r}
6 \text{ tons 784 lbs.} \\
2{,}000\overline{)12{,}784} \text{ lbs.} \\
12{,}000 \\
\hline
784 \text{ lbs.}
\end{array}
$$
Divide by 2,000 since 2,000 pounds = 1 ton.

11
595 tons
+ 6 tons 784 lbs.
601 tons 784 lbs.

2.125

$$
\begin{array}{r}
\overset{1}{} \qquad \overset{1}{}\overset{2}{} \\
\text{42 lbs.} \qquad \text{15 oz.} \\
\times\ 25 \\
\hline
210 \quad 75 \\
84 \quad 30 \\
\hline
\text{1,050 lbs. 375 oz.}
\end{array}
$$

$$
\begin{array}{r}
23 \text{ lbs. 7 oz.} \\
16\overline{)375} \text{ oz.} \\
32 \\
\hline
55 \\
48 \\
\hline
7 \text{ oz.}
\end{array}
$$
Divide by 16 since 16 ounces = 1 pound.

1,050 lbs.
+ 23 lbs. 7 oz.
1,073 lbs. 7 oz.

2.126

$$
\begin{array}{r}
\overset{1}{}\overset{0}{} \\
\text{34 min.} \qquad \text{27 sec.} \\
\times\ 21 \\
\hline
^{1}34 \quad 27 \\
68 \quad 54 \\
\hline
\text{714 min. 567 sec.}
\end{array}
$$

$$
\begin{array}{r}
9 \text{ min. 27 sec.} \\
60\overline{)567} \\
540 \\
\hline
27 \text{ sec.}
\end{array}
$$
Divide by 60 since 60 seconds = 1 minute.

1
714 min.
+ 9 min. 27 sec.
723 min. 27 sec.

2.127

$$
\begin{array}{r}
\text{61 lbs.} \qquad \text{11 oz.} \\
\times\ 23 \\
\hline
^{1}183 \quad 33 \\
122 \quad 22 \\
\hline
\text{1,403 lbs. 253 oz.}
\end{array}
$$

$$
\begin{array}{r}
15 \text{ lbs. 13 oz.} \\
16\overline{)253} \text{ oz.} \\
16 \\
\hline
93 \\
80 \\
\hline
13 \text{ oz.}
\end{array}
$$
Divide by 16 since 16 ounces = 1 pound.

1,403 lbs.
+ 15 lbs. 13 oz.
1,418 lbs. 13 oz.

2.128

$$
\begin{array}{rr}
52 \text{ ft.} & 7 \text{ in.} \\
& \times\ 31 \\
\hline
52 & 217 \\
1560 & \\
\hline
1{,}612 \text{ ft.} & 217 \text{ in.}
\end{array}
$$

$$
\begin{array}{l}
18\ \text{ ft. 1 in.} \\
12)\overline{217}\ \text{in.} \qquad \text{Divide by 12} \\
\quad\underline{12}\qquad\qquad\ \text{since 12 inches} \\
\quad\ 97\qquad\qquad = 1 \text{ foot.} \\
\quad\ \underline{96} \\
\qquad 1\ \text{in.}
\end{array}
$$

$$
\begin{array}{l}
\ \ ^{1} \\
1{,}612 \text{ ft.} \\
+\ \ 18 \text{ ft. 1 in.} \\
\hline
1{,}630 \text{ ft. 1 in.}
\end{array}
$$

2.129

$$
\begin{array}{l}
\ \ ^{2} \\
\ \ ^{2} \\
37 \text{ hrs.} \quad 31 \text{ min.} \\
\qquad \times\ \ 33 \\
\hline
111 \qquad 93 \\
111 \qquad {}^{1}93 \\
\hline
1{,}221 \text{ hrs. } 1{,}023 \text{ min.}
\end{array}
$$

$$
\begin{array}{l}
17 \text{ hrs. 3 min.} \\
60)\overline{1{,}023}\ \text{min.} \quad \text{Divide by 60} \\
\quad\underline{60}\qquad\qquad\ \text{since 60} \\
\quad 423\qquad\qquad \text{minutes} = 1 \\
\quad \underline{420}\qquad\qquad \text{hour.} \\
\qquad 3 \text{ min.}
\end{array}
$$

$$
\begin{array}{l}
1{,}221 \text{ hrs.} \\
+\ 17 \text{ hrs. } \ 3 \text{ min.} \\
\hline
1{,}238 \text{ hrs. } \ 3 \text{ min.}
\end{array}
$$

2.130

$$
\begin{array}{l}
\ \ ^{2} \\
\ \ ^{1} \\
27 \text{ yds.} \quad 1 \text{ ft.} \\
\qquad \times\ 42 \\
\hline
54 \qquad 42 \\
1^{1}08 \\
\hline
1{,}134 \text{ yds. } 42 \text{ ft.}
\end{array}
$$

$$
\begin{array}{l}
14 \text{ yds. 0 ft.} \\
3)\overline{42}\ \text{ft.} \qquad \text{Divide by 3 since} \\
\ \underline{3}\qquad\qquad\ 3 \text{ feet} = 1 \text{ yard.} \\
\ 12 \\
\ \underline{12} \\
\ \ 0 \text{ ft.}
\end{array}
$$

$$
\begin{array}{l}
1{,}134 \text{ yds.} \\
+\ 14 \text{ yds.} \\
\hline
1{,}148 \text{ yds.}
\end{array}
$$

2.131

$$
\begin{array}{l}
\qquad\qquad\quad ^{2} \\
\qquad\qquad\quad ^{2} \\
32 \text{ tons} \quad 750 \text{ lbs.} \\
\qquad\qquad \times\ 44 \\
\hline
{}^{1}128 \qquad 3000 \\
128 \qquad 3000 \\
\hline
1{,}408 \text{ tons} \quad 33{,}000 \text{ lbs.}
\end{array}
$$

$$
\begin{array}{l}
\qquad\quad 16 \text{ tons 1,000 lbs.} \\
2{,}000)\overline{33{,}000}\ \text{lbs.} \quad \text{Divide by} \\
\quad\ \underline{2000}\qquad\qquad 2{,}000 \text{ since} \\
\quad 13000\qquad\qquad 2{,}000 \text{ pounds} \\
\quad \underline{12000}\qquad\qquad = 1 \text{ ton.} \\
\qquad 1000 \text{ lbs.}
\end{array}
$$

$$
\begin{array}{l}
\quad ^{1} \\
1{,}408 \text{ tons} \\
+\ \ 16 \text{ tons 1,000 lbs.} \\
\hline
1{,}424 \text{ tons 1,000 lbs.}
\end{array}
$$

2.132

$$
\begin{array}{l}
71 \text{ lbs.} \quad 5 \text{ oz.} \\
\qquad \times\ 61 \\
\hline
71 \qquad 305 \\
4^{1}26 \\
\hline
4{,}331 \text{ lbs.} \quad 305 \text{ oz.}
\end{array}
$$

$$
\begin{array}{l}
19 \text{ lbs. 1 oz.} \\
16)\overline{305}\ \text{oz.} \qquad \text{Divide by 16} \\
\quad\underline{16}\qquad\qquad\ \text{since 16} \\
\ 145\qquad\qquad \text{ounces} = 1 \\
\ \underline{144}\qquad\qquad \text{pound.} \\
\qquad 1 \text{ oz.}
\end{array}
$$

$$
\begin{array}{l}
\ ^{1} \\
4{,}331 \text{ lbs.} \\
+\ 19 \text{ lbs. 1 oz.} \\
\hline
4{,}350 \text{ lbs. 1 oz.}
\end{array}
$$

2.133

$$
\begin{array}{l}
\ \ ^{1}\qquad\quad ^{3} \\
\ \ ^{0}\qquad\quad ^{2} \\
92 \text{ hrs.} \quad 15 \text{ min.} \\
\qquad \times\ 64 \\
\hline
368 \qquad 60 \\
552 \qquad 90 \\
\hline
5{,}888 \text{ hrs.} \quad 960 \text{ min.}
\end{array}
$$

$$
\begin{array}{l}
16 \text{ hrs. 0 min.} \\
60)\overline{960}\ \text{min.} \qquad \text{Divide by 60} \\
\quad\underline{60}\qquad\qquad\ \text{since 60} \\
\ 360\qquad\qquad \text{minutes} = 1 \\
\ \underline{360}\qquad\qquad \text{hour.} \\
\quad\ 0
\end{array}
$$

$$
\begin{array}{l}
\ ^{11} \\
5{,}888 \text{ hrs.} \\
+\ 16 \text{ hrs.} \\
\hline
5{,}904 \text{ hrs.}
\end{array}
$$

2.134
```
      91 ft.        3 in.
                   x  50
      00           150
     455
    4,550 ft.      150 in.
```

```
            12 ft. 6 in.
     12)150 in.   Divide by 12
        12        since 12 inches
        30        = 1 foot.
        24
         6 in.
```

```
    4,550 ft.
  + 12 ft. 6 in.
    4,562 ft. 6 in.
```

2.135
```
       4
       1
      77 yds.       2 ft.
                  x 62
     154          124
     462
    4,774 yds.  124 ft.
```

```
           41 yds.  1 ft.
     3)124 ft.   Divide by 3
        12       since 3 feet
        04       = 1 yard.
         3
         1 ft.
```

```
       1
    4,774 yds.
  + 41 yds. 1 ft.
    4,815 yds. 1 ft.
```

2.136
```
     1        1
     5        4
    17 tons   350 lbs.    8 oz.
                        x 28
    136     2800      224
    34       700
    476 tons 9,800 lbs.  224 oz.
```

```
            14 lbs. 0 oz.
    16)224 oz.   Divide by 16
       16        since 16 ounces
       64        = 1 pound.
       64
        0 oz.
```

```
    476 tons 9,800 lbs.
              + 14 lbs.
    476 tons  9,814 lbs.
```

2.136 cont.
```
              4  tons 1,814 lbs.
    2,000)9,814 lbs.    Divide by
          8000          2,000
          1814 lbs.  since 2,000
                     pounds = 1
                     ton.
```

```
      1
    476 tons
  +   4 tons 1,814 lbs.
    480 tons  1,814 lbs.
```

2.137
```
                  3
    22 hrs.    38 min.     30 sec.
                        x  41
     22        38         30
    ¹88       152        120
    902 hrs.  1,558 min.  1,230 sec.
```

```
          20 min. 30 sec.
    60)1,230 sec.    Divide by 60
       120           since 60
        30           seconds = 1
        00           minute.
        30 sec.
```

```
    902 hrs.  1,558 min.
              + 20 min. 30 sec.
    902 hrs.  1,578 min. 30 sec.
```

```
          26 hrs. 18 min.
    60)1,578 min.    Divide by 60
       120           since 60
       378           minute = 1
       360           hour
        18 min.
```

```
    902 hrs.               30 sec.
  +26 hrs.  18 min.
    928 hrs.  18 min.      30 sec.
```

2.138
```
         7 ft. 8 in.
    3)23 ft.
      21
       2 ft. = 24 in.   Multiply
              24        by 12
               0        since 12
                        inches
                        = 1 foot.
```

141

2.139

$$\begin{array}{r} 14\text{ hrs.} \quad 24\text{ min.} \\ 7\overline{)100\text{ hrs.}} \quad 48\text{ min.} \\ \underline{7} \\ 30 \\ \underline{28} \\ 2\text{ hrs.} = \underline{120\text{ min.}} \\ 168\text{ min.} \\ \underline{14} \\ 28 \\ \underline{28} \\ 0 \end{array}$$

Multiply by 60 since
60 minutes = 1 hour.

2.140

$$\begin{array}{r} 4\text{ lbs.} \quad 7\text{ oz.} \\ 5\overline{)22\text{ lbs.}} \quad 3\text{ oz.} \\ \underline{20} \\ 2\text{ lbs.} = \underline{32\text{ oz.}} \\ 35\text{ oz.} \\ \underline{35} \\ 0 \end{array}$$

Multiply by 16 since
16 ounces = 1 pound.

2.141

$$\begin{array}{r} 14\text{ yds.} \quad 2\text{ ft.} \quad R\ 1\text{ ft.} \\ 12\overline{)176\text{ yds.}} \quad 1\text{ ft.} \\ \underline{12} \\ 56 \\ \underline{48} \\ 8\text{ yds} = \underline{24\text{ ft.}} \\ 25\text{ ft.} \\ \underline{24} \\ 1\text{ ft.} = 12\text{ in.} \\ \underline{12} \\ 0 \end{array}$$

Multiply by 3 since
3 feet = 1 yard.
Multiply by 12 since
12 inches = 1 foot.

2.142

$$\begin{array}{r} 3\text{ min.} \quad 30\text{ sec. } R2\text{ sec.} \\ 7\overline{)24\text{ min.}} \quad 32\text{ sec.} \\ \underline{21} \\ 3\text{ min.} = \ ^1180\text{ sec.} \\ 212\text{ sec.} \\ \underline{21} \\ \overline{02\text{ sec.}} \end{array}$$

Multiply by 60 since
60 seconds = 1 minute.

2.143

$$\begin{array}{r} 12\text{ miles } 1{,}466\text{ yds. } R4\text{ yds.} \\ 6\overline{)77\text{ miles}} \quad \text{(or 12} \\ \underline{6} \quad \text{miles} \\ 17 \quad R5\text{ miles)} \\ \underline{12} \\ 5\text{ miles} = 8{,}800\text{ yds.} \\ \underline{6} \\ 28 \\ \underline{24} \\ 40 \\ \underline{36} \\ 40 \\ \underline{36} \\ 4\text{ yds.} \end{array}$$

Multiply by 1,760 since
1,760 yards = 1 mile.

2.144

$$\begin{array}{r} 55\text{ yds.} \\ 8\overline{)440\text{ yds.}} \\ \underline{40} \\ 40 \\ \underline{40} \\ 0 \end{array}$$

2.145

$$\begin{array}{r} 8\text{ lbs.} \quad 5\text{ oz.} \\ 9\overline{)74\text{ lbs.}} \quad 13\text{ oz} \\ \underline{72} \\ 2\text{ lbs.} = \underline{32\text{ oz.}} \\ 45\text{ oz.} \\ \underline{45} \\ 0 \end{array}$$

Multiply by 16 since
16 ounces = 1 pound.

2.146

$$\begin{array}{r} 35\text{ hrs. } 17\text{ min. } R5\text{ min.} \\ 27\overline{)952\text{ hrs.}} \quad 44\text{ min.} \\ \underline{81} \\ 142 \\ \underline{135} \\ 7\text{ hrs.} = \underline{420\text{ min.}} \\ 464\text{ min.} \\ \underline{27} \\ 194 \\ \underline{189} \\ 5\text{ min.} \end{array}$$

Multiply by 60 since
60 minutes = 1 hour.

2.147

$$
\begin{array}{r}
18\text{ ft.} \qquad 7\text{ in R26 in.}\\
35\overline{)652\text{ ft.}} \qquad 7\text{ in.}\\
35\phantom{2\text{ ft.}}\\
\overline{302}\phantom{\text{ ft.}}\\
280\phantom{\text{ ft.}}\\
\overline{22}\text{ ft.} \quad = \quad 2^{1}64\text{ in.}\\
271\\
\overline{271}\\
245\\
\overline{26}\text{ in.}
\end{array}
$$

Multiply by 12 since
12 inches = 1 foot.

2.148

$$
\begin{array}{r}
18\text{ miles} \quad 515\text{ yds.} \quad R17\text{ yds.}\\
41\overline{)750\text{ miles}} \quad 12\text{ yds.}\\
41\phantom{0\text{ miles}}\\
\overline{340}\phantom{\text{ miles}}\\
328\phantom{\text{ miles}}\\
\overline{12}\text{ miles} = 21{,}120\text{ yds.}\\
21{,}132\text{ yds.}\\
205\\
\overline{63}\\
41\\
\overline{222}\\
205\\
\overline{17}\text{ yds.}
\end{array}
$$

Multiply by 1,760 since
1,760 yards = 1 mile.

2.149

$$
\begin{array}{r}
13\text{ lbs.} \qquad 9\text{ oz.} \quad R49\text{ oz.}\\
70\overline{)952\text{ lbs.}} \qquad 7\text{ oz.}\\
70\phantom{0\text{ lbs.}}\\
\overline{252}\phantom{\text{ lbs.}}\\
210\phantom{\text{ lbs.}}\\
\overline{42}\text{ lbs.} = 672\text{ oz.}\\
679\text{ oz.}\\
630\\
\overline{49}\text{ oz.}
\end{array}
$$

Multiply by 16 since
16 ounces = 1 pound.

2.150

$$
\begin{array}{r}
19\text{ tons} \qquad 1{,}565\text{ lbs.}\\
8\overline{)156\text{ tons}} \qquad 4{,}520\text{ lbs.}\\
8\phantom{00\text{ tons}}\\
\overline{76}\phantom{0\text{ tons}}\\
72\phantom{0\text{ tons}}\\
\overline{4}\text{ tons} = 8{,}000\text{ lbs.}\\
12{,}520\text{ lbs.}\\
8\\
\overline{45}\\
40\\
\overline{52}\\
48\\
\overline{40}\\
40\\
\overline{0}
\end{array}
$$

Multiply by 2,000 since
2,000 pounds = 1 ton.

2.151

$$
\begin{array}{r}
18\text{ gals.} \qquad 2\text{ qts.} \quad R7\text{ qts.}\\
16\overline{)297\text{ gals.}} \qquad 3\text{ qts.}\\
16\phantom{0\text{ gals.}}\\
\overline{137}\phantom{\text{ gals.}}\\
128\phantom{\text{ gals.}}\\
\overline{9}\text{ gals.} = 36\text{ qts.}\\
39\text{ qts.}\\
32\\
\overline{7}\text{ qts.}
\end{array}
$$

Multiply by 4 since
4 quarts = 1 gallon.

2.152

$$
\begin{array}{r}
5\text{ miles} \qquad 4{,}292\text{ ft.}\\
6\overline{)34\text{ miles}} \qquad 4{,}632\text{ ft.}\\
30\phantom{0\text{ miles}}\\
\overline{4}\text{ miles} = 21{,}120\text{ ft.}\\
25{,}752\text{ ft.}\\
24\\
\overline{17}\\
12\\
\overline{55}\\
54\\
\overline{12}\\
12\\
\overline{0}
\end{array}
$$

Multiply by 5,280 since
5,280 feet = 1 mile.

2.153

$$
\begin{array}{r}
2\text{ gals.} \quad 1\text{ qt.} \quad 0\text{ pts. R 31 pts.}\\
32\overline{)75\text{ gals.}} \quad 3\text{ qts.} \quad 1\text{ pt.} \quad \text{pts.}\\
64\phantom{0\text{ gals.}}\\
\overline{11}\text{ gals.} = 44\text{ qts.}\\
47\text{ qts.}\\
32\\
\overline{15}\text{ qts.} = 30\text{ pts.}\\
31\text{ pts.}
\end{array}
$$

Multiply by 4 since
4 quarts = 1 gallon.
Multiply by 2 since
2 pints = 1 quart.

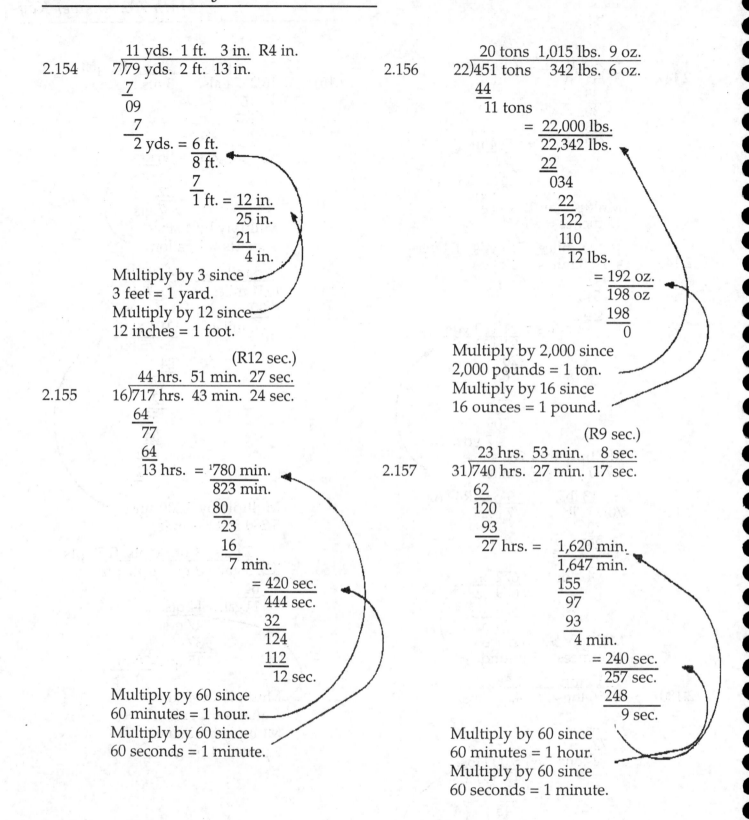

2.154

$$\begin{array}{r} 11 \text{ yds. } 1 \text{ ft. } 3 \text{ in. } \text{R4 in.} \\ 7\overline{)79 \text{ yds. } 2 \text{ ft. } 13 \text{ in.}} \end{array}$$

$$\underline{7}$$
$$09$$
$$\underline{7}$$
$$2 \text{ yds.} = \underline{6 \text{ ft.}}$$
$$8 \text{ ft.}$$
$$\underline{7}$$
$$1 \text{ ft.} = \underline{12 \text{ in.}}$$
$$25 \text{ in.}$$
$$\underline{21}$$
$$4 \text{ in.}$$

Multiply by 3 since
3 feet = 1 yard.
Multiply by 12 since
12 inches = 1 foot.

2.155

(R12 sec.)
$$\begin{array}{r} 44 \text{ hrs. } 51 \text{ min. } 27 \text{ sec.} \\ 16\overline{)717 \text{ hrs. } 43 \text{ min. } 24 \text{ sec.}} \end{array}$$
$$\underline{64}$$
$$77$$
$$\underline{64}$$
$$13 \text{ hrs.} = {}^{1}780 \text{ min.}$$
$$823 \text{ min.}$$
$$\underline{80}$$
$$23$$
$$\underline{16}$$
$$7 \text{ min.}$$
$$= \underline{420 \text{ sec.}}$$
$$444 \text{ sec.}$$
$$\underline{32}$$
$$124$$
$$\underline{112}$$
$$12 \text{ sec.}$$

Multiply by 60 since
60 minutes = 1 hour.
Multiply by 60 since
60 seconds = 1 minute.

2.156

$$\begin{array}{r} 20 \text{ tons } 1,015 \text{ lbs. } 9 \text{ oz.} \\ 22\overline{)451 \text{ tons } 342 \text{ lbs. } 6 \text{ oz.}} \end{array}$$
$$\underline{44}$$
$$11 \text{ tons}$$
$$= \underline{22,000 \text{ lbs.}}$$
$$22,342 \text{ lbs.}$$
$$\underline{22}$$
$$034$$
$$\underline{22}$$
$$122$$
$$\underline{110}$$
$$12 \text{ lbs.}$$
$$= \underline{192 \text{ oz.}}$$
$$198 \text{ oz}$$
$$\underline{198}$$
$$0$$

Multiply by 2,000 since
2,000 pounds = 1 ton.
Multiply by 16 since
16 ounces = 1 pound.

2.157

(R9 sec.)
$$\begin{array}{r} 23 \text{ hrs. } 53 \text{ min. } 8 \text{ sec.} \\ 31\overline{)740 \text{ hrs. } 27 \text{ min. } 17 \text{ sec.}} \end{array}$$
$$\underline{62}$$
$$120$$
$$\underline{93}$$
$$27 \text{ hrs.} = \underline{1,620 \text{ min.}}$$
$$1,647 \text{ min.}$$
$$\underline{155}$$
$$97$$
$$\underline{93}$$
$$4 \text{ min.}$$
$$= \underline{240 \text{ sec.}}$$
$$257 \text{ sec.}$$
$$\underline{248}$$
$$9 \text{ sec.}$$

Multiply by 60 since
60 minutes = 1 hour.
Multiply by 60 since
60 seconds = 1 minute.

2.158

```
                    R4 sq. in.
        12 sq. ft.      113 sq. in.
      7)89 sq. ft.  75 sq. in.
        7
       ──
        19
        14
       ──
         5 sq. ft.  =  720 sq. in.
                        795 sq. in.
                        7
                       ──
                        09
                         7
                        ──
                        25
                        21
                       ──
                         4 sq. in.
```

Multiply by 144 since
144 square inches
= 1 square foot.

2.159

Enter	Display	Comment
C (maybe twice)	0	Clearing calc.
457	457	First factor
x	457	Multiply instruction
322	322	Second factor
x	147, 154	First product, 2nd instruction
217	217	Third factor
=	31,932,418	The answer

```
 12            1  1
 11            0  0
 11           351 32
 457         147,154
x 322         x 217
 914        1¹03¹0¹078
¹914        1 47 1 54
1¹371       ¹2 94 3 08
───────     ──────────
147,154      31,932,418
```

2.160

Enter	Display	Comment
C (maybe twice)	0	Clearing calc.
6,523	6,523	First quantity
x	6,523	Multiply instruction
256	256	Second quantity
÷	1,669,888	Product, Second instruction
32	32	Third quantity
=	52,184	The answer

```
   1          52,184              8
  2 1 1    32)1,669,888  or   32)256
  3 1 1        160               256
  6,523        69               ───
  x  256       64                 0
 ¹39138       ──
 32615         58              4 12
 13046         32              6,523
 ───────       ──              x  8
 1,669,888     268             ──────
               256             52,184
               ───
               128
               128
               ───
                 0
```

2.161

Enter	Display	Comment
C (maybe twice)	0	Clearing calc.
221,116	221,116	First quantity
÷	221,116	Divide instruction
742	742	Second quantity
x	298	Quotient, Second instruction
527	527	Third quantity
=	157,046	The answer

```
                         44
         298             11
    742)221,116          65
        148 4           298
        7271           x 527
        6678           2¹086
        5936            596
        5936          1¹490
        ────          ───────
           0          157,046
```

2.162

Enter	Display	Comment
C maybe twice)	0	Clearing calc.
725	725	First factor
x	725	Multiply instruction
194	194	Second factor
x	140,650	First product, Second instruction
314	314	Third factor
=	44,164,100	The answer

```
  24              1 11
  12              0 00
 725              1 22
x 194           140,650
¹2900           x 314
¹6525         ¹5¹6¹2600
 725          ¹14 0 650
140,650       421 9 50
             44,164,100
```

2.163

Enter	Display	Comment
C maybe twice)	0	Clearing calc.
287,616	287,616	First quantity
x	287,616	Multiply instruction
247	247	Second quantity
÷	71,041,152	Product, Second instruction
642	642	Third quantity
=	110,656	The answer

```
 111 1              110,656
 332 2        642)71,041,152
 65414              642
287,616             684
 x 247              642
¹20¹1¹3312          4211
¹1150464            3852
575232              3595
71,041,152          3210
                    3852
                    3852
                       0
```

2.164

Enter	Display	Comment
C maybe twice)	0	Clearing calc.
2,446,115	2,446,115	First quantity
÷	2,446,115	Divide instruction
1,645	1,645	Second quantity
x	1,487	Quotient, Second instruction
452	452	Third quantity
=	672,124	The answer

```
            1,487        1 3 2
1,645)2,446,115          2 4 3
      1645              0 1 1
      8011              1,487
      6580             x 452
     14311           ²2¹974
     13160           ¹74 35
     11515           ¹594 8
     11515           672,124
         0
```

2.165

Enter	Display	Comment
C maybe twice)	0	Clearing calc.
1,451	1,451	First factor
x	1,451	Muptiply instruction
24	24	Second factor
x	34,824	First product, Second instruction
16	16	Third factor
=	557,184	The answer

```
   1             24 12
  1 2            34,824
 1,451           x 16
 x 24          2¹0¹8944
 5804           34824
¹2902          557,184
 34,824
```

2.166

Enter	Display	Comment
C maybe twice)	0	Clearing calc.
4,352	4,352	First quantity
x	4,352	Multiply instruction
1,475	1,475	Second quantity
+	6,419,200	Product, Second instruction
452,527	452,527	Third quantity
=	6,871,727	The answer

$$
\begin{array}{r}
1\,2 \\
2\,3\,1 \\
1\,2\,1 \\
4,352 \\
\times\ 1,475 \\
\hline
2^11^1760 \\
{}^130\ 4\ 64 \\
{}^1174\ 0\ 8 \\
4352 \\
\hline
6,419,200
\end{array}
\qquad
\begin{array}{r}
1 \\
6,419,200 \\
+\ 452,527 \\
\hline
6,871,727
\end{array}
$$

2.167

Enter	Display	Comment
C maybe twice)	0	Clearing calc.
76,673,776	76,672,776	First quantity
÷	76,672.776	Divide instruction
1,256	1,256	Second quantity
-	61,046	Quotient, Second instruction
16,523	16,523	Third quantity
=	44,523	The answer

$$
\begin{array}{r}
61,046 \\
1,256)\overline{76,673,776} \\
7536 \\
\hline
1313 \\
1256 \\
\hline
5777 \\
5024 \\
\hline
7536 \\
7536 \\
\hline
0
\end{array}
\qquad
\begin{array}{r}
510_1 \\
61,046 \\
-\ 16,523 \\
\hline
44,523
\end{array}
$$

2.168

Enter	Display	Comment
C maybe twice)	0	Clearing calc.
12	12	First factor
x	12	Multiply instruction
27	27	Second factor
=	324	The answer

$$
\begin{array}{r}
1 \\
12 \\
\times\ 27 \\
\hline
84 \\
{}^124 \\
\hline
324
\end{array}\ \text{alarms}
$$

2.169

Enter	Display	Comment
C maybe twice)	0	Clearing calc.
324	324	First factor
x	324	Multiply instruction
4.95	4.95	Second factor
=	1,603.8	The answer

$$
\begin{array}{r}
1 \\
23 \\
12 \\
324 \\
\times\ \$4.95 \\
\hline
{}^11620 \\
{}^22916 \\
1296 \\
\hline
\$1,603.80
\end{array}
$$

Note: Most calculators do not show a 0 at the end of an answer. You must remember to write it for money amounts when necessary.

I.   SECTION ONE

1.1 $\overleftrightarrow{RB}$

1.2 $\overleftrightarrow{UV}$

1.3 $\overleftrightarrow{PQ}$

1.4 $\overleftrightarrow{SW}$

1.5 $\overleftrightarrow{EF}$

1.6   line

1.7   not line (not straight)

1.8   lines

1.9   not line (not straight)

1.10   lines

1.11   a. $2\frac{5}{8}$
       b. 6.7

1.12   a. $2\frac{1}{4}$
       b. 5.6

1.13   a. $1\frac{1}{8}$
       b. 2.9

1.14   a. $2\frac{1}{4}$
       b. 5.6

1.15   a. $\frac{3}{4}$
       b. 2

1.16   a. $\frac{5}{8}$
       b. 1.6

1.17   a. $5\frac{1}{8}$
       b. 13

1.18   a. $2\frac{3}{4}$
       b. 7.0

1.19   a. $1\frac{3}{4}$
       b. 4.4

1.20   a. $2\frac{5}{8}$
       b. 6.7

1.21   A ●————————————● B

1.22   C ●————————● D

1.23   E ●——————————————● F

1.24   G ●———————————————————————● H

1.25   I ●————————————————● J

1.26   K ●——————————————● L

1.27   M ●————————————————————● N

149

1.28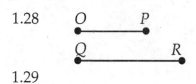

1.29

1.30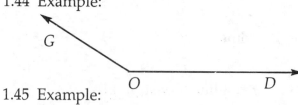

S ————————————————————————— T

1.31 a. ∠ RST
b. ∠ S

1.32 a ∠ WXY
b. ∠ X

1.33 a. ∠ ABC
b. ∠ B

1.34 a. ∠ ABC
b. ∠ B

1.35 a. ∠ ABC OR ∠ CBA
b. ∠ B

1.36 a. ∠ MOP OR ∠ POM
b. ∠ O

1.37 ∠ ADB, ∠ BDC, ∠ ADC

1.38 ∠ TOS or ∠ SOT, ∠ SOU or ∠ UOS,
∠ UOR or ∠ ROU, ∠ ROT or ∠ TOR,
∠ TOU or ∠ UOT, ∠ ROS or ∠ SOR

1.39 ∠ EGF or ∠ FGE, ∠ GFE or ∠ EFG,
∠ FEG or ∠ GEF

1.40 ∠ MTA or ∠ ATM, ∠ MTH or ∠ HTM,
∠ ATH or ∠ HTA

1.41 Example

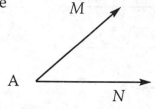

1.42 Example:

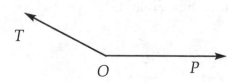

1.43 Example:

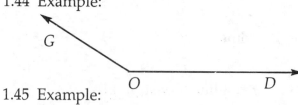

1.44 Example:

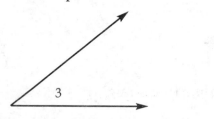

1.45 Example:

1.46 Example:

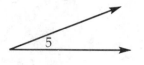

1.47 Example:

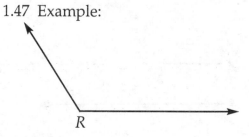

1.48 Example:

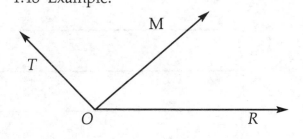

150

1.49 Example

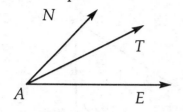

1.50 Example:

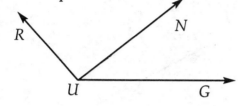

1.51 $\angle AOB = 40°$

1.52 $\angle AOC = 65°$

1.53 $\angle AOD = 70°$

1.54 $\angle AOE = 117°$

1.55 $\angle AOF = 132°$

1.56 $\angle AOG = 161°$

1.57 $\angle BOC = 25°$

1.58 $\angle BOD = 30°$

1.59 $\angle BOE = 77°$

1.60 $\angle BOF = 92°$

1.61 $\angle BOG = 121°$

1.62 $\angle COD = 5°$

1.63 $\angle COE = 52°$

1.64 $\angle COF = 67°$

1.65 $\angle COG = 96°$

1.66 $\angle DOE = 47°$

1.67 $\angle DOF = 62°$

1.68 $\angle DOG = 91°$

1.69 $\angle EOF = 15°$

1.70 $\angle EOG = 44°$

1.71 $\angle FOG = 30°$

1.72

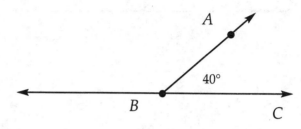

1.73

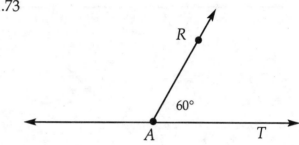

1.74

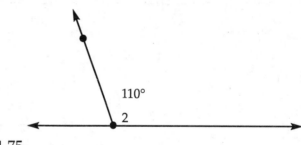

1.75

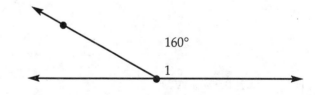

1.76

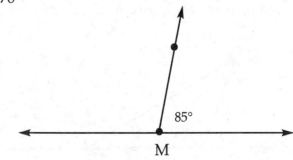

1.77

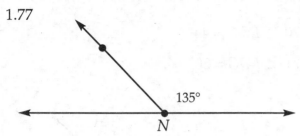

1.78

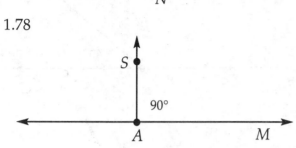

1.79

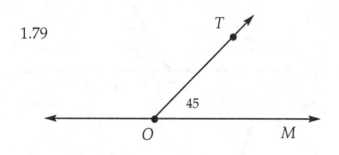

1.80

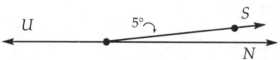

II.   SECTION TWO
NOTE: To conserve space not all of the
drawings in the remainder of the 703 keys are
the correct size. Use the samples given to
evaluate correct procedure.

2.1

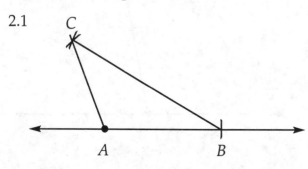

2.2

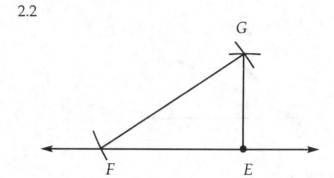

2.3

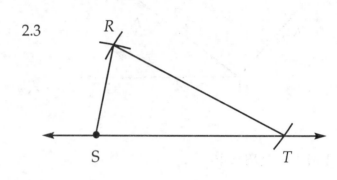

2.4

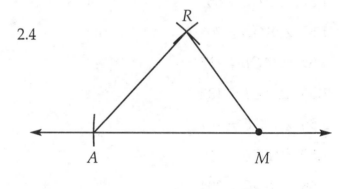

2.5

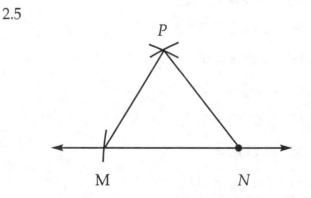

2.6   a. $4\dfrac{7}{8}$

b. 12.3

2.7　a.　$4\frac{5}{8}$

　　b.　11.7

2.8　a.　$5\frac{1}{8}$

　　b.　13.2

2.9　a.　$3\frac{1}{2}$

　　b.　8.8

2.10　a.　$11\frac{3}{4}$

　　b.　29.7

2.11　$P = 6\frac{3}{4}$ inches

　　$P = 17$ cm

　　$\angle A = 40°$

　　$\angle B = 45°$

　　$\angle C = 95°$

　　Sum of $\angle$'s $= 180°$

2.12　$P = 6\frac{3}{8}$ inches

　　$P = 16.2$ cm

　　$\angle R = 25°$

　　$\angle S = 140°$

　　$\angle T = 15°$

　　Sum of $\angle$'s $= 180°$

2.13　$P = 8\frac{3}{8}$ inches

　　$P = 21.4$ cm

　　$\angle A = 90°$

　　$\angle B = 35°$

　　$\angle C = 55°$

　　Sum of $\angle$'s $= 180°$

2.14　8 ft. + 12 ft. + 17 ft. = 37 ft.

2.15　30′ + 30′ + 45′ = 105′

2.16　Area is approximately 12

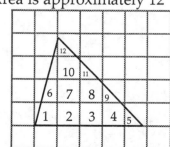

2.17　Area is approximately 14

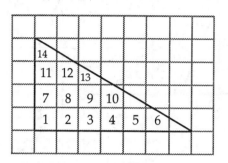

2.18　Area is approximately 14

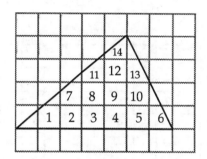

2.19　Area is approximately 36

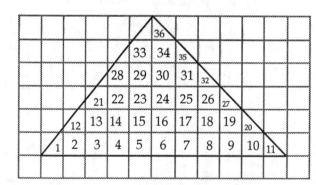

2.20　Area is approximately 33

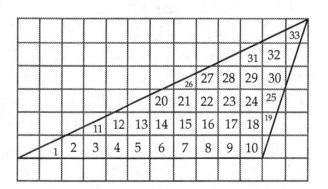

2.21 Area is approximately 38

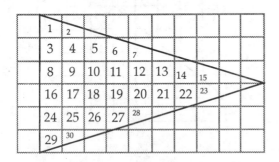

2.22 Area is approximately 30

2.23 Area is approximately 107

2.24

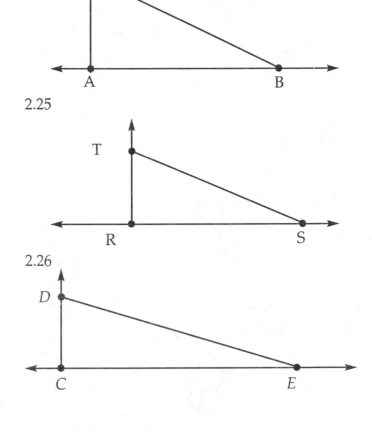

Not all of the drawings are the correct size. Use the samples given to evaluate correct procedure.

2.25

2.26

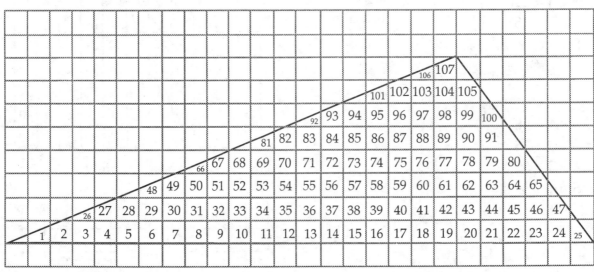

2.27   Not all of the drawings are the correct size. Use the samples given to evaluate correct procedure.

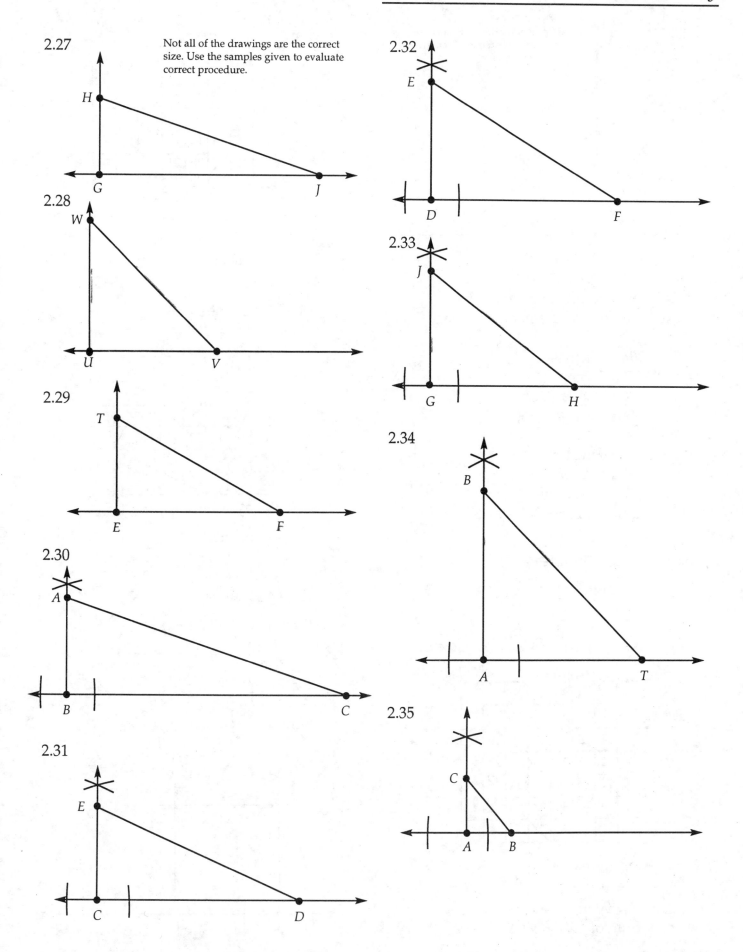

2.28

2.29

2.30

2.31

2.32

2.33

2.34

2.35

2.36

Not all of the drawings are the correct size. Use the samples given to evaluate correct procedure.

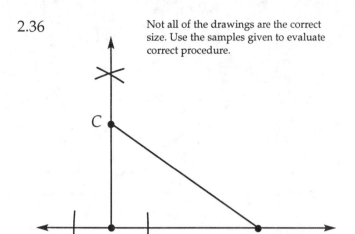

III.   SECTION THREE

NOTE: Due to printing variations, these measurements may vary.

3.1 $P = 3$ inches
$P = 7.6$ cm
$A = 9$ sq. units
Sum of $\angle$'s $= 360°$

3.2 $P = 6$ inches
$P = 15.2$ cm
$A = 36$ sq. units
Sum of $\angle$'s $= 360°$

3.3 $P = 5\frac{1}{2}$ inches
$P = 14$ cm
$A = 40$ sq. units
Sum of $\angle$'s $= 360°$

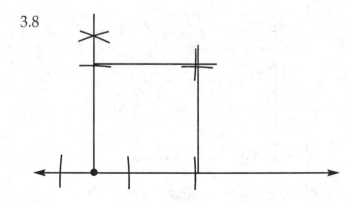

3.4 $P = 4$ inches
$P = 10.4$ cm
$A = 24$ sq. units
Sum of $\angle$'s $= 360°$

3.4 cont.

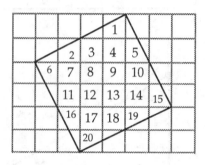

3.5 $P = 4\frac{1}{4}$ inches
$P = 10.8$ cm
$A \doteq 20$ sq. units
Sum of $\angle$'s $= 360°$

3.6 $P = 7\frac{3}{4}$ inches
$P = 20$ cm
$A = 64$ sq. units
Sum of $\angle$'s $= 360°$

3.7 $P = 1$ inches
$P = 2.4$ cm
$A = 1$ sq. unit
Sum of $\angle$'s $= 360°$

3.8

Not all of the drawings are the correct size. Use the samples given to evaluate correct procedure.

3.9

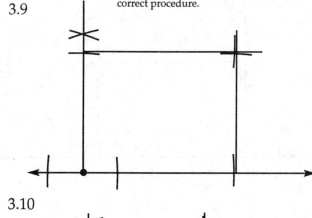

3.10

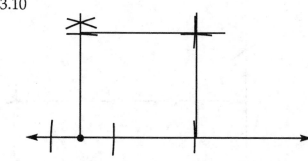

3.11

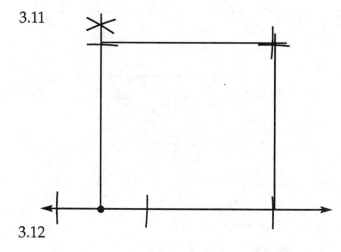

3.12

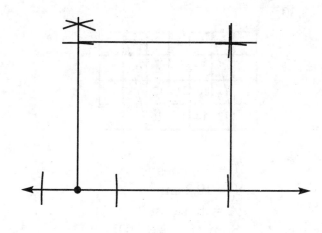

3.13

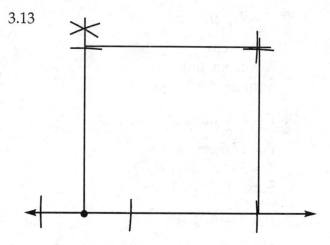

3.14 $P = 4$ inches
        $\therefore$ each side = 1 inch

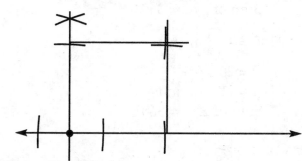

3.15 $P = 3$ inches
        $P = 7.6$ cm
        $A = 8$ sq. units
        Sum of $\angle$'s = 360°

3.16 $P = 5\frac{7}{8}$ inches
        $P = 15$ cm
        $A = 32$ sq. units
        Sum of $\angle$'s = 360°

3.17 $P = 5\frac{3}{8}$ inches
        $P = 13.7$ cm
        $A = 28$ sq. units
        Sum of $\angle$'s = 360°

3.18 $P = 6\frac{3}{4}$ inches
        $P = 16.9$ cm
        $A = 48$ sq. units
        Sum of $\angle$'s = 360°

3.19 $P = 6$ inches
        $P = 15.1$ cm
        $A = 11$ sq. units
        Sum of $\angle$'s = 360°

Not all of the drawings are the correct size. Use the samples given to evaluate correct procedure.

3.20 $P = 7\frac{1}{8}$ inches

$P = 18$ cm

$A = 42$ sq. units

Sum of $\angle$'s $= 360°$

3.21 $P = 6\frac{7}{8}$ inches

$P = 17.6$ cm

$A = 33$ sq. units

Sum of $\angle$'s $= 360°$

3.22 $P = 5\frac{3}{8}$ inches

$P = 13.8$ cm

$A = 30$ sq. units

Sum of $\angle$'s $= 360°$

3.23 $P = 11\frac{1}{4}$ inches

$P = 28.6$ cm

$A = 112$ sq. units

Sum of $\angle$'s $= 360°$

3.24

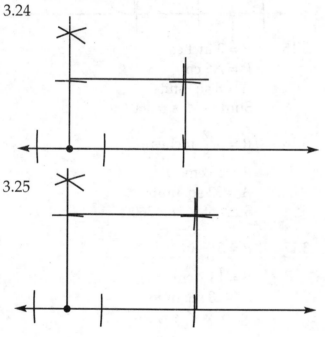

3.25

3.26

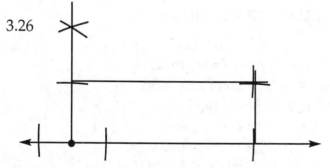

3.27

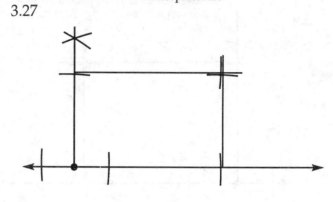

3.28

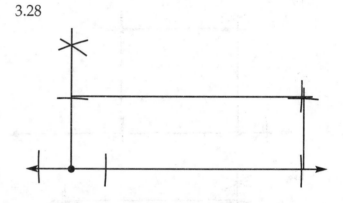

3.29

		1	2	3	4	5
	6	7	8	9	10	
11	12	13	14	15		

$P = 4\frac{1}{4}$ inches

$P = 11$ cm

$A \doteq 15$ sq. units

Sum of $\angle$'s $= 360°$

3.30

	1	2	3	4	
	5	6	7	8	
9	10	11	12		
13	14	15	16		

$P = 4$ inches

$P = 10.2$ cm

$A \doteq 16$ sq. units

Sum of $\angle$'s $= 360°$

3.31

$P = 4\frac{3}{4}$ inches

$P = 12.2$ cm

$A \doteq 21$ sq. units

Sum of $\angle$'s = 360°

3.32

$P = 6\frac{3}{8}$ inches

$P = 16.4$ cm

$A \doteq 30$ sq. units

Sum of $\angle$'s = 360°

3.33

$P = 7\frac{1}{4}$ inches

$P = 18.6$ cm

$A \doteq 54$ sq. units

Sum of $\angle$'s = 360°

3.34

$P = 7$ inches

$P = 17.4$ cm

$A \doteq 45$ sq. units

Sum of $\angle$'s = 360°

3.35

$P = 9\frac{1}{2}$ inches

$P = 24$ cm

$A \doteq 89$ sq. units

Sum of $\angle$'s = 360°

3.36

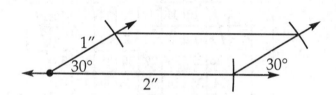

$P = 8\frac{5}{8}$ inches

$P = 21.6$ cm

$A \doteq 75$ sq. units

Sum of $\angle$'s = 360°

3.37

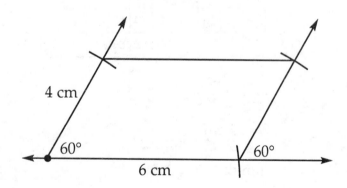

Not all of the drawings are the correct size. Use the samples given to evaluate correct procedure.

3.38

3.39

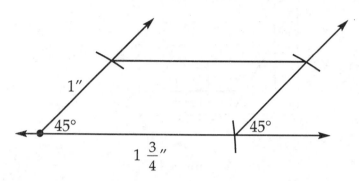

3.40

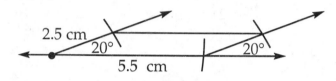

3.41

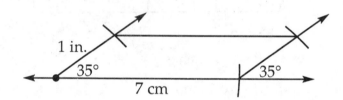

3.42

3.43

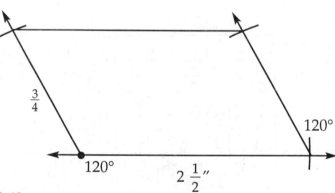

3.43 cont.

$$P = 4\frac{5}{8} \text{ inches}$$
$$P = 11.8 \text{ cm}$$
$$A \doteq 20 \text{ sq. units}$$
Sum of $\angle$'s = 360°

3.44

(figure: trapezoid on grid with 24 squares)

$$P = 5\frac{1}{8} \text{ inches}$$
$$P = 13.0 \text{ cm}$$
$$A \doteq 24 \text{ sq. units}$$
Sum of $\angle$'s = 360°

3.45

(figure: trapezoid on grid with 30 squares)

$$P = 5\frac{3}{8} \text{ inches}$$
$$P = 13.7 \text{ cm}$$
$$A \doteq 30 \text{ sq. units}$$
Sum of $\angle$'s = 360°

3.46

(figure: shape on grid with 15 squares)

$$P = 4\frac{1}{16} \text{ inches}$$
$$P = 10.4 \text{ cm}$$
$$A \doteq 15 \text{ sq. units}$$
Sum of $\angle$'s = 360°

3.47

(figure: trapezoid on grid with 49 squares)

$$P = 7\frac{7}{8} \text{ inches}$$
$$P = 19.9 \text{ cm}$$
$$A \doteq 49 \text{ sq. units}$$
Sum of $\angle$'s = 360°

3.48

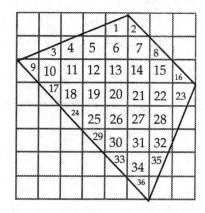

$$P = 5\frac{5}{8} \text{ inches}$$
$$P = 14.6 \text{ cm}$$
$$A \doteq 36 \text{ sq. units}$$
Sum of $\angle$'s = 360°

3.49

(figure: parallelogram on grid with 108 squares)

3.49 cont.

$$P = 10\frac{7}{8} \text{ inches}$$

$P = 27.5$ cm

$A \doteq 108$ sq. units

Sum of $\angle$'s $= 360°$

3.50

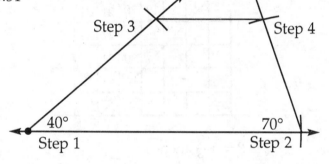

$$P = 7\frac{3}{16} \text{ inches}$$

$P = 18.4$ cm

$A \doteq 35$ sq. units

Sum of $\angle$'s $= 360°$

3.51

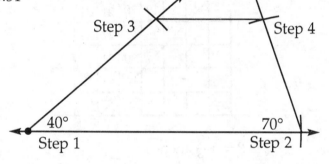

3.52

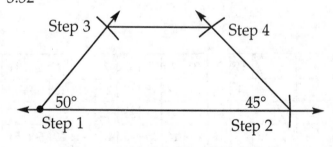

IV. SECTION FOUR

4.1 through 4.6

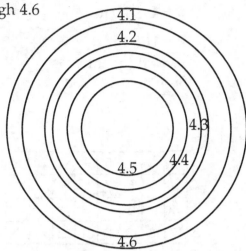

4.1

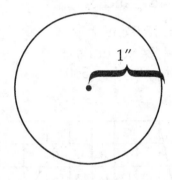

4.2

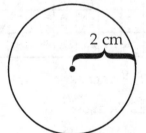

4.3

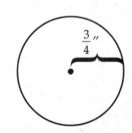

4.4

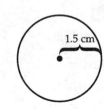

162

4.5

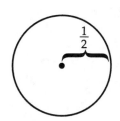

4.6

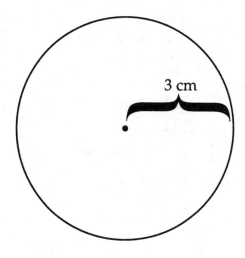

4.7 $C = \pi d$
$d = 2$

$C = 3\frac{1}{7} \times 2$

$C = \frac{22}{7} \times \frac{2}{1}$

$C = \frac{44}{7}$

$C = 6\frac{2}{7}''$

4.8 $C = \pi d$
$d = 14$

$C = 3\frac{1}{7} \times 14$

$C = \frac{22}{{}_1\cancel{7}} \times \frac{\cancel{14}^{\,2}}{1}$

$C = \frac{44}{1}$

$C = 44$ ft.

4.9 $C = \pi d$
$d = 7$

$C = 3\frac{1}{7} \times 7$

$C = \frac{22}{{}_1\cancel{7}} \times \frac{\cancel{7}^{\,1}}{1}$

$C = 22$ cm

4.10 $C = \pi d$
$d = 2r$
$r = 5$
$d = 2 \times 5 = 10$

$C = 3\frac{1}{7} \times 10$

$C = \frac{22}{7} \times \frac{10}{1}$

$C = \frac{220}{7}$

$C = 31\frac{3}{7}$ cm

4.11 $C = \pi d$
$d = 2r$
$r = 14$
$d = 2 \times 14 = 28$

$C = 3\frac{1}{7} \times 28$

$C = \frac{22}{{}_1\cancel{7}} \times \frac{\cancel{28}^{\,4}}{1}$

$C = 88$ cm

4.12 $C = \pi d$
$C = 3.14 \times 6$
$C = 18.84''$

4.13 $C = \pi d$
$C = 3.14 \times 10$
$C = 31.4''$

4.14 $C = \pi d$
$d = 2r$
$r = 4.5$
$d = 2 \times 4.5 = 9$
$C = 3.14 \times 9$
$C = 28.26$ cm

4.15 $C = \pi d$
$d = 2r$
$r = 1.6$
$d = 2 \times 1.6 = 3.2$
$C = 3.14 \times 3.2$
$C = 10.048$ cm

4.16 $C = \pi d$
$C = 3.14 \times 100$
$C = 314$ ft.

4.17 $A = \pi r^2$

$A = 3\frac{1}{7} \times 2 \times 2$

$A = \frac{22}{7} \times \frac{4}{1}$

$A = \frac{88}{7}$

$A = 12\frac{4}{7}$ sq. in.

4.18 $A = \pi r^2$

$A = 3\frac{1}{7} \times 3 \times 3$

$A = \frac{22}{7} \times \frac{9}{1}$

$A = \frac{198}{7}$

$A = 28\frac{2}{7}$ sq. cm

4.19 $A = \pi r^2$

$A = 3\frac{1}{7} \times 7 \times 7$

$A = \frac{22}{7} \times \frac{49}{1}$

$A = 154$ sq. ft.

4.20 $A = \pi r^2$

$A = 3\frac{1}{7} \times \frac{1}{2} \times \frac{1}{2}$

$A = \frac{22}{7} \times \frac{1}{4}$

$A = \frac{11}{14}$ sq. in.

4.21 $A = \pi r^2$

$d = 2r$

$r = \frac{d}{2}$

$2r = 10$

$r = 5$

$A = 3\frac{1}{7} \times 5 \times 5$

$A = \frac{22}{7} \times \frac{25}{1}$

$A = \frac{550}{7}$

$A = 78\frac{4}{7}$ sq. in.

4.22 $A = \pi r^2$
$A = 3.14 \times 4 \times 4$
$A = 3.14 \times 16$
$A = 50.24$ sq. in.

4.23 $A = \pi r^2$
$A = 3.14 \times 1.2 \times 1.2$
$A = 3.14 \times 1.44$
$A = 4.5216$ sq. cm

4.24 $A = \pi r^2$
$A = 3.14 \times 6 \times 6$
$A = 3.14 \times 36$
$A = 113.04$ sq. ft.

4.25 $A = \pi r^2$
$d = 2r$
$r = \frac{d}{2}$
$2r = d$
$2r = 10$
$r = 5$
$A = 3.14 \times 5 \times 5$

$A = \frac{22}{7} \times \frac{25}{1}$

$A = 3.14 \times 25$
$A = 78.5$ sq. in.

4.26

$P = 6 \dfrac{1}{8}$ inches

$P = 15.6$ cm

$A \doteq 45$ sq. units

Sum of $\angle$'s $= 720°$

4.27

$P = 6 \dfrac{3}{8}$ inches

$P = 16.4$ cm

$A \doteq 49$ sq. units

Sum of $\angle$'s $= 720°$

4.28

$P = 7 \dfrac{4}{8} = 7 \dfrac{1}{2}$ inches

$P = 19.0$ cm

$A \doteq 63$ sq. units

Sum of $\angle$'s $= 720°$

4.29

$P = 5 \dfrac{4}{8} = 5 \dfrac{1}{2}$ inches

$P = 14.2$ cm

$A \doteq 36$ sq. units

Sum of $\angle$'s $= 720°$

4.30

$P = 10 \dfrac{1}{4}$ inches

$P = 26$ cm

$A \doteq 132$ sq. units

Sum of 's $= 720°$

4.31

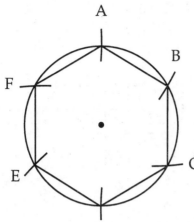

4.35

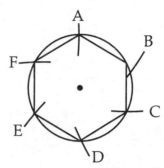

4.32

4.36

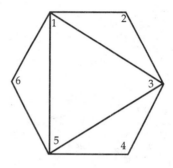

One large triangle and
three smaller triangles
are formed.

4.33

4.37

4.34

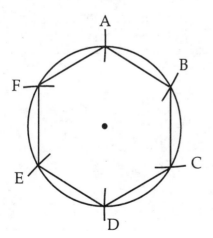

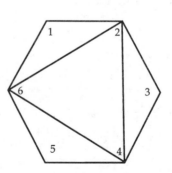

One large triangle
and three smaller
triangles are formed.

4.38

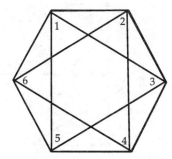

Twelve small triangles and
one smaller hexagon are formed.

4.40

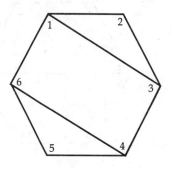

Two triangles and one
rectangle are formed.

4.39

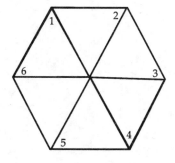

Six triangles that are
all the same size are
formed.

4.41

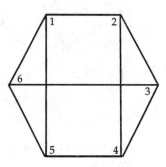

Four right triangles and
two rectangles are formed.

4.42

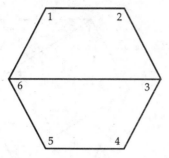

Two trapezoids are formed.

4.44

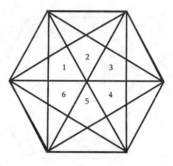

6 nonoverlapping quadrilaterals are formed.

4.43

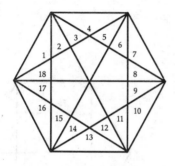

18 nonoverlapping triangles are formed.

I. SECTION ONE

1.1 $\frac{1}{2}$; 1 tells how many parts; 2 tells the size of each part

1.2 $\frac{4}{5}$; 4 tells how many parts; 5 tells the size of each part

1.3 $\frac{4}{9}$

1.4 $\frac{4}{15}$

1.5 $\frac{4}{9}$ is larger than $\frac{4}{15}$ because ninths are larger than fifteenths.

1.6 $\frac{4}{15}$ is smaller than $\frac{4}{9}$ because fifteenths are smaller than ninths.

1.7   Examples:

$$\frac{3}{7} = \frac{3 \times 2}{7 \times 2} = \frac{6}{14}$$

$$\frac{3}{7} = \frac{3 \times 3}{7 \times 3} = \frac{9}{21}$$

$$\frac{3}{7} = \frac{3 \times 4}{7 \times 4} = \frac{12}{28}$$

1.8   Examples:

$$\frac{11}{14} = \frac{11 \times 2}{14 \times 2} = \frac{22}{28}$$

$$\frac{11}{14} = \frac{11 \times 3}{14 \times 3} = \frac{33}{42}$$

$$\frac{11}{14} = \frac{11 \times 4}{14 \times 4} = \frac{44}{56}$$

1.9   more than can be counted; an infinite amount

1.10 $\frac{10}{12} = \frac{2 \times 5}{2 \times 6} = \frac{5}{6}$

1.11 $\frac{16}{24} = \frac{8 \times 2}{8 \times 3} = \frac{2}{3}$

1.12   Reduce each fraction to lowest terms:

$$\frac{1}{2} = \frac{1}{2}$$

$$\frac{2}{4} = \frac{2 \times 1}{2 \times 2} = \frac{1}{2}$$

$$\frac{3}{6} = \frac{3 \times 1}{3 \times 2} = \frac{1}{2}$$

$$\frac{4}{9} = \frac{4}{9}$$

$$\frac{5}{10} = \frac{5 \times 1}{5 \times 2} = \frac{1}{2}$$

$$\frac{6}{12} = \frac{6 \times 1}{6 \times 2} = \frac{1}{2}$$

$\frac{4}{9}$ is not equivalent to the other fractions.

1.13 $\frac{4}{6} = \frac{2 \times 2}{2 \times 3} = \frac{2}{3}$

1.14 $\frac{5}{10} = \frac{5 \times 1}{5 \times 2} = \frac{1}{2}$

1.15 $\frac{8}{12} = \frac{4 \times 2}{4 \times 3} = \frac{2}{3}$

1.16 $\frac{10}{15} = \frac{5 \times 2}{5 \times 3} = \frac{2}{3}$

1.17 $\frac{16}{28} = \frac{4 \times 4}{4 \times 7} = \frac{4}{7}$

1.18 $\frac{20}{50} = \frac{10 \times 2}{10 \times 5} = \frac{2}{5}$

1.19 $\frac{13}{39} = \frac{13 \times 1}{13 \times 3} = \frac{1}{3}$

1.20 $\frac{9}{27} = \frac{9 \times 1}{9 \times 3} = \frac{1}{3}$

1.21 $\dfrac{50}{60} = \dfrac{10 \times 5}{10 \times 6} = \dfrac{5}{6}$

1.22 $\dfrac{42}{56} = \dfrac{14 \times 3}{14 \times 4} = \dfrac{3}{4}$

1.23 $16 \div 4 = 4$

$\dfrac{1}{4} = \dfrac{1 \times 4}{4 \times 4} = \dfrac{4}{16}$

1.24 $24 \div 6 = 4$

$\dfrac{5}{6} = \dfrac{5 \times 4}{6 \times 4} = \dfrac{20}{24}$

1.25   Examples:

$\dfrac{2}{5} = \dfrac{2 \times 2}{5 \times 2} = \dfrac{4}{10}$

$\dfrac{2}{5} = \dfrac{2 \times 3}{5 \times 3} = \dfrac{6}{15}$

$\dfrac{2}{5} = \dfrac{2 \times 4}{5 \times 4} = \dfrac{8}{20}$

$\dfrac{2}{5} = \dfrac{2 \times 5}{5 \times 5} = \dfrac{10}{25}$

$\dfrac{2}{5} = \dfrac{2 \times 6}{5 \times 6} = \dfrac{12}{30}$

$\dfrac{2}{5} = \dfrac{2 \times 7}{5 \times 7} = \dfrac{14}{35}$

1.26   Examples:

$\dfrac{5}{9} = \dfrac{5 \times 2}{9 \times 2} = \dfrac{10}{18}$

$\dfrac{5}{9} = \dfrac{5 \times 3}{9 \times 3} = \dfrac{15}{27}$

$\dfrac{5}{9} = \dfrac{5 \times 4}{9 \times 4} = \dfrac{20}{36}$

$\dfrac{5}{9} = \dfrac{5 \times 5}{9 \times 5} = \dfrac{25}{45}$

$\dfrac{5}{9} = \dfrac{5 \times 6}{9 \times 6} = \dfrac{30}{54}$

$\dfrac{5}{9} = \dfrac{5 \times 7}{9 \times 7} = \dfrac{35}{63}$

1.27 $15 \div 3 = 5$

$\dfrac{3}{7} = \dfrac{3 \times 5}{7 \times 5} = \dfrac{15}{35}$

1.28 $28 \div 7 = 4$

$\dfrac{7}{11} = \dfrac{7 \times 4}{11 \times 4} = \dfrac{28}{44}$

1.29   more than can be counted; an infinite number

1.30   Examples:

$\dfrac{2}{3} = \dfrac{2 \times 2}{3 \times 2} = \dfrac{4}{6}$

$\dfrac{2}{3} = \dfrac{2 \times 3}{3 \times 3} = \dfrac{6}{9}$

$\dfrac{2}{3} = \dfrac{2 \times 4}{3 \times 4} = \dfrac{8}{12}$

$\dfrac{2}{3} = \dfrac{2 \times 5}{3 \times 5} = \dfrac{10}{15}$

$\dfrac{2}{3} = \dfrac{2 \times 6}{3 \times 6} = \dfrac{12}{18}$

1.31   dividend

1.32   divisor

1.33 $\dfrac{11}{4} = 11 \div 4 = 2\,\dfrac{3}{4}$

1.34 $\dfrac{25}{7} = 25 \div 7 = 3\,\dfrac{4}{7}$

1.35 $\dfrac{33}{8} = 33 \div 8 = 4\,\dfrac{1}{8}$

1.36 $\dfrac{50}{11} = 50 \div 11 = 4\,\dfrac{6}{11}$

1.37 $\dfrac{26}{3}$

1.38 $\dfrac{41}{5}$

1.39 $\dfrac{4}{1}$

1.40 $\dfrac{11}{1}$

1.41 $\dfrac{6}{1} = 6 \div 1 = 6$

1.42 $\dfrac{32}{1} = 32 \div 1 = 32$

1.43 $\dfrac{12}{4} = 12 \div 4 = 3$

1.44 $\dfrac{36}{9} = 36 \div 9 = 4$

1.45 Examples:

$$\frac{4}{2} = \frac{{}^1\cancel{2} \times 2}{{}_1\cancel{2} \times 1} = \frac{2}{1}$$

$$\frac{4}{2} = \frac{4 \times 2}{2 \times 2} = \frac{8}{4}$$

$$\frac{4}{2} = \frac{4 \times 3}{2 \times 3} = \frac{12}{6}$$

1.46 Examples:

$$\frac{28}{7} = \frac{{}^1\cancel{7} \times 4}{{}_1\cancel{7} \times 1} = \frac{4}{1}$$

$$\frac{28}{7} = \frac{28 \times 2}{7 \times 2} = \frac{56}{14}$$

$$\frac{28}{7} = \frac{28 \times 3}{7 \times 3} = \frac{84}{21}$$

1.47 $\frac{7}{2} = 7 \div 2 = 3\frac{1}{2}$

1.48 $\frac{13}{4} = 13 \div 4 = 3\frac{1}{4}$

1.49 $\frac{31}{5} = 31 \div 5 = 6\frac{1}{5}$

1.50 $\frac{53}{6} = 53 \div 6 = 8\frac{5}{6}$

1.51 $\frac{62}{4} = 62 \div 4 = 15\frac{2}{4} = 15\frac{1}{2}$

1.52 $\frac{56}{21} = 56 \div 21 = 2\frac{14}{21} = 2\frac{2}{3}$

1.53

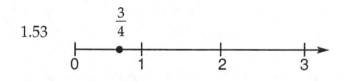

1.54

1.55

1.56

1.57 $\frac{1}{2}, 1\frac{1}{4}, 2\frac{1}{2}$

1.58 $\frac{2}{3}, 1\frac{1}{2}, 2\frac{3}{4}$

1.59 more than can be counted

1.60 more than can be counted

1.61 Multiply 2 x 1 = 2 and add 1; $1\frac{1}{2} = \frac{3}{2}$.

1.62 Multiply 5 x 2 = 10 and add 2; $2\frac{2}{5} = \frac{12}{5}$.

1.63 Multiply 7 x 3 = 21 and add 4; $3\frac{4}{7} = \frac{25}{7}$.

1.64 Multiply 8 x 4 = 32 and add 1; $4\frac{1}{8} = \frac{33}{8}$.

1.65 Multiply 5 x 5 = 25 and add 3; $5\frac{3}{5} = \frac{28}{5}$.

1.66 Multiply 9 x 6 = 54 and add 7; $6\frac{7}{9} = \frac{61}{9}$.

1.67 Halves are larger than fourths, so $\frac{1}{2}$ is larger than $\frac{1}{4}$.

1.68 Fifths are larger than ninths, so $\frac{2}{5}$ is larger than $\frac{2}{9}$.

1.69 Fourths are larger than tenths, so $\frac{3}{4}$ is larger than $\frac{3}{10}$.

1.70 Fifths are larger than sevenths, so $\frac{4}{5}$ is larger than $\frac{4}{7}$.

1.71 Both fractions are in sevenths. 3 is smaller than 5, so $\frac{3}{7}$ is smaller than $\frac{5}{7}$.

1.72 Both fractions are in thirteenths. 5 is larger than 3, so $\frac{5}{13}$ is larger than $\frac{3}{13}$.

1.73 $2\frac{2}{3}$

1.74 $\frac{1}{10}$

1.75 The larger the denominator is, the smaller the fractional parts are.

$\frac{1}{10}, \frac{1}{8}, \frac{1}{7}, \frac{1}{5}, \frac{1}{4}, \frac{1}{3}, \frac{1}{2}$

1.76 The larger the denominator is, the smaller the fractional parts are.

$5\frac{1}{2}, 4\frac{1}{2}, 3\frac{2}{5}, 1\frac{2}{3}, \frac{2}{3}, \frac{2}{5}, \frac{1}{10}$

II. SECTION TWO

2.1 five and two tenths

2.2 eight and nine tenths

2.3 thirteen and thirty-two hundredths

2.4 forty-five and twenty-seven hundredths

2.5 10.7

2.6 15.24

2.7 one hundred seventy-three and seven hundred eighty-four thousandths

2.8 268.425

2.9 0.145

2.10 0.0456

2.11 5.45

2.12 127.009

2.13 $3 \times 10 + 5 \times 1 + \frac{8 \times 1}{10} + \frac{7 \times 1}{100} =$

$30 + 5 + \frac{8}{10} + \frac{7}{100} = 35.87$

2.14 $5 \times 100 + 2 \times 10 +$ $7 \times 1 + \frac{6 \times 1}{10} + \frac{5 \times 1}{100} + \frac{9 \times 1}{1,000} =$

$500 + 20 + 7 + \frac{6}{10} + \frac{5}{100} + \frac{9}{1,000}$

$= 527.659$

2.15 15 is larger than 1.5

2.16 5.4 is larger than 0.54

2.17 0.003, 0.03, 0.3, 3, 30

2.18 450, 45, 4.5, 0.45, 0.045, 0.0045

2.19 0.35 is smaller

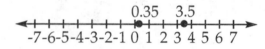

2.20 13 is larger

2.21 $\frac{1}{5} = \frac{1 \times 20}{5 \times 20} = \frac{20}{100} = 0.2$

2.22 $\frac{3}{4} = \frac{3 \times 25}{4 \times 25} = \frac{75}{100} = 0.75$

2.23 $\frac{4}{25} = \frac{4 \times 4}{25 \times 4} = \frac{16}{100} = 0.16$

2.24 $\frac{13}{20} = \frac{13 \times 5}{20 \times 5} = \frac{65}{100} = 0.65$

2.25 $\frac{6}{10} = 0.6$

2.26 $\frac{111}{200} = \frac{111 \div 2}{200 \div 2} = \frac{55.5}{100} = 0.555$ or

2.26 cont. $\frac{111 \times 5}{200 \times 5} = \frac{55.5}{1,000} = 0.555$

2.27 $0.75 = \frac{75}{100} = \frac{75 \div 25}{100 \div 25} = \frac{3}{4}$

2.28 $0.35 = \frac{35}{100} = \frac{35 \div 5}{100 \div 5} = \frac{7}{20}$

2.29 $0.033 = \frac{33}{1,000}$

2.30 $0.7338 = \frac{7,338}{10,000} = \frac{7,338 \div 2}{10,000 \div 2} = \frac{3,669}{5,000}$

2.31 $2.25 = 2\frac{25}{100} = 2\frac{25 \div 25}{100 \div 25} = 2\frac{1}{4}$

2.32 $25.456 = 25\frac{456}{1,000} = 25\frac{456 \div 8}{1,000 \div 8} = 25\frac{57}{125}$

2.33 $0.66\frac{2}{3}$

2.34 0.75

2.35 $\frac{1}{2} = 0.5$
$5\frac{1}{2} = 5.5$

2.36 $\frac{8}{9} = 0.88\frac{8}{9}$
$345\frac{8}{9} = 345.88\frac{8}{9}$

2.37 $\frac{1}{4}$

2.38 $\frac{2}{9}$

2.39 $0.875 = \frac{7}{8}$
$6.875 = 6\frac{7}{8}$

2.40 $0.16\frac{2}{3} = \frac{1}{6}$
$947.16\frac{2}{3} = 947\frac{1}{6}$

2.41 $\frac{1}{8} = 0.125$
$\frac{1}{16} = \frac{1}{8} \times \frac{1}{2}$
$\frac{1}{16} = 0.125 \times 0.5 = 0.0625$

2.42 $\frac{1}{6} = 0.16\frac{2}{3}$
$\frac{1}{12} = \frac{1}{6} \times \frac{1}{2}$
$\frac{1}{12} = 0.016\frac{2}{3} \times 0.5 = 0.083\frac{1}{3}$

2.43 $6\% = 6 \text{ per cent} = \frac{6}{100} = 0.06$

2.44 $18\% = 18 \text{ per cent} = \frac{18}{100} = 0.18$

2.45 $\frac{34}{100} = 34\%$

2.46 $\frac{98}{100} = 98\%$

2.47 $\frac{21}{100} = 21\%$

2.48 $\frac{56}{100} = 56\%$

2.49 $\frac{71}{100} = 71\%$

2.50 Examples:
a. 100ths
b. per 100
c. out of 100
d. divided by 100

For Problems 2.51 through 2.56, move the decimal point two places to the left and drop the % sign

2.51 0.18

2.52 0.24

2.53 0.72

2.54 0.07

173

2.55   0.123

2.56   0.479

For Problems 2.57 through 2.62, move the decimal point two places to the right and add the % sign.

2.57   26%

2.58   82%

2.59   4%

2.60   0.8%

2.61   45.9%

2.62   11.1%

2.63 $\dfrac{21}{100}$

2.64 $\dfrac{63}{100}$

2.65 $\dfrac{42}{100} = \dfrac{42 \div 2}{100 \div 2} = \dfrac{21}{50}$

2.66 $\dfrac{85}{100} = \dfrac{85 \div 5}{100 \div 5} = \dfrac{17}{20}$

2.67 $\dfrac{25}{100} = \dfrac{25 \div 25}{100 \div 25} = \dfrac{1}{4}$

2.68 $\dfrac{12.5}{100} = \dfrac{12.5 \times 10}{100 \times 10} =$

$\dfrac{125}{1,000} = \dfrac{125 \div 125}{1,000 \div 125} =$

$\dfrac{1}{8}$

2.69 $\dfrac{1}{2} = 0.5 = 50\%$

2.70 $\dfrac{1}{6} = 0.16 \dfrac{2}{3} = 16 \dfrac{2}{3}\%$

2.71 $\dfrac{3}{8} = 0.375 = 37.5\%$

2.72 $\dfrac{4}{9} = 0.44 \dfrac{4}{9} = 44 \dfrac{4}{9}\%$

2.73 $\dfrac{3}{10} = 0.3 = 30\%$

2.74 $\dfrac{1}{3} = 0.33 \dfrac{1}{3} = 33 \dfrac{1}{3}\%$

III. SECTION THREE

3.1 $\dfrac{7}{11}$ or 7:11

3.2 $\dfrac{23}{37}$ or 23:37

3.3 $\dfrac{14}{28} = \dfrac{14 \div 14}{28 \div 14} = \dfrac{1}{2}$ or 1:2

3.4 $\dfrac{48}{72} = \dfrac{48 \div 24}{72 \div 24} = \dfrac{2}{3}$ or 2:3

3.5 $\dfrac{37}{55}$ or 37:55

3.6 $\dfrac{36}{56} = \dfrac{36 \div 4}{56 \div 4} = \dfrac{9}{14}$ or 9:14

3.7 $\dfrac{500}{800} = \dfrac{500 \div 100}{800 \div 100} = \dfrac{5}{8}$
or 5:8

3.8 $\dfrac{12}{36} = \dfrac{12 \div 12}{36 \div 12} = \dfrac{1}{3}$ or 1:3

3.9   Convert 5 dimes to nickels.
5 dimes = 10 nickels
$\dfrac{5}{10} = \dfrac{5 \div 5}{10 \div 5} = \dfrac{1}{2}$ or 1:2

3.10   Convert 2 lbs. to oz.
2 x 16 = 32 ozs.
$\dfrac{27}{32}$ or 27:32

3.11   Convert 2 feet to inches
2 x 12 = 24 inches
$\dfrac{24}{27} = \dfrac{24 \div 3}{27 \div 3} = \dfrac{8}{9}$ or 8:9

3.12   Convert $1.25 to quarters.
$1.25 = 5 quarters
$\dfrac{5}{8}$ or 5:8

3.13   a. 3:4
b. 2:3
c. 5:7
d. 34:45

3.14   must have the same units

3.15   The means are 8 and 6, and the extremes are 3 and 16. The rule states that 8 x 6 = 3 x 16; 48 = 48.

3.16   5:6 = 10:12
The means are 6 and 10, and the extremes are 5 and 12. The rule states that 6 x 10 = 5 x 12; 60 = 60.

3.17 $\dfrac{2}{7} = \dfrac{6}{21}$

3.18 $\dfrac{23}{46} = \dfrac{12}{24}$

3.19   1:2 = 8:16

3.20   4:9 = 12:27

3.21   5 x 8 = 4 x ?
     40 = 4 x ?
     ? = 10

3.22   5:16 = 10:?
     16 x 10 = 5 x ?
     160 = 5 x ?
     ? = 32

3.23   2:3 = ?:9
     3 x ? = 2 x 9
     3 x ? = 18
     ? = 6

3.24   6:? = 12:14
     ? x 12 = 6 x 14
     ? x 12 = 84
     ? = 7

3.25   ? x 9 = 2:18
     9 x 2 = ? x 18
     18 = ? x 18
     ? = 1

3.26 $\overline{AB}:\overline{DE} = \overline{BC}:\overline{EF} = \overline{AC}:\overline{DF}$ OR

$$\frac{\overline{AB}}{\overline{DE}} = \frac{\overline{BC}}{\overline{EF}} = \frac{\overline{AC}}{\overline{DF}}$$

3.27 $\overline{EF}:\overline{HI} = \overline{FG}:\overline{IJ} = \overline{EG}:\overline{HJ}$ OR

$$\frac{\overline{EF}}{\overline{HI}} = \frac{\overline{FG}}{\overline{IJ}} = \frac{\overline{EG}}{\overline{HJ}}$$

3.28 $\overline{AB}:\overline{FG} = \overline{BC}:\overline{GH} = \overline{CD}:\overline{HI} = \overline{DE}:\overline{IJ} = \overline{AE}:\overline{FJ}$ OR

$$\frac{\overline{AB}}{\overline{FG}} = \frac{\overline{BC}}{\overline{GH}} = \frac{\overline{CD}}{\overline{HI}} = \frac{\overline{DE}}{\overline{IJ}} = \frac{\overline{AE}}{\overline{FJ}}$$

3.29 $\overline{AB}:\overline{GH} = \overline{BC}:\overline{HI} = \overline{CD}:\overline{IJ} = \overline{DE}:\overline{JK} = \overline{EF}:\overline{KL} = \overline{AF}:\overline{GL}$ OR

$$\frac{\overline{AB}}{\overline{GH}} = \frac{\overline{BC}}{\overline{HI}} = \frac{\overline{CD}}{\overline{IJ}} = \frac{\overline{DE}}{\overline{JK}} = \frac{\overline{EF}}{\overline{KL}} = \frac{\overline{AF}}{\overline{GL}}$$

3.30 $\overline{PQ}:\overline{ST} = \overline{PR}:\overline{SU}$
     $6:3 = 4:\overline{SU}$
     $3 \times 4 = 6 \times \overline{SU}$
     $12 = 6 \times \overline{SU}$
     $\overline{SU} = 2$

3.31 a. $\overline{AB}:\overline{EF} = \overline{CD}:\overline{GH}$
     $2:6 = 6:\overline{GH}$
     $6 \times 6 = 2 \times \overline{GH}$
     $36 = 2 \times \overline{GH}$
     $\overline{GH} = 18$

     b. $\overline{AB}:\overline{EF} = \overline{AD}:\overline{EH}$
     $2:6 = 8:\overline{EH}$
     $6 \times 8 = 2 \times \overline{EH}$
     $48 = 2 \times \overline{EH}$
     $\overline{EH} = 24$

3.32   40 feet:
     1:10 = ?:40
     10 x ? = 1 x 40
     10 x ? = 40
     ? = 4 inches
     50 feet:
     1:10 = ?:50
     10 x ? = 1 x 50
     10 x ? = 50
     ? = 5 inches
     60 feet:
     1:10 = ?:60
     10 x ? = 1 x 60
     10 x ? = 60
     ? = 6 inches

3.33　12 yards:
　　　　1:2 = ?:12
　　　　2 x ? = 1 x 12
　　　　2 x ? = 12
　　　　　? = 6 inches
　　　14 yards:
　　　　1:2 = ?:14
　　　　2 x ? = 1 x 14
　　　　2 x ? = 14
　　　　　? = 7 inches
　　　16 yards:
　　　　1:2 = ?:16
　　　　2 x ? = 1 x 16
　　　　2 x ? = 16
　　　　　? = 8 inches
　　　18 yards:
　　　　1:2 = ?:18
　　　　2 x ? = 1 x 18
　　　　2 x ? = 18
　　　　　? = 9 inches

3.34　300 miles:
　　　　1:10 = ?:300
　　　　10 x ? = 1 x 300
　　　　10 x ? = 300
　　　　　? = 30 cm
　　　150 miles:
　　　　1:10 = ?:150
　　　　10 x ? = 1 x 150
　　　　10 x ? = 150
　　　　　? = 15 cm
　　　60 miles:
　　　　1:10 = ?:60
　　　　10 x ? = 1 x 60
　　　　10 x ? = 60
　　　　　? = 6 cm
　　　170 miles:
　　　　1:10 = ?:170
　　　　10 x ? = 1 x 170
　　　　10 x ? = 170
　　　　　? = 17 cm

3.35　80 miles:
　　　　1:40 = ?:80
　　　　40 x ? = 1 x 80
　　　　40 x ? = 80
　　　　　? = 2 inches

120 miles:
　　　1:40 = ?:120
　　　40 x ? = 1 x 120
　　　40 x ? = 120
　　　　? = 3 inches
160 miles:
　　　1:40 = ?:160
　　　40 x ? = 1 x 160
　　　40 x ? = 160
　　　　? = 4 inches
240 miles:
　　　1:40 = ?:240
　　　40 x ? = 1 x 240
　　　40 x ? = 240
　　　　? = 6 inches

3.36　4 cm:
　　　1:10 = 4:?
　　　10 x 4 = 1 x ?
　　　40 = 1 x ?
　　　? = 40 feet

　　　5 cm:
　　　1:10 = 5:?
　　　10 x 5 = 1 x ?
　　　50 = 1 x ?
　　　? = 50 feet

　　　6 cm:
　　　1:10 = 6:?
　　　10 x 6 = 1 x ?
　　　60 = 1 x ?
　　　? = 60 feet

3.37　3 inches:
　　　1:50 = 3:?
　　　50 x 3 = 1 x ?
　　　150 = 1 x ?
　　　? = 150 miles
　　　7 inches:
　　　1:50 = 7:?
　　　50 x 7 = 1 x ?
　　　350 = 1 x ?
　　　? = 350 miles
　　　10 inches:
　　　1:50 = 10:?
　　　50 x 10 = 1 x ?
　　　500 = 1 x ?
　　　? = 500 miles

3.38   1 g = 1, 000 mg

3.39   1 mm = 0.0001 dm

3.40   1 dl = 0.001 hl

3.41   1 km = 1,000,000 mm

3.42   1 mm = 0.1 cm
        4 mm = 4 x 0.1 = 0.4 cm

3.43   1 km = 1,000 m
        7 km = 7 x 1,000 = 7,000m

3.44   1 g = 100 cg
        56 g = 56 x 100 = 5,600 cg

3.45   1 hl = 1, 000 dl
        47 hl = 47 x 1,000 = 47,000 dl

3.46   1 L = 1,000 ml
        4.7 L = 4.7 x 1,000 = 4,700.0 ml

3.47   1 mm = 0.000001 km
        34.29 mm = 34.29 x 0.000001
        = 0.00003429 km

3.48   1 m = 0.01 hm
        0.04 m = 0.04 x 0.01
        = 0.0004hm

3.49   1 dm = 0.0001 km
        2,348 dm = 2,348 x 0.0001
        = 0.2348 km

3.50   1 kg = 100,000 cg
        0.00045 kg = 0.00045 x 100,000
        = 45 cg

3.51   1 mm = 0.01 dm
        0.00318 mm = 0.00318 x 0.01
        = 0.0000318 dm

3.52   1 dl = 0.0001 kl
        40,000 dl = 40,000 x 0.0001
        = 4 kl

3.53 $\frac{9}{3}$ = 9 ÷ 3 = 3 or 3.0

3.54 $\frac{23}{5}$ = 23 ÷ 5 = 4.6 or 4.60

3.55 $\frac{40}{4}$ = 40 ÷ 4 = 10 or 10.0

3.56 $\frac{1}{8}$ = 1 ÷ 8 = 0.125 or 0.1250

3.57 $\frac{3}{5}$ = 3 ÷ 5 = 0.6 or 0.60

3.58 $8\frac{5}{8}$ = 8 + 5 ÷ 8 = 8 + 0.625
        = 8.625 or 8.6250

3.59 $0.25 = \frac{25}{100} = \frac{25 \div 25}{100 \div 25} = \frac{1}{4}$

3.60 $6.375 = 6\frac{375}{1,000} = 6\frac{375 \div 125}{1,000 \div 125} = 6\frac{3}{8}$

3.61 $\frac{2}{3} = 0.66$ $\frac{2}{3} = 0.666 \ldots = 0.\overline{6}$

3.62 $\frac{4}{9} = 0.44$ $\frac{4}{9} = 0.444 \ldots = 0.\overline{4}$

3.63 $1\frac{1}{6} = 1.16$ $\frac{2}{3} = 1.1666 \ldots = 1.1\overline{6}$

3.64 $3\frac{5}{6} = 3.83$ $\frac{1}{3} = 3.8333 \ldots = 3.8\overline{3}$

3.65 $0.\overline{54}$

3.66 $78.48\overline{43}$

3.67   191.8

3.68   52.88

3.69   1,349.85

3.70   47.02

3.71   23.21

3.72   1,096.05

The solutions to Problems 3.73 through 3.78 are shown for students who do not have access to calculators.

3.73
$$
\begin{array}{r}
835 \\
\underline{349} \\
7515 \\
3340 \\
\underline{2505} \\
2{,}914.15
\end{array}
$$

3.74
$$
\begin{array}{r}
421.88 \\
\underline{8.451} \\
42188 \\
210940 \\
168752 \\
\underline{337504} \\
3{,}565.30788
\end{array}
$$

3.75
$$
\begin{array}{r}
1.1118 \\
\underline{457.2} \\
22236 \\
77826 \\
55590 \\
\underline{44472} \\
508.31496
\end{array}
$$

3.76
$$
\begin{array}{r}
12.378378 \\
37)\overline{458.000000} \\
\underline{37} \\
88 \\
\underline{74} \\
140 \\
\underline{111} \\
290 \\
\underline{259} \\
310 \\
\underline{296} \\
140 \\
\underline{111} \\
290 \\
\underline{259} \\
310 \\
\underline{296} \\
14
\end{array}
$$

3.77
$$
\begin{array}{r}
12.5719148 \\
235)\overline{2{,}954.4000000} \\
\underline{235} \\
604 \\
\underline{470} \\
1344 \\
\underline{1175} \\
1690 \\
\underline{1645} \\
450 \\
\underline{235} \\
2150 \\
\underline{2115} \\
350 \\
\underline{235} \\
1150 \\
\underline{940} \\
2100 \\
\underline{1880} \\
220
\end{array}
$$

3.78
$$
\begin{array}{r}
10.150098 \\
27{,}482)\overline{278{,}945.000000} \\
\underline{27482} \\
41250 \\
\underline{27482} \\
137680 \\
\underline{137410} \\
270000 \\
\underline{247338} \\
226620 \\
\underline{219856} \\
6764
\end{array}
$$

3.79
$$
\frac{2}{7} = 2 \div 7 =
$$
$$
\begin{array}{r}
0.28571428 \\
7)\overline{2.00000000} \\
\underline{1\ 4} \\
60 \\
\underline{56} \\
40 \\
\underline{35} \\
50 \\
\underline{49} \\
10 \\
\underline{7} \\
30 \\
\underline{28} \\
20 \\
\underline{14} \\
60 \\
\underline{56} \\
4
\end{array}
$$

3.80 $\frac{4}{13} = 4 \div 13 =$

$$
\begin{array}{r}
0.3076923 \\
13\overline{)4.0000000} \\
\underline{39} \\
100 \\
\underline{91} \\
90 \\
\underline{78} \\
120 \\
\underline{117} \\
30 \\
\underline{26} \\
40 \\
\underline{39} \\
1
\end{array}
$$

3.81 $\frac{59}{17} = 59 \div 17 =$

$$
\begin{array}{r}
3.4705882 \\
17\overline{)59.0000000} \\
\underline{51} \\
80 \\
\underline{68} \\
120 \\
\underline{119} \\
100 \\
\underline{85} \\
150 \\
\underline{136} \\
140 \\
\underline{136} \\
40 \\
\underline{34} \\
6
\end{array}
$$

For Problems 3.82 through 3.87, move the decimal point two places to the left and drop the % sign.

3.82   0.008

3.83   0.0023

3.84   0.0006

3.85   0.00305

3.86   0.000005

3.87   0.009998

For Problems 3.88 through 3.92, move the decimal point two places to the right and add the % sign.

3.88   0.8%

3.89   0.018%

3.90   0.0005%

3.91   0.4447%

3.92   11.1001%

3.93 $0.005 = 0.5\% = \frac{1}{2}\%$

3.94 $\frac{3}{4} \times \frac{1}{100} = 0.75 \times 0.01 = 0.0075$

For Problems 3.95 through 3.101, move the decimal point two places to the left and drop the % sign.

3.95   2.00 or 2

3.96   4.00 or 4

3.97   1.20 or 1.2

3.98   8.50 or 8.5

3.99   4.78

3.100   7.141

3.101   20.00 or 20

For Problems 3.102 through 3.106, move the decimal point to the right two places and add the % sign.

3.102   500%

3.103   710%

3.104   776%

3.105   1,000%

3.106   2,345%

3.107   5¢ of $1.00 =
5 out of 100 = 5%

3.108   1¢ of $1.00 =
1 out of 100 = 1%

3.109 $60 - 45 = 15$

$\dfrac{15}{60} = \dfrac{15 \div 15}{60 \div 15} = \dfrac{1}{4} = 0.25 =$

25%

3.110 $300 - 150 = 150$

$\dfrac{150}{300} = \dfrac{150 \div 150}{300 \div 150} = \dfrac{1}{2}$

$= 0.5 = 50\%$

3.111 $\dfrac{150}{500} = \dfrac{150 \div 50}{500 \div 50} = \dfrac{3}{10} =$

$0.3 = 30\%$

3.112 $\dfrac{50}{1,000} = \dfrac{50 \div 50}{1,000 \div 50} = \dfrac{1}{20}$

$= 0.05 = 5\%$

3.113 $\dfrac{200}{1,000} = \dfrac{200 \div 200}{1,000 \div 200} = \dfrac{1}{5} = 0.2$

$= 20\%$

I. SECTION ONE

1.1   1, 2, 3, and 4

1.2   14, 18 and 22

1.3   *x, y,* and *z*

1.4   blue, green, and red

1.5   5, 10, 15, 20, through 100

1.6   the element is 0

1.7   {4, 6, 10}

1.8   {tires, oil}

1.9   {eyes, nose, throat}

1.10   {5, toy, can, 10}

1.11   {Matthew, Mark, Luke, John}

1.12   {Genesis, Exodus, Leviticus, Numbers, Deuteronomy}

1.13 through 1.20   Examples:

1.13   {the whole numbers between 0 and 11}

1.14   {the colors of the United States flag}

1.15   {the even numbers between 0 and 16}

1.16   {the vowels}

1.17   {the letters of the alphabet}

1.18   {the days of the week beginning with S} or {days of the weekend}

1.19   {the multiples of five between 0 and 105}

1.20   {the operations of arithmetic}

1.21   {Saturday, Sunday}

1.22   {a, b, c, d, e, f, g, h, i, j}

1.23   {1, 2, 3, 4, . . ., 50}

1.24   {2 John, 3 John, Jude, Revelation}

1.25   Example: {Peter, John, Matthew, Thomas, Andrew}

1.26   {10, 20, 30, 40, 50, 60, 70, 80, 90, 100}

1.27   {5,280 ft.}

1.28   {5, 7, 9, 11, 13, 15, 17}

1.29   finite

1.30   finite

1.31   finite

1.32   infinite

1.33   finite

1.34   infinite

1.35   finite

1.36   infinite

1.37   proper

1.38   improper

1.39   proper

1.40   improper

1.41   {1, 2}, {1}, {2}, ϕ

1.42   {x, y, z}, {x, y}, {y, z}, {x, z}, {x}, {y}, {z}, ϕ

1.43   {Matthew}, ϕ

1.44   {I, V, X, D}; {I, V, X}; {V, X, D}; {I, X, D}; {I, V, D}; {I, V}; {V, X}; {X, D}; {I, X}; {I, D}; {V, D}; {I}; {V}; {X}; {D}; ϕ

1.45   Set *A* is a subset of set *B*.

1.46   Set {1, 2} is a subset of set {1, 2, 3, 4} or set C is a proper subset of set *D*.

1.47   Set *H* is not a subset of set *J*.

1.48   {21; 119; 2,319}

1.49   {74; 51; 119; 3,559}

1.50   {72; 130; 837; 17,052; 47,322}

1.51   {6, 20, 13, 25}

1.52   {6, 15, 24}

1.53   {4, 9, 16, 25}

1.54   {3}

1.55   {10}

1.56   {2}

1.57   {6}

1.58   {7, 8, 9, 10, . . .}

1.59   {0, 1, 2, 3, 4, 5, 6, 7}

1.60   {4, 5, 6, 7, . . .}

1.61   {0, 1, 2, 3}

1.62   Segment *AB* is the set of points between *A* and *B*.

1.63 $\overleftrightarrow{PQ}$ is the set of points on the whole line.

1.64   {R, A, B, C, D, G, S}

1.65   {A, S, R, B, P, Q, C, Y, X}

1.66 through 1.73   Examples:

1.66   {set of chairs}, {set of lamps}, {set of electric appliances}

1.67   {set of canned vegetables}, {set of ice cream}, {set of meat}

1.68   {set of airplanes}, {set of gasoline trucks}, {set of pilots}

1.69   {set of pews}, {set of hymn books}, {set of visitor cards}

1.70   {set of police cars}, {set of firemen}, {set of traffic signals}

1.71   {the set of fans}, {the set of referees}, {the set of players}

1.72   {set of classes each period}, {set of students}, {set of teachers}

1.73   {set of nails}, {set of tools}, {set of paints}

1.74   {7, 8, 9}

1.75   {A, C, E}

1.76   empty set (ϕ)

1.77   {15}

1.78   {a, e, i, o, u}

1.79   {10, 20, 30, 40, 50, . . .}

1.80   {2}

1.81   φ

1.82   {5}

1.83   {1, 3}

1.84   φ

1.85   φ

1.86   φ

1.87   {a, b, c}

1.88   {1, 2, 4, 5, 7, 9, 10, 14}

1.89   {2, 4, 6, 8, 100, 101, 102, 103}

1.90   {I, II, III, A, B, C, D}

1.91   {a, b, c, d, e, f, g}

1.92   {Monday, Wednesday, Friday, Tuesday, Thursday, Saturday}

1.93   {1, 3, 5, a, b, c}

1.94   {a, b, c, red, white, blue}

1.95   {red, white, blue, 1, 2, 3, 4}

1.96   {1, 3, 5, red, white, blue}

1.97   {1, 3, 5, a, b, c, red, white, blue}

1.98   {a, b, c, red, white, blue, 1, 2, 3, 4}

1.99   {1, 2, 3, 4, 5, a, b, c, red, white, blue}

1.100   a.

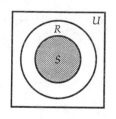 or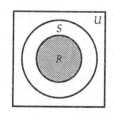

b.

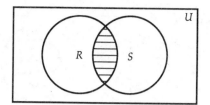

c.

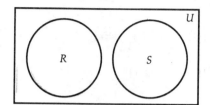

1.101   a.

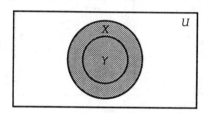

b.

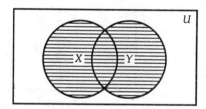

c.

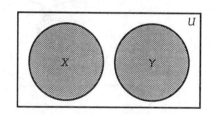

1.102   The complement of *P* is all elements in the universal set except the elements in set *P*.

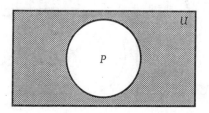

1.103
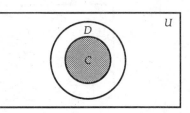

1.104 $10 + 4 = 14$

1.105 $4 + 7 = 11$

1.106   4

1.107 $10 + 4 + 7 = 21$

1.108
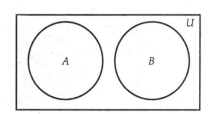

1.109
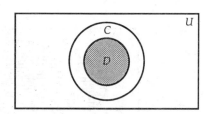

1.110
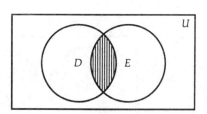

1.111
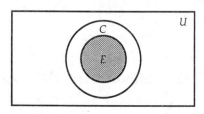

1.112
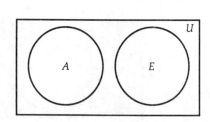

II. SECTION TWO

2.1 $70 + 2 =$
∧∧∧∧∧∧∧||

2.2 $100 + 20 + 6 =$
Ϛ∧∧||||||

2.3 $3,000 + 400 + 10 + 2 =$
ƷƷƷϚϚϚ∧||

2.4 $10,000 + 4,000 + 40 + 1 =$
ℓƷƷƷƷ∧∧∧∧|

2.5 $200,000 + 600 + 10 + 3 =$
ℛℛϚϚϚϚϚϚ∧|||

2.6 $2,000,000 + 100,000 + 10,000 +$
$4,000 + 200 + 10 + 7 =$
ℛℛℛ ℓƷƷƷƷϚϚ ∧||||||||

2.7 $1,000,000 + 100,000 + 10,000 +$
$1,000 + 100 + 10 + 1 =$
ℛℛ ℓƷϚ∧|

2.8 $30 + 5 = 35$

2.9 $200 + 50 + 2 = 252$

2.10 $1,000 + 100 + 20 + 6 = 1,126$

2.11 $3,000,000 + 100,000 + 20,000 =$
$3,120,000$

2.12 $300,000 + 10,000 + 2,000 + 200 =$
$312,200$

2.13 $1,000,000 + 100,000 + 60,000 + 200 +$
$50 + 2 = 1,160,252$

2.14 $1,000,000 + 100,000 + 10,000 + 1,000 +$
$100 + 10 + 1 = 1,111,111$

2.15 $50 + 5 + 1 =$
⌐△⌐|

2.16   50 + 30 + 5 + 3 =

2.17   100 + 40 + 2 =

2.18   500 + 50 + 5 =

2.19   1,000 + 200 + 30 + 2 =

2.20   5,000 + 1,000 + 500 + 200 + 50 + 4 =

2.21   10,000 + 3,000 + 300 + 40 + 5 + 1 =

2.22   30,000 + 5,000 + 2,000 + 500 + 400 + 50 + 30 + 5 + 2 =

2.23   50,000 =

    or MMMMM

2.24   30 + 2 = 32

2.25   50 + 30 + 5 + 4 = 89

2.26   300 + 20 + 5 + 1 = 326

2.27   500 + 100 + 50 + 10 + 5 + 3 = 668

2.28   1,000 + 200 + 50 + 5 + 2 = 1,257

2.29   5,000 + 500 + 50 + 5 = 5,555

2.30   20,000 + 2,000 + 200 + 20 + 2 = 22,222

2.31   30,000 + 5,000 + 1,000 + 500 + 100 + 50 + 10 = 36,660

2.32   50,000 (the Greek symbol means 5 x 10,000)

2.33   10 + 2 = XII

2.34   20 + 9 = XXIX

2.35   40 + 5 + 2 = XLVII

2.36   50 + 30 + 4 = LXXXIV

2.37   90 + 9 = XCIX

2.38   100 + 40 + 4 = CXLIV

2.39   200 + 30 + 5 + 1 = CCXXXVI

2.40   500 + 100 + 50 + 10 + 9 = DCLXIX

2.41   1,000 + 400 + 90 + 4 = MCDXCIV

2.42   3,000 + 900 + 9 = MMMCMIX

2.43   5 + 3 = 8

2.44   30 + 4 = 34

2.45   50 + 10 + 9 = 69

2.46   90 + 4 = 94

2.47   200 + 50 + 20 + 5 + 2 = 277

2.48   400 + 40 + 4 = 444

2.49   1,000 + 500 + 10 + 1 = 1,511

2.50   1,000 + 900 + 40 + 2 = 1,942

2.51   1,000 + 400 + 40 + 9 = 1,449

2.52   3,000 + 500 + 200 + 50 + 20 + 5 = 3,775

2.53

2.54

2.55

2.56   1468

2.57   56,384

2.58   611,485

2.59   1,416,385

2.60   8,888,888

2.61   112

2.62   334

2.63   896

2.64   2,255

2.65   76,321

2.66   416,885

2.67   3,215,874

2.68   5,662,483

2.69   5,000

2.70   0.007

2.71   20

2.72   0.0009

2.73   50 is 1,000 times larger than 0.05.

2.74   2,000 is 1,000 times larger than 2.

2.75   0.006 is $\frac{1}{100}$ (or 0.01) as large as 0.6.

2.76   0.06 is $\frac{1}{100,000}$ (or 0.00001) as large as 6,000.

2.77   80 is 10,000 times larger than 0.008

2.78   Move the decimal point to the right 2 places to multiply by 100.
$$0.6 \times 100 = 60$$

2.79   Move the decimal point to the left 1 place to multiply by $\frac{1}{10}$.
$$0.07 \times \frac{1}{10} = 0.007$$

2.80   Move the decimal point to the right 4 places to multiply by 10,000.
$$0.0009 \times 10,000 = 9$$

2.81   Move the decimal point to the left 3 places to multiply by $\frac{1}{1,000}$.
$$400 \times \frac{1}{1,000} = 0.4$$

2.82   4000.004
     123 456
     0.004 is the number.

2.83   5000.05
     543 21
     5,000 is the number.

2.84 $500 = 5 \times 100 = 5 \times 10^2$

2.85 $3,000 = 3 \times 1,000 = 3 \times 10^3$

2.86 $6,000,000 = 6 \times 1,000,000 = 6 \times 10^6$

2.87 $10,000 = 10^4$

2.88 $0.03 = 3 \times 0.01 = 3 \times 10^{-2}$

2.89 $0.000009 = 9 \times 0.000001 = 9 \times 10^{-6}$

2.90 $0.0004 = 4 \times 0.0001 = 4 \times 10^{-4}$

2.91 $0.0000000001 = 10^{-10}$

2.92 10^{12}

2.93 $5 \times 10^2 = 5 \times 100 = 500$

2.94 $6 \times 10^4 = 6 \times 10,000 = 60,000$

2.95 $2 \times 10^{-2} = 2 \times 0.01 = 0.02$

2.96 $7 \times 10^{-4} = 7 \times 0.0001 - 0.0007$

2.97   1,000,000

2.98   0.000001

2.99   1, 2, 3, 4, 5, 6, 7, 8, 9, 10, 11, 12, 13, 14

2.100   1, 2, 3, 4, 5, 6, 7, 8, 9, 10, 11, 12, 13, 14, 15, 16, 17, 18, 19, 20, 21

2.101   1, 2, 3, 4, 5, 6, 7, 8

2.102   {1, 2, 3, 4, 5, 6, 7, 8, 9, 0}

2.103

1: $1\overline{)1}$
    $\underline{1}$
    0
    $1 = 1_2$

2: $2\overline{)2}$ $1\overline{)0}$
    $\underline{2}$ $\underline{0}$
    0 0
    $2 = 10_2$

3: $2\overline{)3}$ $1\overline{)1}$
    $\underline{2}$ $\underline{1}$
    1 0
    $3 = 11_2$

4: $4\overline{)4}$ $2\overline{)0}$ $1\overline{)0}$
    $\underline{4}$ $\underline{0}$ $\underline{0}$
    0 0 0
    $4 = 100_2$

5: $4\overline{)5}$ $2\overline{)1}$ $1\overline{)1}$
    $\underline{4}$ $\underline{0}$ $\underline{1}$
    1 1 0
    $5 = 101_2$

2.104   The solutions to the first five numerals are in Problem 2.103.
     $1 = 1_2$
     $2 = 10_2$
     $3 = 11_2$
     $4 = 100_2$
     $5 = 101_2$

6: $4\overline{)6}$ $2\overline{)2}$ $1\overline{)0}$
    $\underline{4}$ $\underline{2}$ $\underline{0}$
    2 0 0
    $6 = 110_2$

7: $4\overline{)7}$ $2\overline{)3}$ $1\overline{)1}$
    $\underline{4}$ $\underline{2}$ $\underline{1}$
    3 1 0
    $7 = 111_2$

8: $8\overline{)8}$ $4\overline{)0}$ $2\overline{)0}$ $1\overline{)0}$
    $\underline{8}$ $\underline{0}$ $\underline{0}$ $\underline{0}$
    0 0 0 0
    $8 = 1000_2$

9: $8\overline{)9}$ $4\overline{)1}$ $2\overline{)1}$ $1\overline{)1}$
    $\underline{8}$ $\underline{0}$ $\underline{0}$ $\underline{1}$
    1 1 1 0
    $9 = 1001_2$

10: $8\overline{)10}$ $4\overline{)2}$ $2\overline{)2}$ $1\overline{)0}$
    $\underline{8}$ $\underline{0}$ $\underline{2}$ $\underline{0}$
    2 2 0 0
    $10 = 1010_2$

11: $8\overline{)11}$ $4\overline{)3}$ $2\overline{)3}$ $1\overline{)1}$
    $\underline{8}$ $\underline{0}$ $\underline{2}$ $\underline{1}$
    3 3 1 0
    $11 = 1011_2$

12: $8\overline{)12}$ $4\overline{)4}$ $2\overline{)0}$ $1\overline{)0}$
    $\underline{8}$ $\underline{4}$ $\underline{0}$ $\underline{0}$
    4 0 0 0
    $12 = 1100_2$

2.105   The solutions to the first twelve numerals are in Problems 2.103 and 2.104.

$$1 = 1_2$$
$$2 = 10_2$$
$$3 = 11_2$$
$$4 = 100_2$$
$$5 = 101_2$$
$$6 = 110_2$$
$$7 = 111_2$$
$$8 = 1000_2$$
$$9 = 1001_2$$
$$10 = 1010_2$$
$$11 = 1011_2$$
$$12 = 1100_2$$

13:
$$8\overline{)13} \quad 4\overline{)5} \quad 2\overline{)1} \quad 1\overline{)1}$$
quotients 1 1 0 1
$$\underline{8} \quad \underline{4} \quad \underline{0} \quad \underline{1}$$
$$5 \quad 1 \quad 1 \quad 0$$
$$13 = 1101_2$$

14:
$$8\overline{)14} \quad 4\overline{)6} \quad 2\overline{)2} \quad 1\overline{)0}$$
quotients 1 1 1 0
$$\underline{8} \quad \underline{4} \quad \underline{2} \quad \underline{0}$$
$$6 \quad 2 \quad 0 \quad 0$$
$$14 = 1110_2$$

15:
$$8\overline{)15} \quad 4\overline{)7} \quad 2\overline{)3} \quad 1\overline{)1}$$
quotients 1 1 1 1
$$\underline{8} \quad \underline{4} \quad \underline{2} \quad \underline{1}$$
$$7 \quad 3 \quad 1 \quad 0$$
$$15 = 1111_2$$

16:
$$16\overline{)16} \quad 8\overline{)0} \quad 4\overline{)0} \quad 2\overline{)0} \quad 1\overline{)0}$$
quotients 1 0 0 0 0
$$\underline{16} \quad \underline{0} \quad \underline{0} \quad \underline{0} \quad \underline{0}$$
$$0 \quad 0 \quad 0 \quad 0 \quad 0$$
$$16 = 10000_2$$

17:
$$16\overline{)17} \quad 8\overline{)1} \quad 4\overline{)1} \quad 2\overline{)1} \quad 1\overline{)1}$$
quotients 1 0 0 0 1
$$\underline{16} \quad \underline{0} \quad \underline{0} \quad \underline{0} \quad \underline{1}$$
$$1 \quad 1 \quad 1 \quad 1 \quad 0$$
$$17 = 10001_2$$

18:
$$16\overline{)18} \quad 8\overline{)2} \quad 4\overline{)2} \quad 2\overline{)2} \quad 1\overline{)0}$$
quotients 1 0 0 1 0
$$\underline{16} \quad \underline{0} \quad \underline{0} \quad \underline{2} \quad \underline{0}$$
$$2 \quad 2 \quad 2 \quad 0 \quad 0$$
$$18 = 10010_2$$

2.106   The solutions to the first eighteen numerals are in Problems 2.103 through 2.105.

$$1 = 1_2$$
$$2 = 10_2$$
$$3 = 11_2$$
$$4 = 100_2$$
$$5 = 101_2$$
$$6 = 110_2$$
$$7 = 111_2$$
$$8 = 1000_2$$
$$9 = 1001_2$$
$$10 = 1010_2$$
$$11 = 1011_2$$
$$12 = 1100_2$$
$$13 = 1101_2$$
$$14 = 1110_2$$
$$15 = 1111_2$$
$$16 = 10000_2$$
$$17 = 10001_2$$
$$18 = 10010_2$$

19:
$$16\overline{)19} \quad 8\overline{)3} \quad 4\overline{)3} \quad 2\overline{)3} \quad 1\overline{)1}$$
quotients 1 0 0 1 1
$$\underline{16} \quad \underline{0} \quad \underline{0} \quad \underline{2} \quad \underline{1}$$
$$3 \quad 3 \quad 3 \quad 1 \quad 0$$
$$19 = 10011_2$$

20:
$$16\overline{)20} \quad 8\overline{)4} \quad 4\overline{)4} \quad 2\overline{)0} \quad 1\overline{)0}$$
quotients 1 0 1 0 0
$$\underline{16} \quad \underline{0} \quad \underline{4} \quad \underline{0} \quad \underline{0}$$
$$4 \quad 4 \quad 0 \quad 0 \quad 0$$
$$20 = 10000_2$$

2.107
$$4\overline{)6} \quad 2\overline{)2} \quad 1\overline{)0}$$
quotients 1 1 0
$$\underline{4} \quad \underline{2} \quad \underline{0}$$
$$2 \quad 0 \quad 0$$
$$6 = 110_2$$

2.108

$$8\overline{)12}^{\,1} \qquad 4\overline{)4}^{\,1} \qquad 2\overline{)0}^{\,0} \qquad 1\overline{)0}^{\,0}$$
$$\underline{8} \qquad\quad \underline{4} \qquad\quad \underline{0} \qquad\quad \underline{0}$$
$$4 \qquad\qquad 0 \qquad\qquad 0 \qquad\qquad 0$$
$$12 = 1100_2$$

2.109

$$8\overline{)15}^{\,1} \qquad 4\overline{)7}^{\,1} \qquad 2\overline{)3}^{\,1} \qquad 1\overline{)1}^{\,1}$$
$$\underline{8} \qquad\quad \underline{4} \qquad\quad \underline{2} \qquad\quad \underline{1}$$
$$7 \qquad\qquad 3 \qquad\qquad 1 \qquad\qquad 0$$
$$15 = 1111_2$$

2.110

$$16\overline{)17}^{\,1} \quad 8\overline{)1}^{\,0} \quad 4\overline{)1}^{\,0} \quad 2\overline{)1}^{\,0} \quad 1\overline{)1}^{\,1}$$
$$\underline{16} \qquad \underline{0} \qquad \underline{0} \qquad \underline{0} \qquad \underline{1}$$
$$1 \qquad\quad 1 \qquad\quad 1 \qquad\quad 1 \qquad\quad 0$$
$$17 = 10001_2$$

2.111

$$16\overline{)24}^{\,1} \quad 8\overline{)8}^{\,1} \quad 4\overline{)0}^{\,0} \quad 2\overline{)0}^{\,0} \quad 1\overline{)0}^{\,0}$$
$$\underline{16} \qquad \underline{8} \qquad \underline{0} \qquad \underline{0} \qquad \underline{0}$$
$$8 \qquad\quad 0 \qquad\quad 0 \qquad\quad 0 \qquad\quad 0$$
$$24 = 11000_2$$

2.112

$$16\overline{)31}^{\,1} \quad 8\overline{)15}^{\,1} \quad 4\overline{)7}^{\,1} \quad 2\overline{)3}^{\,1} \quad 1\overline{)1}^{\,1}$$
$$\underline{16} \qquad \underline{8} \qquad \underline{4} \qquad \underline{2} \qquad \underline{1}$$
$$15 \qquad 7 \qquad 3 \qquad 1 \qquad 0$$
$$31 = 11111_2$$

2.113

$$32\overline{)45}^{\,1} \qquad 16\overline{)13}^{\,0} \qquad 8\overline{)13}^{\,1}$$
$$\underline{32} \qquad\quad \underline{0} \qquad\quad \underline{8}$$
$$13 \qquad\qquad 13 \qquad\qquad 5$$

$$4\overline{)5}^{\,1} \qquad 2\overline{)1}^{\,0} \qquad 1\overline{)1}^{\,1}$$
$$\underline{4} \qquad\quad \underline{0} \qquad\quad \underline{1}$$
$$1 \qquad\qquad 1 \qquad\qquad 0$$
$$45 = 101101_2$$

2.114

$$32\overline{)50}^{\,1} \qquad 16\overline{)18}^{\,1} \qquad 8\overline{)2}^{\,0}$$
$$\underline{32} \qquad\quad \underline{16} \qquad\quad \underline{0}$$
$$18 \qquad\qquad 2 \qquad\qquad 2$$

2.115 (continued)

$$4\overline{)2}^{\,0} \qquad 2\overline{)2}^{\,1} \qquad 1\overline{)0}^{\,0}$$
$$\underline{0} \qquad\quad \underline{2} \qquad\quad \underline{0}$$
$$2 \qquad\qquad 0 \qquad\qquad 0$$
$$50 = 110010_2$$

2.115

$$64\overline{)100}^{\,1} \quad 32\overline{)36}^{\,1} \quad 16\overline{)4}^{\,0} \quad 8\overline{)4}^{\,0}$$
$$\underline{64} \qquad \underline{32} \qquad \underline{0} \qquad \underline{0}$$
$$36 \qquad\quad 4 \qquad\quad 4 \qquad\quad 4$$

$$4\overline{)4}^{\,1} \qquad 2\overline{)0}^{\,0} \qquad 1\overline{)0}^{\,0}$$
$$\underline{4} \qquad\quad \underline{0} \qquad\quad \underline{0}$$
$$0 \qquad\qquad 0 \qquad\qquad 0$$
$$100 = 1100100_2$$

2.116

$$128\overline{)128}^{\,1} \quad 64\overline{)0}^{\,0} \quad 32\overline{)0}^{\,0} \quad 16\overline{)0}^{\,0}$$
$$\underline{128} \qquad \underline{0} \qquad \underline{0} \qquad \underline{0}$$
$$0 \qquad\quad 0 \qquad\quad 0 \qquad\quad 0$$

$$8\overline{)0}^{\,0} \qquad 4\overline{)0}^{\,0} \qquad 2\overline{)0}^{\,0} \qquad 1\overline{)0}^{\,0}$$
$$\underline{0} \qquad\quad \underline{0} \qquad\quad \underline{0} \qquad\quad \underline{0}$$
$$0 \qquad\qquad 0 \qquad\qquad 0 \qquad\qquad 0$$
$$128 = 10000000_2$$

2.117 $\quad 11_2 = 2^1 + 1 = 2 + 1 = 3$

2.118 $\quad 1000_2 = 2^3 = 8$

2.119 $\quad 11100_2 = 2^4 + 2^3 + 2^2$
$$= 16 + 8 + 4$$
$$= 28$$

2.120 $\quad 111111_2 = 2^5 + 2^4 + 2^3 + 2^2 + 2^1 + 1$
$$= 32 + 16 + 8 + 4 + 2 + 1$$
$$= 63$$

2.121 $\quad 10011_2 = 2^4 + 2^1 + 1$
$$= 16 + 2 + 1$$
$$= 19$$

2.122 $\quad 100110_2 = 2^5 + 2^2 + 2^1$
$$= 32 + 4 + 2$$
$$= 38$$

2.123 $1000000_2 = 2^6$
$= 64$

2.124 $111_2 = 2^2 + 2^1 + 1$
$= 4 + 2 + 1$
$= 7$

2.125 $1010101_2 = 2^6 + 2^4 + 2^2 + 1$
$= 64 + 16 + 4 + 1$
$= 85$

2.126 $11001100_2 = 2^7 + 2^6 + 2^3 + 2^2$
$= 128 + 64 + 8 + 4$
$= 204$

2.127   1: $1\overline{)1}$
$\underline{1}$
0

$1 = 1_5$

2:    $1\overline{)2}$
$\underline{2}$
0

$2 = 2_5$

3:    $1\overline{)3}$
$\underline{3}$
0

$3 = 3_5$

4:    $1\overline{)4}$
$\underline{4}$
0

$4 = 4_5$

5:    $5\overline{)5}$ $1\overline{)0}$
$\underline{5}$ $\underline{0}$
0 0

$5 = 10_5$

6:    $5\overline{)6}$ $1\overline{)1}$
$\underline{5}$ $\underline{1}$
1 0

$6 = 11_5$

7:    $5\overline{)7}$ $1\overline{)2}$
$\underline{5}$ $\underline{2}$
2 0

$7 = 12_5$

2.128   The solutions for the first seven
numerals are in Problem 2.127.

$1 = 1_5$
$2 = 2_5$
$3 = 3_5$
$4 = 4_5$
$5 = 10_5$
$6 = 11_5$
$7 = 12_5$

8:    $5\overline{)8}$ $1\overline{)3}$
$\underline{5}$ $\underline{3}$
3 0

$8 = 13_5$

9:    $5\overline{)9}$ $1\overline{)4}$
$\underline{5}$ $\underline{4}$
4 0

$9 = 14_5$

10:    $5\overline{)10}$ $1\overline{)0}$
$\underline{10}$ $\underline{0}$
0 0

$10 = 20_5$

2.129   The solutions for the first ten
numerals are in Problems 2.127 and
2.128.

$1 = 1_5$
$2 = 2_5$
$3 = 3_5$
$4 = 4_5$
$5 = 10_5$
$6 = 11_5$
$7 = 12_5$
$8 = 13_5$
$9 = 14_5$
$10 = 20_5$

11:
$$5\overline{)11} \quad 1\overline{)1}$$
$$\underline{10} \qquad \underline{1}$$
$$1 \qquad\; 0$$
$$11 = 21_5$$

12:
$$5\overline{)12} \quad 1\overline{)2}$$
$$\underline{10} \qquad \underline{2}$$
$$2 \qquad\; 0$$
$$12 = 22_5$$

13:
$$5\overline{)13} \quad 1\overline{)3}$$
$$\underline{10} \qquad \underline{3}$$
$$3 \qquad\; 0$$
$$13 = 23_5$$

14:
$$5\overline{)14} \quad 1\overline{)4}$$
$$\underline{10} \qquad \underline{4}$$
$$4 \qquad\; 0$$
$$14 = 24_5$$

15:
$$5\overline{)15} \quad 1\overline{)0}$$
$$\underline{15} \qquad \underline{0}$$
$$0 \qquad\; 0$$
$$15 = 30_5$$

16:
$$5\overline{)16} \quad 1\overline{)1}$$
$$\underline{15} \qquad \underline{1}$$
$$1 \qquad\; 0$$
$$16 = 31_5$$

17:
$$5\overline{)17} \quad 1\overline{)2}$$
$$\underline{15} \qquad \underline{2}$$
$$2 \qquad\; 0$$
$$17 = 32_5$$

18:
$$5\overline{)18} \quad 1\overline{)3}$$
$$\underline{15} \qquad \underline{3}$$
$$3 \qquad\; 0$$
$$18 = 33_5$$

19:
$$5\overline{)19} \quad 1\overline{)4}$$
$$\underline{15} \qquad \underline{4}$$
$$4 \qquad\; 0$$
$$19 = 34_5$$

20:
$$5\overline{)20} \quad 1\overline{)0}$$
$$\underline{20} \qquad \underline{0}$$
$$0 \qquad\; 0$$
$$20 = 40_5$$

2.130 The solutions for the first twenty numerals are in Problems 2.127 through 2.129.

$$1 = 1_5$$
$$2 = 2_5$$
$$3 = 3_5$$
$$4 = 4_5$$
$$5 = 10_5$$
$$6 = 11_5$$
$$7 = 12_5$$
$$8 = 13_5$$
$$9 = 14_5$$
$$10 = 20_5$$
$$11 = 21_5$$
$$12 = 22_5$$
$$13 = 23_5$$
$$14 = 24_5$$
$$15 = 30_5$$
$$16 = 31_5$$
$$17 = 32_5$$
$$18 = 33_5$$
$$19 = 34_5$$
$$20 = 40_5$$

21:
$$5\overline{)21} \quad 1\overline{)1}$$
$$\underline{20} \qquad \underline{1}$$
$$1 \qquad\; 0$$
$$21 = 41_5$$

22:
$$5\overline{)22} \quad 1\overline{)2}$$
$$\underline{20} \qquad \underline{2}$$
$$2 \qquad\; 0$$
$$22 = 42_5$$

23:
$$5\overline{)23} \quad 1\overline{)3}$$
$$\underline{20} \qquad \underline{3}$$
$$3 \qquad\; 0$$
$$23 = 43_5$$

24:
$$5\overline{)24} \quad 1\overline{)4}$$
$$\quad \underline{20} \qquad \underline{4}$$
$$\quad 4 \qquad 0$$

$24 = 44_5$

25:
$$25\overline{)25} \quad 5\overline{)0} \quad 1\overline{)0}$$
$$\quad \underline{25} \qquad \underline{0} \qquad \underline{0}$$
$$\quad 0 \qquad 0 \qquad 0$$

$25 = 100_5$

2.131 {1, 2, 3, 4, 0}

2.132
$$5\overline{)6} \quad 1\overline{)1}$$
$$\quad \underline{5} \qquad \underline{1}$$
$$\quad 1 \qquad 0$$

$6 = 11_5$

2.133
$$5\overline{)9} \quad 1\overline{)4}$$
$$\quad \underline{5} \qquad \underline{4}$$
$$\quad 4 \qquad 0$$

$9 = 14_5$

2.134
$$5\overline{)13} \quad 1\overline{)3}$$
$$\quad \underline{10} \qquad \underline{3}$$
$$\quad 3 \qquad 0$$

$13 = 23_5$

2.135
$$5\overline{)23} \quad 1\overline{)3}$$
$$\quad \underline{20} \qquad \underline{3}$$
$$\quad 3 \qquad 0$$

$23 = 43_5$

2.136
$$25\overline{)27} \quad 5\overline{)2} \quad 1\overline{)2}$$
$$\quad \underline{25} \qquad \underline{0} \qquad \underline{2}$$
$$\quad 2 \qquad 2 \qquad 0$$

$27 = 102_5$

2.137
$$25\overline{)51} \quad 5\overline{)1} \quad 1\overline{)1}$$
$$\quad \underline{50} \qquad \underline{0} \qquad \underline{1}$$
$$\quad 1 \qquad 1 \qquad 0$$

$51 = 201_5$

2.138
$$5\overline{)19} \quad 1\overline{)4}$$
$$\quad \underline{15} \qquad \underline{4}$$
$$\quad 4 \qquad 0$$

$19 = 34_5$

2.139
$$25\overline{)84} \quad 5\overline{)9} \quad 1\overline{)4}$$
$$\quad \underline{75} \qquad \underline{5} \qquad \underline{4}$$
$$\quad 9 \qquad 4 \qquad 0$$

$84 = 314_5$

2.140
$$125\overline{)141} \quad 25\overline{)16} \quad 5\overline{)16} \quad 1\overline{)1}$$
$$\quad \underline{125} \qquad \underline{0} \qquad \underline{15} \qquad \underline{1}$$
$$\quad 16 \qquad 16 \qquad 1 \qquad 0$$

$141 = 1031_5$

2.141
$$125\overline{)228} \quad 25\overline{)103} \quad 5\overline{)3} \quad 1\overline{)3}$$
$$\quad \underline{125} \qquad \underline{100} \qquad \underline{0} \qquad \underline{3}$$
$$\quad 103 \qquad 3 \qquad 3 \qquad 0$$

$228 = 1403_5$

2.142
$21_5 = 2 \times 5^1 = 1$
$= 10 + 1$
$= 11$

2.143
$33_5 = 3 \times 5^1 + 3 \times 1$
$= 15 + 3$
$= 18$

2.144
$40_5 = 4 \times 5^1$
$= 20$

2.145
$44_5 = 4 \times 5^1 + 4 \times 1$
$= 20 + 4$
$= 24$

2.146
$100_5 = 1 \times 5^2$
$= 25$

2.147
$121_5 = 1 \times 5^2 + 2 \times 5^1 + 1 \times 1$
$= 25 + 10 + 1$
$= 36$

2.148 $200_5 = 2 \times 5^2$
$$= 2 \times 25$$
$$= 50$$

2.149 $222_5 = 2 \times 5^2 + 2 \times 5^1 + 2 \times 1$
$$= 2 \times 25 + 2 \times 5 + 2$$
$$= 50 + 10 + 2$$
$$= 62$$

2.150 $333_5 = 3 \times 5^2 + 3 \times 5^1 + 3 \times 1$
$$= 3 \times 25 + 3 \times 5 + 3$$
$$= 75 + 15 + 3$$
$$= 93$$

2.151 $400_5 = 4 \times 5^2$
$$= 4 \times 25$$
$$= 100$$

III. SECTION THREE

3.1 1×14 or 2×7

3.2 1×24 or 2×12 or 3×8 or 4×6

3.3 1×50 or 2×25 or 5×10

3.4 96×1 or 2×48 or 4×24 or 8×12 or 3×32 or 6×16

3.5 1×150 or 2×75 or 3×50 or 5×30 or 6×25 or 10×15

3.6 1×216 or 2×108 or 3×72 or 4×54 or 6×36 or 8×27 or 9×24 or 12×18

3.7   {1, 3, 9, 27}

3.8   {1, 2, 3, 4, 6, 8, 9, 12, 18, 24, 36, 72}

3.9   {1, 2, 4, 5, 10, 20, 25, 50, 100}

3.10   {1, 2, 3, 4, 6, 8, 9, 12, 16, 18, 24, 36, 48, 72, 144}

3.11   {1, 3, 5, 9, 15, 25, 45, 75, 225}

3.12   The factors are $3 \cdot 3 \cdot x \cdot y \cdot z$.

3.13   The factors are $2 \cdot 2 \cdot 3 \cdot p \cdot q \cdot r \cdot s$

3.14   The set of factors is $\{3 \cdot 5 \cdot a \cdot b \cdot c \cdot d\}$.

3.15   The set of factors is $\{2 \cdot 2 \cdot 2 \cdot 2 \cdot 2 \cdot e \cdot f \cdot g\}$.

3.16   2, 3, 5, 7, 11, 13, 17, 19, 23, 29

3.17   31, 37, 41, 43, 47

3.18   79, 83, 89, 97

3.19   19, 23, 29, 31, 37

3.20   4, 6, 8, 9, 10, 12, 14, 15, 16, 18

3.21   30, 32, 33, 34, 35

3.22   18, 20, 21, 22, 24, 25, 26, 27, 28, 30

3.23   54, 55, 56, 57, 58, 60, 62, 63, 64, 65, 66, 68, 69, 70, 72, 74, 75, 76, 77, 78

3.24   6 is divisible by 2: $6 = 2 \times 3$

3.25   9 is divisible by 3: $9 = 3 \times 3$

3.26   14 is divisible by 2: $14 = 2 \times 7$

3.27   18 is divisible by 2: $18 = 2 \times 9$
9 is divisible by 3:
$$18 = 2 \times 3 \times 3$$

3.28   24 is divisible by 2: $24 = 2 \times 12$
12 is divisible by 2:
$$24 = 2 \times 2 \times 6$$
6 is divisible by 2:
$$24 = 2 \times 2 \times 2 \times 3$$

3.29   39 is divisible by 3: $39 = 3 \times 13$

3.30   42 is divisible by 2: $42 = 2 \times 21$
21 is divisible by 3:
$$42 = 2 \times 3 \times 7$$

3.31   54 is divisible by 2: $54 = 2 \times 27$
27 is divisible by 3:
$$54 = 2 \times 3 \times 9$$
9 is divisible by 3:
$$54 = 2 \times 3 \times 3 \times 3$$

3.32   75 is divisible by 3: $75 = 3 \times 25$
        25 is divisible by 5:
          $75 = 3 \times 5 \times 5$

3.33   90 is divisible by 2: $90 = 2 \times 45$
        45 is divisible by 3:
          $90 = 2 \times 3 \times 15$
        15 is divisible by 3:
          $90 = 2 \times 3 \times 3 \times 5$

3.34   100 is divisible by 2: $100 = 2 \times 50$
        50 is divisible by 2:
          $100 = 2 \times 2 \times 25$
        25 is divisible by 5:
          $100 = 2 \times 2 \times 5 \times 5$

3.35   27 is divisible by 3: $27 = 3 \times 9$
        9 is divisible by 3:
          $27 = 3 \times 3 \times 3$
           $= 3^3$

3.36   36 is divisible by 2: $36 = 2 \times 18$
        18 is divisible by 2:
          $36 = 2 \times 2 \times 9$
        9 is divisible by 3:
          $36 = 2 \times 2 \times 3 \times 3$
           $= 2^2 \times 3^2$

3.37   49 is divisible by 7: $49 = 7 \times 7$
                     $= 7^2$

3.38   64 is divisible by 2: $64 = 2 \times 32$
        32 is divisible by 2:
          $64 = 2 \times 2 \times 16$
        16 is divisible by 2:
          $64 = 2 \times 2 \times 2 \times 8$
        8 is divisible by 2:
          $64 = 2 \times 2 \times 2 \times 2 \times 4$
        4 is divisible by 2:
          $64 = 2 \times 2 \times 2 \times 2 \times 2 \times 2$
           $= 2^6$

3.39   72 is divisible by 2: $72 = 2 \times 36$
        36 is divisible by 2:
          $72 = 2 \times 2 \times 18$
        18 is divisible by 2:
          $72 = 2 \times 2 \times 2 \times 9$
        9 is divisible by 3:
          $72 = 2 \times 2 \times 2 \times 3 \times 3$
           $= 2^3 \times 3^2$

3.40   81 is divisible by 3: $81 = 3 \times 27$
        27 is divisible by 3:
          $81 = 3 \times 3 \times 9$
        9 is divisible by 3:
          $81 = 3 \times 3 \times 3 \times 3$
           $= 3^4$

3.41   121 is divisible by 11:
          $121 = 11 \times 11$
           $= 11^2$

3.42   150 is divisible by 2:
          $150 = 2 \times 75$
        75 is divisible by 3:
          $150 = 2 \times 3 \times 25$
        25 is divisible by 5:
          $150 = 2 \times 3 \times 5 \times 5$
           $= 2 \times 3 \times 5^2$

3.43   169 is divisible by 13:
          $169 = 13 \times 13$
           $= 13^2$

3.44   225 is divisible by 3:
          $225 = 3 \times 75$
        75 is divisible by 3:
          $225 = 3 \times 3 \times 25$
        25 is divisible by 5:
          $225 = 3 \times 3 \times 5 \times 5$
           $= 3^2 \times 5^2$

3.45 $12 = 2 \times \boxed{2 \times 3}$
        $30 = \boxed{2 \times 3} \times 5$
    GCF $= 2 \times 3 = 6$

3.46 $24 = \boxed{2 \times 2} \times 2 \times 3$
        $44 = \boxed{2 \times 2} \times 11$
    GCF $= 2 \times 2 = 4$

3.47 $30 = \boxed{2} \times 3 \times \boxed{5}$
        $70 = \boxed{2 \times 5} \times 7$
    GCF $= 2 \times 5 = 10$

3.48 $75 = \boxed{3 \times 5} \times 5$
      $120 = 2 \times 2 \times 2 \times \boxed{3 \times 5}$
    GCF $= 3 \times 5 = 15$

3.49 $81 = \boxed{3 \times 3} \times 3 \times 3$
$144 = 2 \times 2 \times 2 \times 2 \times \boxed{3 \times 3}$
$GCF = 3 \times 3 = 9$

3.50 $125 = \boxed{5 \times 5} \times 5$
$300 = 2 \times 2 \times 3 \times \boxed{5 \times 5}$
$GCF = 5 \times 5 = 25$

3.51 $54 = \boxed{2 \times 3 \times 3 \times 3}$
$108 = 2 \times \boxed{2 \times 3 \times 3 \times 3}$
$GCF = 2 \times 3 \times 3 \times 3 = 54$
$$\frac{54 \div 54}{108 \div 54} = \frac{1}{2}$$

3.52 $39 = 3 \times \boxed{13}$
$65 = 5 \times \boxed{13}$
$GCF = 13$
$$\frac{39 \div 13}{65 \div 13} = \frac{3}{5}$$

3.53 $85 = 5 \times \boxed{17}$
$102 = 2 \times 3 \times \boxed{17}$
$GCF = 17$
$$\frac{85 \div 17}{102 \div 17} = \frac{5}{6}$$

3.54 $38 = 2 \times \boxed{19}$
$95 = 5 \times \boxed{19}$
$GCF = 19$
$$\frac{38 \div 19}{95 \div 19} = \frac{2}{5}$$

3.55 $48 = \boxed{2 \times 2} \times 2 \times 2 \times 3$
$96 = \boxed{2 \times 2} \times 2 \times 2 \times 2 \times 3$
$100 = \boxed{2 \times 2} \times 5 \times 5$
$GCF = 2 \times 2 = 4$

3.56 $15 = 3 \times 5$
$21 = 3 \times 7$
$LCM = 3 \times 5 \times 7 = 105$

3.57 $27 = 3 \times 3 \times 3$
$36 = 2 \times 2 \times 3 \times 3$
$LCM = 3 \times 3 \times 3 \times 2 \times 2 = 108$

3.58 $12 = 2 \times 2 \times 3$
$18 = 2 \times 3 \times 3$
$30 = 2 \times 3 \times 5$
$LCM = 2 \times 2 \times 3 \times 3 \times 5 = 180$

3.59 $42 = 2 \times 3 \times 7$
$91 = 7 \times 13$
$LCM = 2 \times 3 \times 7 \times 13 = 546$

3.60 $21 = 3 \times 7$
$77 = 7 \times 11$
$63 = 3 \times 3 \times 7$
$LCM = 3 \times 3 \times 7 \times 11 = 693$

3.61 $8 = 2 \times 2 \times 2$
$9 = 3 \times 3$
$36 = 2 \times 2 \times 3 \times 3$
$LCM = 2 \times 2 \times 2 \times 3 \times 3 = 72$

3.62 $3 = 3$
$5 = 5$
$6 = 2 \times 3$
$LCM = 2 \times 3 \times 5 = 30$

3.63 $7 = 7$
$12 = 2 \times 2 \times 3$
$21 = 3 \times 7$
$27 = 3 \times 3 \times 3$
$LCM = 2 \times 2 \times 3 \times 3 \times 3 \times 7 = 756$

3.64 $3 = 3$
$9 = 3 \times 3$
$LCD = 3 \times 3 = 9$

3.65 $11 = 11$
$33 = 3 \times 11$
$LCD = 3 \times 11 = 33$

3.66 $12 = 2 \times 2 \times 3$
$21 = 3 \times 7$
$LCD = 2 \times 2 \times 3 \times 7 = 84$

3.67 $15 = 3 \times 5$
$33 = 3 \times 11$
$LCD = 3 \times 5 \times 11 = 165$

3.68 $26 = 2 \times 13$
$39 = 3 \times 13$
$LCD = 2 \times 3 \times 13 = 78$

3.69   Since the units' digit is even, 138 is divisible by 2.

Since the units' digit is neither 5 nor 0; 138 is not divisible by 5.

Since the sum of the digits equals $1 + 3 + 8 = 12$ and 12 is divisible by 3; 138 is divisible by 3.

3.70   Since the units' digit is not even, 417 is not divisible by 2.

Since the units' digit is neither 5 nor 0; 417 is not divisible by 5.

Since the sum of the digits equals $4 + 1 + 7 = 12$ and 12 is divisible by 3; 417 is divisible by 3.

3.71   Since the units' digit is not even, 6,315 is not divisible by 2.

Since the units' digit is 5; 6,315 is divisible by 5.

Since the sum of the digits equals $6 + 3 + 1 + 5 = 15$ and 15 is divisible by 3; 6,315 is divisible by 3.

3.72   Since the units' digit is even, 12,052 is divisible by 2.

Since the units' digit is neither 5 nor 0; 12,052 is not divisible by 5.

Since the sum of the digits equals $1 + 2 + 0 + 5 + 2 = 10$ and 10 is not divisible by 3; 12,052 is not divisible by 3.

3.73   Since the units' digit is not even, 2,146,115 is not divisible by 2.

Since the units' digit is 5; 2,146,115 is divisible by 5.

Since the sum of the digits equals $2 + 1 + 4 + 6 + 1 + 1 + 5 = 20$ and 20 is not divisible by 3; 2,146,115 is not divisible by 3.

3.74   Since the units' digit is even, 3,186,212 is divisible by 2.

Since the units' digit is neither 5 nor 0; 3,186,212 is not divisible by 5.

Since the sum of the digits equals $3 + 1 + 8 + 6 + 2 + 1 + 2 = 23$ and 23 is not divisible by 3; 3,186,212 is not divisible by 3.

3.75   Since the units' digit is even, 128,600 is divisible by 2.

Since the units' digit is 0, then 128,600 is divisible by 5.

Since the sum of the digits equals $1 + 2 + 8 + 6 + 0 + 0 = 17$ and 17 is not divisible by 3; 128,600 is not divisible by 3.

3.76   Since the units' digit is even, 3,166 is divisible by 2.

Since the units' digit is neither 5 nor 0; 3,166 is not divisible by 5.

Since the sum of the digits equals $3 + 1 + 6 + 6 = 16$ and 16 is not divisible by 3; 3,166 is not divisible by 3.

3.77   Since the units' digit is not even, 76,145 is not divisible by 2.

Since the units' digit is 5; 76,145 is divisible by 5.

Since the sum of the digits equals $7 + 6 + 1 + 4 + 5 = 23$ and 23 is not divisible by 3; 76,145 is not divisible by 3.

3.78   Since the units' digit is even, 476,100 is divisible by 2.

Since the units' digit is 0; 476,100 is divisible by 5.

Since the sum of the digits equals $4 + 7 + 6 + 1 + 0 + 0 = 18$ and 18 is divisible by 3; 476,100 is divisible by 3.

3.79 Since the units' digit is even, 437,386 is divisible by 2.

Since the units' digit is neither 5 nor 0; 437,386 is not divisible by 5.

Since the sum of the digits equals $4 + 3 + 7 + 3 + 8 + 6 = 31$ and 31 is not divisible by 3; 437,386 is not divisible by 3.

3.80 Since the units' digit is even, 627,266 is divisible by 2.

Since the units' digit is neither 5 nor 0; 627,266 is not divisible by 5.

Since the sum of the digits equals $6 + 2 + 7 + 2 + 6 + 6 = 29$ and 29 is not divisible by 3; 627,266 is not divisible by 3.

3.81 Since the units' digit is not even, 563,565 is not divisible by 2.

Since the units' digit is 5; 563,565 is divisible by 5.

Since the sum of the digits equals $5 + 6 + 3 + 5 + 6 + 5 = 30$ and 30 is divisible by 3; 563,565 is divisible by 3.

3.82 Since the units' digit is not even, 174,331 is not divisible by 2.

Since the units' digit is neither 5 nor 0; 175,331 is not divisible by 5.

Since the sum of the digits equals $1 + 7 + 5 + 3 + 3 + 1 = 20$ and 20 is not divisible by 3; 175,331 is not divisible by 3.

3.83 Since the units' digit is not even, 123,429 is not divisible by 2.

Since the units' digit is neither 5 nor 0; 123,429 is not divisible by 5.

Since the sum of the digits equals $1 + 2 + 3 + 4 + 2 + 9 = 21$ and 21 is divisible by 3; 123,429 is divisible by 3.

3.84 Since the units' digit is even, 79,246 is divisible by 2.

Since the units' digit is neither 5 nor 0; 79,246 is not divisible by 5.

Since the sum of the digits equals $7 + 9 + 2 + 4 + 6 = 28$ and 28 is not divisible by 3; 79,246 is not divisible by 3.

3.85 Since the units' digit is even, 46,474 is divisible by 2

Since the units' digit is neither 5 nor 0; 46,474 is not divisible by 5.

Since the sum of the digits equals $4 + 6 + 4 + 7 + 4 = 25$ and 25 is not divisible by 3; 46,474 is not divisible by 3.

3.86 Since the units' digit is even, 943,512 is divisible by 2.

Since the units' digit is neither 5 nor 0; 943,512 is not divisible by 5.

Since the sum of the digits equals $9 + 4 + 3 + 5 + 1 + 2 = 24$ and 24 is divisible by 3; 943,512 is divisible by 3.

3.87 Since the units' digit is not even, 187,335 is not divisible by 2.

Since the units' digit is 5; 187,335 is divisible by 5.

Since the sum of the digits equals $1 + 8 + 7 + 3 + 3 + 5 = 27$ and 27 is divisible by 3; 187,335 is divisible by 3.

3.88 Since the units' digit is not even, 150,475 is not divisible by 2.

Since the units' digit is 5; 150,475 is divisible by 5.

Since the sum of the digits equals $1 + 5 + 0 + 4 + 7 + 5 = 22$ and 22 is not divisible by 3; 150,475 is not divisible by 3.

3.89 Since the units' digit is even, 61,844 is divisible by 2.

Since the units' digit is neither 5 nor 0; 61,844 is not divisible by 5.

Since the sum of the digits equals 6 + 1 + 8 + 4 + 4 = 23 and 23 is not divisible by 3; 61,844 is not divisible by 3.

3.90 Since the units' digit is even, 52,430 is divisible by 2.

Since the units' digit is 0; 52,430 is divisible by 5.

Since the sum of the digits equals 5 + 2 + 4 + 3 + 0 = 14 and 14 is not divisible by 3; 52,430 is not divisible by 3.

3.91 Since the units' digit is not even, 6,713 is not divisible by 2.

Since the units' digit is neither 5 nor 0; 6,713 is not divisible by 5.

Since the sum of the digits equals 6 + 7 + 1 + 3 = 17 and 17 is not divisible by 3; 6,713 is not divisible by 3.

3.92 Since the units' digit is not even, 4,511 is not divisible by 2.

Since the units' digit is neither 5 nor 0; 4,511 is not divisible by 5.

Since the sum of the digits equals 4 + 5 + 1 + 1 = 11 and 11 is not divisible by 3; 4,511 is not divisible by 3.

3.93 Since the units' digit is even, 1,800 is divisible by 2.

Since the units' digit is 0; 1,800 is divisible by 5.

Since the sum of the digits equals 1 + 8 + 0 + 0 = 9 and 9 is divisible by 3; 1,800 is divisible by 3.

3.94 Since the units' digit is even, 7,390 is divisible by 2.

Since the units' digit is 0; 7,390 is divisible by 5.

Since the sum of the digits equals 7 + 3 + 9 + 0 = 19 and 19 is not divisible by 3; 7,390 is not divisible by 3.

3.95 Since the units' digit is not even, 9,779 is not divisible by 2.

Since the units' digit is neither 5 nor 0; 9,779 is not divisible by 5.

Since the sum of the digits equals 9 + 7 + 7 + 9 = 32 and 32 is not divisible by 3; 9,779 is not divisible by 3.

3.96 Since the units' digit is even, 7,060 is divisible by 2.

Since the units' digit is 0; 7,060 is divisible by 5.

Since the sum of the digits equals 7 + 0 + 6 + 0 = 13 and 13 is not divisible by 3; 7,060 is not divisible by 3.

3.97 Since the units' digit is not even, 8,525 is not divisible by 2.

Since the units' digit is 5; 8,525 is divisible by 5.

Since the sum of the digits equals 8 + 5 + 2 + 5 = 20 and 20 is not divisible by 3; 8,525 is not divisible by 3.

3.98 Since the units' digit is not even, 2,119 is not divisible by 2.

Since the units' digit is neither 5 nor 0; 2,119 is not divisible by 5.

Since the sum of the digits equals 2 + 1 + 1 + 9 = 13 and 13 is not divisible by 3; 2,119 is not divisible by 3.

3.99   Since the units' digit is even,
2,020,202 is divisible by 2.

Since the units' digit is neither 5 nor
0; 2,020,202 is not divisible by 5.

Since the sum of the digits equals
$2 + 0 + 2 + 0 + 2 + 0 + 2 = 8$ and 8 is
not divisible by 3; 2,020,202 is not
divisible by 3.

3.100   Since the units' digit is even,
8,100,000 is divisible by 2.

Since the units' digit is 0; 8,100,000 is
divisible by 5.

Since the sum of the digits equals
$8 + 1 + 0 + 0 + 0 + 0 + 0 = 9$ and 9 is
divisible by 3; 8,100,000 is divisible
by 3.

I.   SECTION ONE

1.1 $\dfrac{5}{8}$

1.2 $\dfrac{10}{11}$

1.3 $\dfrac{14}{17}$

1.4 $\dfrac{1}{6}$

1.5 $\dfrac{7}{19}$

1.6 $\dfrac{3}{23}$

1.7 $\dfrac{7}{9}$

1.8 $\dfrac{13}{15}$

1.9 $\dfrac{5}{29}$

1.10 $\dfrac{2}{5}$

1.11 $\dfrac{2}{4} = \dfrac{2 \div 2}{4 \div 2} = \dfrac{1}{2}$

1.12 $\dfrac{8}{12} = \dfrac{8 \div 4}{12 \div 4} = \dfrac{2}{3}$

1.13 $\dfrac{8}{16} = \dfrac{8 \div 8}{16 \div 8} = \dfrac{1}{2}$

1.14 $\dfrac{5}{10} = \dfrac{5 \div 5}{10 \div 5} = \dfrac{1}{2}$

1.15 $\dfrac{12}{18} = \dfrac{12 \div 6}{18 \div 6} = \dfrac{2}{3}$

1.16 $\dfrac{3}{9} = \dfrac{3 \div 3}{9 \div 3} = \dfrac{1}{3}$

1.17 $\dfrac{9}{15} = \dfrac{9 \div 3}{15 \div 3} = \dfrac{3}{5}$

1.18 $\dfrac{16}{18} = \dfrac{16 \div 2}{18 \div 2} = \dfrac{8}{9}$

1.19 $\dfrac{13}{26} = \dfrac{13 \div 13}{26 \div 13} = \dfrac{1}{2}$

1.20 $\dfrac{10}{16} = \dfrac{10 \div 2}{16 \div 2} = \dfrac{5}{8}$

1.21 $\dfrac{16}{20} = \dfrac{16 \div 4}{20 \div 4} = \dfrac{4}{5}$

1.22 $\dfrac{18}{24} = \dfrac{18 \div 6}{24 \div 6} = \dfrac{3}{4}$

1.23 $\dfrac{3}{12} = \dfrac{3 \div 3}{12 \div 3} = \dfrac{1}{4}$

1.24 $\dfrac{24}{30} = \dfrac{24 \div 6}{30 \div 6} = \dfrac{4}{5}$

1.25 $\dfrac{11}{8} = 8\overline{)11} = 1\dfrac{3}{8}$
$$\underline{8}$$
$$3$$

1.26 $\dfrac{4}{3} = 3\overline{)4} = 1\dfrac{1}{3}$
$$\underline{3}$$
$$1$$

1.27 $\dfrac{41}{16} = 16\overline{)41} = 2\dfrac{9}{16}$
$$\underline{32}$$
$$9$$

1.28 $\dfrac{19}{6} = 6\overline{)19} = 3\dfrac{1}{6}$
$$\underline{18}$$
$$1$$

1.29 $\dfrac{16}{9} = 9\overline{)16} = 1\dfrac{7}{9}$
$$\underline{9}$$
$$7$$

1.30 $\dfrac{22}{3} = 3\overline{)22} = 7\dfrac{1}{3}$
$$\underline{21}$$
$$1$$

1.31 $\dfrac{16}{5} = 5\overline{)16} = 3\dfrac{1}{5}$
$$\underline{15}$$
$$1$$

1.32 $\dfrac{44}{17} = 17\overline{)44} = 2\dfrac{10}{17}$
$$\underline{34}$$
$$10$$

1.33 $\dfrac{42}{11} = 11\overline{)42} = 3\dfrac{9}{11}$
$$\underline{33}$$
$$9$$

1.34 $\dfrac{43}{12} = \quad 12\overline{)43} = 3\dfrac{7}{12}$

$$\begin{array}{r} 3 \\ 12\overline{)43} \\ \underline{36} \\ 7 \end{array}$$

1.35 $\dfrac{25}{2} = \quad 2\overline{)25} = 12\dfrac{1}{2}$

$$\begin{array}{r} 12 \\ 2\overline{)25} \\ \underline{2} \\ 05 \\ \underline{4} \\ 1 \end{array}$$

1.36 $\dfrac{41}{15} = \quad 15\overline{)41} = 2\dfrac{11}{15}$

$$\begin{array}{r} 2 \\ 15\overline{)41} \\ \underline{30} \\ 11 \end{array}$$

1.37 $\dfrac{13}{7} = \quad 7\overline{)13} = 1\dfrac{6}{7}$

$$\begin{array}{r} 1 \\ 7\overline{)13} \\ \underline{7} \\ 6 \end{array}$$

1.38 $\dfrac{71}{20} = \quad 20\overline{)71} = 3\dfrac{11}{20}$

$$\begin{array}{r} 3 \\ 20\overline{)71} \\ \underline{60} \\ 11 \end{array}$$

1.39 $\dfrac{10}{6} = \dfrac{10 \div 2}{6 \div 2} = \dfrac{5}{3} = \quad 3\overline{)5} = 1\dfrac{2}{3}$

$$\begin{array}{r} 1 \\ 3\overline{)5} \\ \underline{3} \\ 2 \end{array}$$

1.40 $\dfrac{12}{10} = \dfrac{12 \div 2}{10 \div 2} = \dfrac{6}{5} = \quad 5\overline{)6} = 1\dfrac{1}{5}$

$$\begin{array}{r} 1 \\ 5\overline{)6} \\ \underline{5} \\ 1 \end{array}$$

1.41 $\dfrac{42}{16} = \dfrac{42 \div 2}{16 \div 2} = \dfrac{21}{8} = 8\overline{)21} = 2\dfrac{5}{8}$

$$\begin{array}{r} 2 \\ 8\overline{)21} \\ \underline{16} \\ 5 \end{array}$$

1.42 $\dfrac{87}{21} = \dfrac{87 \div 3}{21 \div 3} = \dfrac{29}{7} = 7\overline{)29} = 4\dfrac{1}{7}$

$$\begin{array}{r} 4 \\ 7\overline{)29} \\ \underline{28} \\ 1 \end{array}$$

1.43 $\dfrac{40}{12} = \dfrac{40 \div 4}{12 \div 4} = \dfrac{10}{3} = 3\overline{)10} = 3\dfrac{1}{3}$

$$\begin{array}{r} 3 \\ 3\overline{)10} \\ \underline{9} \\ 1 \end{array}$$

1.44 $\dfrac{24}{15} = \dfrac{24 \div 3}{15 \div 3} = \dfrac{8}{5} = \quad 5\overline{)8} = 1\dfrac{3}{5}$

$$\begin{array}{r} 1 \\ 5\overline{)8} \\ \underline{5} \\ 3 \end{array}$$

1.45 $\dfrac{26}{4} = \dfrac{26 \div 2}{4 \div 2} = \dfrac{13}{2} = 2\overline{)13} = 6\dfrac{1}{2}$

$$\begin{array}{r} 6 \\ 2\overline{)13} \\ \underline{12} \\ 1 \end{array}$$

1.46 $\dfrac{45}{20} = \dfrac{45 \div 5}{20 \div 5} = \dfrac{9}{4} = \quad 4\overline{)9} = 2\dfrac{1}{4}$

$$\begin{array}{r} 2 \\ 4\overline{)9} \\ \underline{8} \\ 1 \end{array}$$

1.47 $\dfrac{39}{6} = \dfrac{39 \div 3}{6 \div 3} = \dfrac{13}{2} = 2\overline{)13} = 6\dfrac{1}{2}$

$$\begin{array}{r} 6 \\ 2\overline{)13} \\ \underline{12} \\ 1 \end{array}$$

1.48 $\dfrac{55}{10} = \dfrac{55 \div 5}{10 \div 5} = \dfrac{11}{2} = 2\overline{)11} = 5\dfrac{1}{2}$

$$\begin{array}{r} 5 \\ 2\overline{)11} \\ \underline{10} \\ 1 \end{array}$$

1.49 $\dfrac{85}{25} = \dfrac{85 \div 5}{25 \div 5} = \dfrac{17}{5} = 5\overline{)17} = 3\dfrac{2}{5}$

$$\begin{array}{r} 3 \\ 5\overline{)17} \\ \underline{15} \\ 2 \end{array}$$

1.50 $\dfrac{64}{24} = \dfrac{64 \div 8}{24 \div 8} = \dfrac{8}{3} = \quad 3\overline{)8} = 2\dfrac{2}{3}$

$$\begin{array}{r} 2 \\ 3\overline{)8} \\ \underline{6} \\ 2 \end{array}$$

1.51 $\dfrac{46}{8} = \dfrac{46 \div 2}{8 \div 2} = \dfrac{23}{4} = 4\overline{)23} = 5\dfrac{3}{4}$

$$\begin{array}{r} 5 \\ 4\overline{)23} \\ \underline{20} \\ 3 \end{array}$$

1.52 $\dfrac{24}{9} = \dfrac{24 \div 3}{9 \div 3} = \dfrac{8}{3} = \quad 3\overline{)8} = 2\dfrac{2}{3}$

$$\begin{array}{r} 2 \\ 3\overline{)8} \\ \underline{6} \\ 2 \end{array}$$

1.53 $\dfrac{30}{8} = \dfrac{30 \div 2}{8 \div 2} = \dfrac{15}{4} = 4\overline{)15} = 3\dfrac{3}{4}$

$$\begin{array}{r} 3 \\ 4\overline{)15} \\ \underline{12} \\ 3 \end{array}$$

1.54 $\dfrac{39}{27} = \dfrac{39 \div 3}{27 \div 3} = \dfrac{13}{9} = 9\overline{)13} = 1\dfrac{4}{9}$
$\underline{9}$
4

1.55 $\dfrac{7}{9}$

1.56 $\dfrac{3}{12} = \dfrac{3 \div 3}{12 \div 3} = \dfrac{1}{4}$

1.57 $\dfrac{25}{7} = \dfrac{3}{7\overline{)25}} = 3\dfrac{4}{7}$
$\underline{21}$
4

1.58 $\dfrac{16}{26} = \dfrac{16 \div 2}{26 \div 2} = \dfrac{8}{13}$

1.59 $\dfrac{19}{10} = \dfrac{1}{10\overline{)19}} = 1\dfrac{9}{10}$
$\underline{10}$
9

1.60 $\dfrac{49}{35} = \dfrac{49 \div 7}{35 \div 7} = \dfrac{7}{5} = \dfrac{1}{5\overline{)7}} = 1\dfrac{2}{5}$
$\underline{5}$
2

1.61 $9\dfrac{3}{5}$

1.62 $18\dfrac{7}{9}$

1.63 $5\dfrac{11}{13}$

1.64 $22\dfrac{12}{22} = 22\dfrac{12 \div 2}{22 \div 2} = 22\dfrac{6}{11}$

1.65 $43\dfrac{12}{16} = 43\dfrac{12 \div 4}{16 \div 4} = 43\dfrac{3}{4}$

1.66 $61\dfrac{8}{12} = 61\dfrac{8 \div 4}{12 \div 4} = 61\dfrac{2}{3}$

1.67 $\dfrac{14}{35} = \dfrac{14 \div 7}{35 \div 7} = \dfrac{2}{5}$

1.68 $8\dfrac{8}{15}$

1.69 $34\dfrac{17}{21}$

1.70 $43\dfrac{5}{10} = 43\dfrac{5 \div 5}{10 \div 5} = 43\dfrac{1}{2}$

1.71 $39\dfrac{8}{28} = 39\dfrac{8 \div 4}{28 \div 4} = 39\dfrac{2}{7}$

1.72 $4\dfrac{24}{32} = 4\dfrac{24 \div 8}{32 \div 8} = 4\dfrac{3}{4}$

1.73 $15\dfrac{7}{8}$

1.74 $78\dfrac{8}{12} = 78\dfrac{8 \div 4}{12 \div 4} = 78\dfrac{2}{3}$

1.75 $22\dfrac{20}{24} = 22\dfrac{20 \div 4}{24 \div 4} = 22\dfrac{5}{6}$

1.76 $28\dfrac{9}{18} = 28\dfrac{9 \div 9}{18 \div 9} = 28\dfrac{1}{2}$

1.77 $21\dfrac{35}{49} = 21\dfrac{35 \div 7}{49 \div 7} = 21\dfrac{5}{7}$

1.78 $17\dfrac{9}{19}$

1.79 $21\dfrac{2}{2} = 21 + 1 = 22$

1.80 $42\dfrac{15}{11} = 42 + 1\dfrac{4}{11} = 43\dfrac{4}{11}$

1.81 $89\dfrac{26}{16} = 89\dfrac{26 \div 2}{16 \div 2} = 89\dfrac{13}{8} = 89 + 1\dfrac{5}{8} = 90\dfrac{5}{8}$

1.82 $163\dfrac{40}{30} = 163\dfrac{40 \div 10}{30 \div 10} = 163\dfrac{4}{3} = 163 + 1\dfrac{1}{3} = 164\dfrac{1}{3}$

1.83 $93\dfrac{30}{24} = 93\dfrac{30 \div 6}{24 \div 6} = 93\dfrac{5}{4} = 93 + 1\dfrac{1}{4} = 94\dfrac{1}{4}$

1.84 $138\dfrac{7}{5} = 138 + 1\dfrac{2}{5} = 139\dfrac{2}{5}$

1.85 $37\dfrac{20}{15} = 37\dfrac{20 \div 5}{15 \div 5} = 37\dfrac{4}{3} = 37 + 1\dfrac{1}{3} = 38\dfrac{1}{3}$

1.86 $95\dfrac{28}{18} = 95\dfrac{28 \div 2}{18 \div 2} = 95\dfrac{14}{9} = 95 + 1\dfrac{5}{9} = 96\dfrac{5}{9}$

1.87 $39\dfrac{25}{12} = 39 + 2\dfrac{1}{12} = 41\dfrac{1}{12}$

1.88 $\quad 22\frac{14}{6} = 22\frac{14 \div 2}{6 \div 2} = 22\frac{7}{3} = 22 + 2\frac{1}{3} = 24\frac{1}{3}$

1.89
$$14\frac{2}{13} = 13\frac{15}{13}$$
$$-6\frac{9}{13} = \ 6\frac{9}{13}$$
$$\overline{\qquad\quad 7\frac{6}{13}}$$

1.90
$$9\frac{3}{14} = 8\frac{17}{14}$$
$$-\ \frac{9}{14} = \ \ \frac{9}{14}$$
$$\overline{\qquad 8\frac{8}{14} = 8\frac{8 \div 2}{14 \div 2} = 8\frac{4}{7}}$$

1.91
$$7\frac{1}{4} = 6\frac{5}{4}$$
$$-3\frac{3}{4} = 3\frac{3}{4}$$
$$\overline{\qquad 3\frac{2}{4} = 3\frac{2 \div 2}{4 \div 2} = 3\frac{1}{2}}$$

1.92
$$13\frac{2}{7} = 12\frac{9}{7}$$
$$-12\frac{5}{7} = 12\frac{5}{7}$$
$$\overline{\qquad\quad \frac{4}{7}}$$

1.93
$$21\frac{1}{3} = 20\frac{4}{3}$$
$$-3\frac{2}{3} = \ 3\frac{2}{3}$$
$$\overline{\qquad\quad 17\frac{2}{3}}$$

1.94
$$39\frac{8}{12} = 38\frac{20}{12}$$
$$-27\frac{11}{12} = 27\frac{11}{12}$$
$$\overline{\qquad 11\frac{9}{12} = 11\frac{9 \div 3}{12 \div 3} = 11\frac{3}{4}}$$

1.95
$$56\frac{6}{21} = 55\frac{27}{21}$$
$$-22\frac{16}{21} = 22\frac{16}{21}$$
$$\overline{\qquad\quad 33\frac{11}{21}}$$

1.96
$$6\frac{3}{8} = 5\frac{11}{8}$$
$$-2\frac{7}{8} = 2\frac{7}{8}$$
$$\overline{\qquad 3\frac{4}{8} = 3\frac{4 \div 4}{8 \div 4} = 3\frac{1}{2}}$$

1.97
$$25\frac{3}{10} = 24\frac{13}{10}$$
$$-21\frac{8}{10} = 21\frac{8}{10}$$
$$\overline{\qquad 3\frac{5}{10} = 3\frac{5 \div 5}{10 \div 5} = 3\frac{1}{2}}$$

1.98
$$10\frac{1}{9} = 9\frac{10}{9}$$
$$-4\frac{4}{9} = 4\frac{4}{9}$$
$$\overline{\qquad 5\frac{6}{9} = 5\frac{6 \div 3}{9 \div 3} = 5\frac{2}{3}}$$

1.99 $\quad \frac{2}{9} + \frac{4}{9} + \frac{3}{9} = \frac{9}{9} = 1;$

the whole picture has been completed.

1.100
$$\frac{7}{12}$$
$$-\ \frac{3}{12}$$
$$\overline{\ \frac{4}{12} = \frac{4 \div 4}{12 \div 4} = \frac{1}{3}}$$

1.101
$$4\frac{5}{16}$$
$$+9\frac{3}{16}$$
$$\overline{13\frac{8}{16} = 13\frac{8 \div 8}{16 \div 8} = 13\frac{1}{2} \text{ gallons}}$$

1.102
$$62\ \ = 61\frac{3}{3}$$
$$-45\frac{2}{3} = 45\frac{2}{3}$$
$$\overline{\qquad 16\frac{1}{3} \text{ gallons}}$$

1.103
$$2\frac{3}{10} = 1\frac{13}{10}$$
$$-1\frac{7}{10} = 1\frac{7}{10}$$
$$\overline{\qquad \frac{6}{10} = \frac{6 \div 2}{10 \div 2} = \frac{3}{5} \text{ minute}}$$

1.104

$$2\frac{7}{8}$$

$$+\ 1\frac{3}{8}$$

$$3\frac{10}{8} = 3 + 1\frac{2}{8} = 4\frac{2\div 2}{8\div 2} = 4\frac{1}{4} \text{ yards}$$

1.105 On which side does a chicken have the most feathers? On the outside.

e. $\dfrac{16}{10} = \dfrac{16\div 2}{10\div 2} = \dfrac{8}{5} = \;\; \begin{array}{r} 1\frac{3}{5} \\ 5\overline{)8} \\ \underline{5} \\ 3 \end{array}$

h. $\dfrac{11}{20}$

i. $2\dfrac{4}{7}$

n. $\dfrac{18}{21} = \dfrac{18\div 3}{21\div 3} = \dfrac{6}{7}$

o. $13\dfrac{11}{18}$

s. $\dfrac{30}{24} = \dfrac{30\div 6}{24\div 6} = \dfrac{5}{4} = \;\; \begin{array}{r} 1\frac{1}{4} \\ 4\overline{)5} \\ \underline{4} \\ 1 \end{array}$

t. $17\dfrac{13}{8} = 17 + 1\dfrac{5}{8} = 18\dfrac{5}{8}$

u. $\dfrac{2}{9}$

1.106

$$\frac{2}{3} = \frac{2}{3}$$

$$\frac{1}{4} = \frac{1}{2 \times 2}$$

$$\text{LCD} = 2 \times 2 \times 3 = 12$$

1.107

$$\frac{6}{7} = \frac{6}{7}$$

$$\frac{1}{3} = \frac{1}{3}$$

$$\text{LCD} = 3 \times 7 = 21$$

1.108

$$\frac{6}{7} = \frac{6}{7}$$

$$\frac{1}{2} = \frac{1}{2}$$

$$\frac{9}{14} = \frac{9}{2 \times 7}$$

$$\text{LCD} = 2 \times 7 = 14$$

1.109

$$\frac{1}{2} = \frac{1}{2}$$

$$\frac{2}{3} = \frac{2}{3}$$

$$\frac{3}{4} = \frac{3}{2 \times 2}$$

$$\text{LCD} = 2 \times 2 \times 3 = 12$$

1.110

$$\frac{5}{6} = \frac{5}{2 \times 3}$$

$$\frac{1}{2} = \frac{1}{2}$$

$$\text{LCD} = 2 \times 3 = 6$$

1.111

$$\frac{2}{9} = \frac{2}{3 \times 3}$$

$$\frac{1}{6} = \frac{1}{2 \times 3}$$

$$\text{LCD} = 2 \times 3 \times 3 = 18$$

1.112

$$\frac{3}{4} = \frac{3}{2 \times 2}$$

$$\frac{2}{5} = \frac{2}{5}$$

$$\frac{7}{8} = \frac{7}{2 \times 2 \times 2}$$

$$\text{LCD} = 2 \times 2 \times 2 \times 5 = 40$$

1.113

$$\frac{1}{6} = \frac{1}{2 \times 3}$$

$$\frac{1}{8} = \frac{1}{2 \times 2 \times 2}$$

$$\frac{1}{12} = \frac{1}{2 \times 2 \times 3}$$

$$\text{LCD} = 2 \times 2 \times 2 \times 3 = 24$$

1.114

$$\frac{3}{8} = \frac{3}{2 \times 2 \times 2}$$

$$\frac{5}{6} = \frac{5}{2 \times 3}$$

$$\text{LCD} = 2 \times 2 \times 2 \times 3 = 24$$

1.115

$$\frac{3}{7} = \frac{3}{7}$$

$$\frac{2}{9} = \frac{2}{3 \times 3}$$

$$\text{LCD} = 3 \times 3 \times 7 = 63$$

1.116

$$\frac{7}{12} = \frac{7}{2 \times 2 \times 3}$$

$$\frac{1}{3} = \frac{1}{3}$$

$$\frac{5}{16} = \frac{5}{2 \times 2 \times 2 \times 2}$$

$$LCD = 2 \times 2 \times 2 \times 2 \times 3 = 48$$

1.117

$$\frac{4}{5} = \frac{4}{5}$$

$$\frac{2}{3} = \frac{2}{3}$$

$$\frac{1}{15} = \frac{1}{3 \times 5}$$

$$LCD = 3 \times 5 = 15$$

1.118

$$\frac{3}{4} = \frac{3}{2 \times 2}$$

$$\frac{7}{8} = \frac{7}{2 \times 2 \times 2}$$

$$LCD = 2 \times 2 \times 2 = 8$$

$$\frac{3 \times 2}{4 \times 2} = \frac{6}{8}$$

$$\frac{6}{8}, \frac{7}{8}$$

1.119

$$\frac{2}{3} = \frac{2}{3}$$

$$\frac{2}{7} = \frac{2}{7}$$

$$LCD = 3 \times 7 = 21$$

$$\frac{2 \times 7}{3 \times 7} = \frac{14}{21}$$

$$\frac{2 \times 3}{7 \times 3} = \frac{6}{21}$$

$$\frac{14}{21}, \frac{6}{21}$$

1.120

$$\frac{1}{6} = \frac{1}{2 \times 3}$$

$$\frac{5}{8} = \frac{5}{2 \times 2 \times 2}$$

$$LCD = 2 \times 2 \times 2 \times 3 = 24$$

$$\frac{1 \times 4}{6 \times 4} = \frac{4}{24}$$

$$\frac{5 \times 3}{8 \times 3} = \frac{15}{24}$$

$$\frac{4}{24}, \frac{15}{24}$$

1.121

$$\frac{11}{18} = \frac{11}{2 \times 3 \times 3}$$

$$\frac{1}{6} = \frac{1}{2 \times 3}$$

$$LCD = 2 \times 3 \times 3 = 18$$

$$\frac{1 \times 3}{6 \times 3} = \frac{3}{18}$$

$$\frac{11}{18}, \frac{3}{18}$$

1.122

$$\frac{2}{4} = \frac{2}{2 \times 2}$$

$$\frac{3}{16} = \frac{3}{2 \times 2 \times 2 \times 2}$$

$$LCD = 2 \times 2 \times 2 \times 2 = 16$$

$$\frac{2 \times 4}{4 \times 4} = \frac{8}{16}$$

$$\frac{8}{16}, \frac{3}{16}$$

1.123

$$\frac{4}{5} = \frac{4}{5}$$

$$\frac{2}{7} = \frac{2}{7}$$

$$LCD = 5 \times 7 = 35$$

$$\frac{4 \times 7}{5 \times 7} = \frac{28}{35}$$

$$\frac{2 \times 5}{7 \times 5} = \frac{10}{35}$$

$$\frac{28}{35}, \frac{10}{35}$$

1.124

$$\frac{3}{8} = \frac{3}{2 \times 2 \times 2}$$

$$\frac{1}{24} = \frac{1}{2 \times 2 \times 2 \times 3}$$

$$\frac{7}{12} = \frac{7}{2 \times 2 \times 3}$$

$$LCD = 2 \times 2 \times 2 \times 3 = 24$$

$$\frac{3 \times 3}{8 \times 3} = \frac{9}{24}$$

$$\frac{1}{24} = \frac{1}{24}$$

$$\frac{7 \times 2}{12 \times 2} = \frac{14}{24}$$

$$\frac{9}{24}, \frac{1}{24}, \frac{14}{24}$$

1.125

$$\frac{3}{26} = \frac{3}{2 \times 13}$$

$$\frac{1}{4} = \frac{1}{2 \times 2}$$

$$\frac{2}{13} = \frac{2}{13}$$

LCD = 2 x 2 x 13 = 52

$$\frac{3 \times 2}{26 \times 2} = \frac{6}{52}$$

$$\frac{1 \times 13}{4 \times 13} = \frac{13}{52}$$

$$\frac{2 \times 4}{13 \times 4} = \frac{8}{52}$$

$$\frac{6}{52}, \frac{13}{52}, \frac{8}{52}$$

1.126

$$\frac{1}{4} = \frac{1}{2 \times 2}$$

$$\frac{5}{6} = \frac{5}{2 \times 3}$$

$$\frac{4}{9} = \frac{4}{3 \times 3}$$

LCD = 2 x 2 x 3 x 3 = 36

$$\frac{1 \times 9}{4 \times 9} = \frac{9}{36}$$

$$\frac{5 \times 6}{6 \times 6} = \frac{30}{36}$$

$$\frac{4 \times 4}{9 \times 4} = \frac{16}{36}$$

$$\frac{9}{36}, \frac{30}{36}, \frac{16}{36}$$

1.127

$$\frac{3}{5} = \frac{3}{5}$$

$$\frac{1}{4} = \frac{1}{2 \times 2}$$

LCD = 2 x 2 x 5 = 20

$$\frac{3 \times 4}{5 \times 4} = \frac{12}{20}$$

$$+ \frac{1 \times 5}{4 \times 4} = \frac{5}{20}$$

$$\frac{17}{20}$$

1.128

$$\frac{5}{8} = \frac{5}{2 \times 2 \times 2}$$

$$\frac{1}{6} = \frac{1}{2 \times 3}$$

LCD = 2 x 2 x 2 x 3 = 24

$$\frac{5 \times 3}{8 \times 3} = \frac{15}{24}$$

$$+ \frac{1 \times 4}{6 \times 4} = \frac{4}{24}$$

$$\frac{19}{24}$$

1.129

$$\frac{3}{10} = \frac{3}{2 \times 5}$$

$$\frac{11}{20} = \frac{11}{2 \times 2 \times 5}$$

LCD = 2 x 2 x 5 = 20

$$\frac{3 \times 2}{10 \times 2} = \frac{6}{20}$$

$$+ \frac{11}{20} = \frac{11}{20}$$

$$\frac{17}{20}$$

1.130

$$\frac{1}{6} = \frac{1}{2 \times 3}$$

$$\frac{2}{3} = \frac{2}{3}$$

LCD = 2 x 3 = 6

$$\frac{1}{6} = \frac{1}{6}$$

$$+ \frac{2 \times 2}{3 \times 2} = \frac{4}{6}$$

$$\frac{5}{6}$$

1.131

$$\frac{7}{8} = \frac{7}{2 \times 2 \times 2}$$

$$\frac{3}{4} = \frac{3}{2 \times 2}$$

LCD = 2 x 2 x 2 = 8

$$\frac{7}{8} = \frac{7}{8}$$

$$- \frac{3 \times 2}{4 \times 4} = \frac{6}{8}$$

$$\frac{1}{8}$$

1.132

$$\frac{1}{2} = \frac{1}{2}$$

$$\frac{5}{12} = \frac{5}{2 \times 2 \times 3}$$

LCD = 2 x 2 x 3 = 12

1.132 cont.

$$\frac{1 \times 6}{2 \times 6} = \frac{6}{12}$$

$$- \frac{5}{12} = \frac{5}{12}$$

$$\frac{1}{12}$$

1.133

$$\frac{2}{3} = \frac{2}{3}$$

$$\frac{3}{7} = \frac{3}{7}$$

$$LCD = 3 \times 7 = 21$$

$$\frac{2 \times 7}{3 \times 7} = \frac{14}{21}$$

$$- \frac{3 \times 3}{7 \times 3} = \frac{9}{21}$$

$$\frac{5}{21}$$

1.134

$$\frac{1}{4} = \frac{1}{2 \times 2}$$

$$\frac{1}{10} = \frac{1}{2 \times 5}$$

$$LCD = 2 \times 2 \times 5 = 20$$

$$\frac{1 \times 5}{4 \times 5} = \frac{5}{20}$$

$$- \frac{1 \times 2}{10 \times 2} = \frac{2}{20}$$

$$\frac{3}{20}$$

1.135

$$\frac{5}{6} = \frac{5}{2 \times 3}$$

$$\frac{2}{5} = \frac{2}{5}$$

$$LCD = 2 \times 3 \times 5 = 30$$

$$\frac{5 \times 5}{6 \times 5} = \frac{25}{30}$$

$$- \frac{2 \times 6}{5 \times 6} = \frac{12}{30}$$

$$\frac{13}{30}$$

1.136

$$\frac{3}{5} = \frac{3}{5}$$

$$\frac{2}{7} = \frac{2}{7}$$

$$LCD = 5 \times 7 = 35$$

$$\frac{3 \times 7}{5 \times 7} = \frac{21}{35}$$

$$+ \frac{2 \times 5}{7 \times 5} = \frac{10}{35}$$

$$\frac{31}{35}$$

1.137

$$\frac{3}{8} = \frac{3}{2 \times 2 \times 2}$$

$$\frac{1}{6} = \frac{1}{2 \times 3}$$

$$\frac{5}{12} = \frac{5}{2 \times 2 \times 3}$$

$$LCD = 2 \times 2 \times 2 \times 3 = 24$$

$$\frac{3 \times 3}{8 \times 3} = \frac{9}{24}$$

$$\frac{1 \times 4}{6 \times 4} = \frac{4}{24}$$

$$+ \frac{5 \times 2}{12 \times 2} = \frac{10}{24}$$

$$\frac{23}{24}$$

1.138

$$\frac{7}{8} = \frac{7}{2 \times 2 \times 2}$$

$$\frac{3}{14} = \frac{3}{2 \times 7}$$

$$LCD = 2 \times 2 \times 2 \times 7 = 56$$

$$\frac{7 \times 7}{8 \times 7} = \frac{49}{56}$$

$$- \frac{3 \times 4}{14 \times 4} = \frac{12}{56}$$

$$\frac{37}{56}$$

1.139

$$\frac{1}{4} = \frac{1}{2 \times 2}$$

$$\frac{7}{12} = \frac{7}{2 \times 2 \times 3}$$

$$LCD = 2 \times 2 \times 3 = 12$$

$$\frac{1 \times 3}{4 \times 3} = \frac{3}{12}$$

$$+ \frac{7}{12} = \frac{7}{12}$$

$$\frac{10}{12} = \frac{10 \div 2}{12 \div 2} = \frac{5}{6}$$

1.140

$$\frac{5}{21} = \frac{5}{3 \times 7}$$

$$\frac{3}{7} = \frac{3}{7}$$

$$LCD = 3 \times 7 = 21$$

$$\frac{5}{21} = \frac{5}{21}$$

$$+ \frac{3 \times 3}{7 \times 3} = \frac{9}{21}$$

$$\frac{14}{21} = \frac{14 \div 7}{21 \div 7} = \frac{2}{3}$$

$$\frac{2 \times 8}{3 \times 8} = \frac{16}{24}$$

$$- \frac{7}{24} = \frac{7}{24}$$

$$\frac{9}{24} = \frac{9 \div 3}{24 \div 3} = \frac{3}{8}$$

1.141

$$\frac{1}{2} = \frac{1}{2}$$

$$\frac{1}{3} = \frac{1}{3}$$

$$\frac{1}{6} = \frac{1}{2 \times 3}$$

$$\text{LCD} = 2 \times 3 = 6$$

$$\frac{1 \times 3}{2 \times 3} = \frac{3}{6}$$

$$\frac{1 \times 2}{3 \times 2} = \frac{2}{6}$$

$$+ \frac{1}{6} = \frac{1}{6}$$

$$\frac{6}{6} = 1$$

1.142

$$\frac{1}{2} = \frac{1}{2}$$

$$\frac{1}{6} = \frac{1}{2 \times 3}$$

$$\text{LCD} = 2 \times 3 = 6$$

$$\frac{1 \times 3}{2 \times 3} = \frac{3}{6}$$

$$+ \frac{1}{6} = \frac{1}{6}$$

$$\frac{4}{6} = \frac{4 \div 2}{6 \div 2} = \frac{2}{3}$$

1.143

$$\frac{9}{10} = \frac{9}{2 \times 5}$$

$$\frac{3}{20} = \frac{3}{2 \times 2 \times 5}$$

$$\text{LCD} = 2 \times 2 \times 5 = 20$$

$$\frac{9 \times 2}{10 \times 2} = \frac{18}{20}$$

$$- \frac{3}{20} = \frac{3}{20}$$

$$\frac{15}{20} = \frac{15 \div 5}{20 \div 5} = \frac{3}{4}$$

1.144

$$\frac{2}{3} = \frac{2}{3}$$

$$\frac{7}{24} = \frac{7}{2 \times 2 \times 2 \times 3}$$

$$\text{LCD} = 2 \times 2 \times 2. \times 3 = 24$$

1.145

$$\frac{2}{3} = \frac{2}{3}$$

$$\frac{1}{6} = \frac{1}{2 \times 3}$$

$$\text{LCD} = 2 \times 3 = 6$$

$$\frac{2 \times 2}{3 \times 2} = \frac{4}{6}$$

$$- \frac{1}{6} = \frac{1}{6}$$

$$\frac{3}{6} = \frac{3 \div 3}{6 \div 3} = \frac{1}{2}$$

1.146

$$\frac{3}{5} = \frac{3}{5}$$

$$\frac{3}{15} = \frac{3}{3 \times 5}$$

$$\text{LCD} = 3 \times 5 = 15$$

$$\frac{3 \times 3}{5 \times 3} = \frac{9}{15}$$

$$- \frac{3}{15} = \frac{3}{15}$$

$$\frac{6}{15} = \frac{6 \div 3}{15 \div 3} = \frac{2}{5}$$

1.147

$$\frac{3}{4} = \frac{3}{2 \times 2}$$

$$\frac{1}{28} = \frac{1}{2 \times 2 \times 7}$$

$$\text{LCD} = 2 \times 2 \times 7 = 28$$

$$\frac{3 \times 7}{4 \times 7} = \frac{21}{28}$$

$$- \frac{1}{28} = \frac{1}{28}$$

$$\frac{20}{28} = \frac{20 \div 4}{28 \div 4} = \frac{5}{7}$$

1.148

$$\frac{5}{6} = \frac{5}{2 \times 3}$$

$$\frac{1}{10} = \frac{1}{2 \times 5}$$

$$\text{LCD} = 2 \times 3 \times 5 = 30$$

1.148 cont.

$$\frac{5 \times 5}{6 \times 5} = \frac{25}{30}$$

$$+ \frac{1 \times 3}{10 \times 3} = \frac{3}{30}$$

$$\frac{28}{30} = \frac{28 \div 2}{30 \div 2} = \frac{14}{15}$$

1.149

$$\frac{3}{4} = \frac{3}{2 \times 2}$$

$$\frac{1}{20} = \frac{1}{2 \times 2 \times 5}$$

$$LCD = 2 \times 2 \times 5 = 20$$

$$\frac{3 \times 5}{4 \times 5} = \frac{15}{20}$$

$$+ \frac{1}{20} = \frac{1}{20}$$

$$\frac{16}{20} = \frac{16 \div 4}{20 \div 4} = \frac{4}{5}$$

1.150

$$\frac{19}{24} = \frac{19}{2 \times 2 \times 2 \times 3}$$

$$\frac{6}{16} = \frac{6}{2 \times 2 \times 2 \times 2}$$

$$LCD = 2 \times 2 \times 2 \times 2 \times 3 = 48$$

$$\frac{19 \times 2}{24 \times 2} = \frac{38}{48}$$

$$- \frac{6 \times 3}{16 \times 3} = \frac{18}{48}$$

$$\frac{20}{48} = \frac{20 \div 4}{48 \div 4} = \frac{5}{12}$$

1.151

$$\frac{2}{3} = \frac{2}{3}$$

$$\frac{4}{9} = \frac{4}{3 \times 3}$$

$$LCD = 3 \times 3 = 9$$

$$\frac{2 \times 3}{3 \times 3} = \frac{6}{9}$$

$$+ \frac{4}{9} = \frac{4}{9}$$

$$\frac{10}{9} = 1\frac{1}{9}$$

1.152

$$\frac{4}{5} = \frac{4}{5}$$

$$\frac{5}{7} = \frac{5}{7}$$

$$LCD = 5 \times 7 = 35$$

$$\frac{4 \times 7}{5 \times 7} = \frac{28}{35}$$

$$+ \frac{5 \times 5}{7 \times 5} = \frac{25}{35}$$

$$\frac{53}{35} = 1\frac{18}{35}$$

1.153

$$\frac{1}{2} = \frac{1}{2}$$

$$\frac{11}{16} = \frac{11}{2 \times 2 \times 2 \times 2}$$

$$LCD = 2 \times 2 \times 2 \times 2 = 16$$

$$\frac{1 \times 8}{2 \times 8} = \frac{8}{16}$$

$$+ \frac{11}{16} = \frac{11}{16}$$

$$\frac{19}{16} = 1\frac{3}{16}$$

1.154

$$\frac{1}{10} = \frac{1}{2 \times 5}$$

$$\frac{3}{20} = \frac{3}{2 \times 2 \times 5}$$

$$\frac{4}{5} = \frac{4}{5}$$

$$LCD = 2 \times 2 \times 5 = 20$$

$$\frac{1 \times 2}{10 \times 2} = \frac{2}{20}$$

$$\frac{3}{20} = \frac{3}{20}$$

$$+ \frac{4 \times 4}{5 \times 4} = \frac{16}{20}$$

$$\frac{21}{20} = 1\frac{1}{20}$$

1.155

$$\frac{5}{6} = \frac{5}{2 \times 3}$$

$$\frac{7}{8} = \frac{7}{2 \times 2 \times 2}$$

$$LCD = 2 \times 2 \times 2 \times 3 = 24$$

$$\frac{5 \times 4}{6 \times 4} = \frac{20}{24}$$

$$+ \frac{7 \times 3}{8 \times 3} = \frac{21}{24}$$

$$\frac{41}{24} = 1\frac{17}{24}$$

1.156

$$\frac{5}{12} = \frac{5}{2 \times 2 \times 3}$$

$$\frac{5}{6} = \frac{5}{2 \times 3}$$

$$LCD = 2 \times 2 \times 3 = 12$$

$$\frac{5}{12} = \frac{5}{12}$$

$$+\ \frac{5 \times 2}{6 \times 2} = \frac{10}{12}$$

$$\frac{15}{12} = 1\frac{3}{12} = 1\frac{3 \div 3}{12 \div 3} = 1\frac{1}{4}$$

$$\frac{32 \times 2}{5 \times 2} = \frac{64}{10}$$

$$-\ \frac{1}{10} = \frac{1}{10}$$

$$\frac{63}{10} = 6\frac{3}{10}$$

1.157

$$\frac{11}{8} = \frac{11}{2 \times 2 \times 2}$$

$$\frac{1}{6} = \frac{1}{2 \times 3}$$

$$LCD = 2 \times 2 \times 2 \times 3 = 24$$

$$\frac{11 \times 3}{8 \times 3} = \frac{33}{24}$$

$$-\ \frac{1 \times 4}{6 \times 4} = \frac{4}{24}$$

$$\frac{29}{24} = 1\frac{5}{24}$$

1.161

$$\frac{27}{16} = \frac{27}{2 \times 2 \times 2 \times 2}$$

$$\frac{3}{8} = \frac{3}{2 \times 2 \times 2}$$

$$LCD = 2 \times 2 \times 2 \times 2 = 16$$

$$\frac{27}{16} = \frac{27}{16}$$

$$-\ \frac{3 \times 2}{8 \times 2} = \frac{6}{16}$$

$$\frac{21}{16} = 1\frac{5}{16}$$

1.158

$$\frac{12}{7} = \frac{12}{7}$$

$$\frac{1}{3} = \frac{1}{3}$$

$$LCD = 7 \times 3 = 21$$

$$\frac{12 \times 3}{7 \times 3} = \frac{36}{21}$$

$$-\ \frac{1 \times 7}{3 \times 7} = \frac{7}{21}$$

$$\frac{29}{21} = 1\frac{8}{21}$$

1.162

$$\frac{22}{9} = \frac{22}{3 \times 3}$$

$$\frac{3}{4} = \frac{3}{2 \times 2}$$

$$LCD = 2 \times 2 \times 3 \times 3 = 36$$

$$\frac{22 \times 4}{9 \times 4} = \frac{88}{36}$$

$$-\ \frac{3 \times 9}{4 \times 9} = \frac{27}{36}$$

$$\frac{61}{36} = 1\frac{25}{36}$$

1.159

$$\frac{20}{9} = \frac{20}{3 \times 3}$$

$$\frac{2}{27} = \frac{2}{3 \times 3 \times 3}$$

$$LCD = 3 \times 3 \times 3 = 27$$

$$\frac{20 \times 3}{9 \times 3} = \frac{60}{27}$$

$$-\ \frac{2}{27} = \frac{2}{27}$$

$$\frac{58}{27} = 2\frac{4}{27}$$

1.163

$$\frac{8}{3} = \frac{8}{3}$$

$$\frac{2}{18} = \frac{2}{2 \times 3 \times 3}$$

$$LCD = 2 \times 3 \times 3 = 18$$

$$\frac{8 \times 6}{3 \times 6} = \frac{48}{18}$$

$$+\ \frac{2}{18} = \frac{2}{18}$$

$$\frac{50}{18} = \frac{50 \div 2}{18 \div 2} = \frac{25}{9} = 2\frac{7}{9}$$

1.160

$$\frac{32}{5} = \frac{32}{5}$$

$$\frac{1}{10} = \frac{1}{2 \times 5}$$

$$LCD = 2 \times 5 = 10$$

1.164

$$\frac{2}{4} = \frac{2}{2 \times 2}$$

$$\frac{4}{5} = \frac{4}{5}$$

$$\frac{9}{10} = \frac{9}{2 \times 5}$$

$$LCD = 2 \times 2 \times 5 = 20$$

1.164 cont.

$$\frac{2 \times 5}{4 \times 5} = \frac{10}{20}$$

$$\frac{4 \times 4}{5 \times 4} = \frac{16}{20}$$

$$+ \frac{9 \times 2}{10 \times 2} = \frac{18}{20}$$

$$\frac{44}{20} = \frac{44 \div 4}{20 \div 4} = \frac{11}{5} = 2\frac{1}{5}$$

1.165

$$\frac{10}{14} = \frac{10}{2 \times 7}$$

$$\frac{5}{7} = \frac{5}{7}$$

$$LCD = 2 \times 7 = 14$$

$$\frac{10}{14} = \frac{10}{14}$$

$$+ \frac{5 \times 2}{7 \times 2} = \frac{10}{14}$$

$$\frac{20}{14} = \frac{20 \div 2}{14 \div 2} = \frac{10}{7} = 1\frac{3}{7}$$

1.166

$$\frac{5}{6} = \frac{5}{2 \times 3}$$

$$\frac{4}{15} = \frac{4}{3 \times 5}$$

$$\frac{2}{5} = \frac{2}{5}$$

$$LCD = 2 \times 3 \times 5 = 30$$

$$\frac{5 \times 5}{6 \times 5} = \frac{25}{30}$$

$$\frac{4 \times 2}{15 \times 2} = \frac{8}{30}$$

$$+ \frac{2 \times 6}{5 \times 6} = \frac{12}{30}$$

$$\frac{45}{30} = \frac{45 \div 15}{30 \div 15} = \frac{3}{2} = 1\frac{1}{2}$$

1.167

$$\frac{5}{6} = \frac{5}{2 \times 3}$$

$$\frac{5}{12} = \frac{5}{2 \times 2 \times 3}$$

$$LCD = 2 \times 2 \times 3 = 12$$

$$\frac{5 \times 2}{6 \times 2} = \frac{10}{12}$$

$$+ \frac{5}{12} = \frac{5}{12}$$

$$\frac{15}{12} = \frac{15 \div 3}{12 \div 3} = \frac{5}{4} = 1\frac{1}{4}$$

1.168

$$\frac{3}{4} = \frac{3}{2 \times 2}$$

$$\frac{6}{8} = \frac{6}{2 \times 2 \times 2}$$

$$LCD = 2 \times 2 \times 2 = 8$$

$$\frac{3 \times 2}{4 \times 2} = \frac{6}{8}$$

$$+ \frac{6}{8} = \frac{6}{8}$$

$$\frac{12}{8} = \frac{12 \div 4}{8 \div 4} = \frac{3}{2} = 1\frac{1}{2}$$

1.169

$$\frac{20}{6} = \frac{20}{2 \times 3}$$

$$\frac{2}{3} = \frac{2}{3}$$

$$LCD = 2 \times 3 = 6$$

$$\frac{20}{6} = \frac{20}{6}$$

$$- \frac{2 \times 2}{3 \times 2} = \frac{4}{6}$$

$$\frac{16}{6} = \frac{16 \div 2}{6 \div 2} = \frac{8}{3} = 2\frac{2}{3}$$

1.170

$$\frac{33}{5} = \frac{33}{5}$$

$$\frac{1}{10} = \frac{1}{2 \times 5}$$

$$LCD = 2 \times 5 = 10$$

$$\frac{33 \times 2}{5 \times 2} = \frac{66}{10}$$

$$- \frac{1}{10} = \frac{1}{10}$$

$$\frac{65}{10} = \frac{65 \div 5}{10 \div 5} = \frac{13}{2} = 6\frac{1}{2}$$

1.171

$$\frac{11}{6} = \frac{11}{2 \times 3}$$

$$\frac{2}{8} = \frac{2}{2 \times 2 \times 2}$$

$$LCD = 2 \times 2 \times 2 \times 3 = 24$$

$$\frac{11 \times 4}{6 \times 4} = \frac{44}{24}$$

$$- \frac{2 \times 3}{8 \times 3} = \frac{6}{24}$$

$$\frac{38}{24} = \frac{38 \div 2}{24 \div 2} = \frac{19}{12} = 1\frac{7}{12}$$

1.172

$$\frac{16}{6} = \frac{16}{2 \times 3}$$

$$\frac{2}{9} = \frac{2}{3 \times 3}$$

LCD = 2 x 3 x 3 = 18

$$\frac{16 \times 3}{6 \times 3} = \frac{48}{18}$$

$$-\frac{2 \times 2}{9 \times 2} = \frac{4}{18}$$

$$\frac{44}{18} = \frac{44 \div 2}{18 \div 2} = \frac{22}{9} = 2\frac{4}{9}$$

1.173

$$\frac{15}{6} = \frac{15}{2 \times 3}$$

$$\frac{2}{5} = \frac{2}{5}$$

LCD = 2 x 3 x 5 = 30

$$\frac{15 \times 5}{6 \times 5} = \frac{75}{30}$$

$$-\frac{2 \times 6}{5 \times 6} = \frac{12}{30}$$

$$\frac{63}{30} = \frac{63 \div 3}{30 \div 3} = \frac{21}{10} = 2\frac{1}{10}$$

1.174

$$\frac{4}{3} = \frac{4}{3}$$

$$\frac{2}{15} = \frac{2}{3 \times 5}$$

LCD = 3 x 5 = 15

$$\frac{4 \times 5}{3 \times 5} = \frac{20}{15}$$

$$-\frac{2}{15} = \frac{2}{15}$$

$$\frac{18}{15} = \frac{18 \div 3}{15 \div 3} = \frac{6}{5} = 1\frac{1}{5}$$

1.175

LCD = 10

$$6\frac{2}{5} = 6\frac{4}{10}$$

$$+2\frac{3}{10} = 2\frac{3}{10}$$

$$8\frac{7}{10}$$

1.176

LCD = 10

$$4\frac{2}{10} = 4\frac{2}{10}$$

$$9\frac{1}{5} = 9\frac{2}{10}$$

$$+1\frac{1}{2} = 1\frac{5}{10}$$

$$14\frac{9}{10}$$

1.177

LCD = 15

$$7\frac{2}{5} = 7\frac{6}{15}$$

$$+27\frac{1}{3} = 27\frac{5}{15}$$

$$34\frac{11}{15}$$

1.178

LCD = 6

$$3\frac{1}{6} = 3\frac{1}{6}$$

$$+16\frac{2}{3} = 16\frac{4}{6}$$

$$19\frac{5}{6}$$

1.179

LCD = 30

$$45\frac{1}{6} = 45\frac{5}{30}$$

$$+19\frac{2}{5} = 19\frac{12}{30}$$

$$64\frac{17}{30}$$

1.180

LCD = 10

$$9\frac{3}{10} = 9\frac{3}{10}$$

$$-6\frac{1}{5} = 6\frac{2}{10}$$

$$3\frac{1}{10}$$

1.181

LCD = 12

$$11\frac{5}{6} = 11\frac{10}{12}$$

$$-3\frac{5}{12} = 3\frac{5}{12}$$

$$8\frac{5}{12}$$

1.182

$$LCD = 20$$

$$2\frac{7}{10} = 2\frac{14}{20}$$

$$-\ \frac{7}{20} = \ \ \frac{7}{20}$$

$$2\frac{7}{20}$$

1.183

$$LCD = 24$$

$$34\frac{5}{12} = 34\frac{10}{24}$$

$$-\ 16\frac{3}{8} = 16\frac{9}{24}$$

$$18\frac{1}{24}$$

1.184

$$LCD = 14$$

$$15\frac{9}{14} = 15\frac{9}{14}$$

$$-\ 7\frac{3}{7} = \ 7\frac{6}{14}$$

$$8\frac{3}{14}$$

1.185

$$LCD = 18$$

$$29\frac{1}{6} = 29\frac{3}{18}$$

$$+\ 13\frac{3}{9} = 13\frac{6}{18}$$

$$42\frac{9}{18} = 42\frac{9 \div 9}{18 \div 9} = 42\frac{1}{2}$$

1.186

$$LCD = 30$$

$$7\frac{1}{10} = \ 7\frac{3}{30}$$

$$+\ 23\frac{5}{6} = 23\frac{25}{30}$$

$$30\frac{28}{30} = 30\ \frac{28 \div 2}{30 \div 2} = 30\frac{14}{15}$$

1.187

$$LCD = 18$$

$$12\frac{2}{9} = 12\frac{4}{18}$$

$$+\ 21\frac{5}{18} = 21\frac{5}{18}$$

$$33\frac{9}{18} = 33\frac{9 \div 9}{18 \div 9} = 33\frac{1}{2}$$

1.188

$$LCD = 20$$

$$42\frac{1}{20} = 42\frac{1}{20}$$

$$+\ 51\frac{3}{4} = 51\frac{15}{20}$$

$$93\frac{16}{20} = 93\frac{16 \div 4}{20 \div 4} = 93\frac{4}{5}$$

1.189

$$LCD = 12$$

$$19\frac{5}{12} = 19\frac{5}{12}$$

$$+\ 14\frac{1}{4} = 14\frac{3}{12}$$

$$33\frac{8}{12} = 33\frac{8 \div 4}{12 \div 4} = 33\frac{2}{3}$$

1.190

$$LCD = 21$$

$$4\frac{3}{7} = 4\frac{9}{21}$$

$$+\ \frac{5}{21} = \ \ \frac{5}{21}$$

$$4\frac{14}{21} = 4\frac{14 \div 7}{21 \div 7} = 4\ \frac{2}{3}$$

1.191

$$LCD = 14$$

$$12\frac{1}{2} = 12\frac{7}{14}$$

$$-\ 7\frac{3}{14} = \ 7\frac{3}{14}$$

$$5\frac{4}{14} = 5\frac{4 \div 2}{14 \div 2} = 5\frac{2}{7}$$

1.192

$$LCD = 28$$

$$45\frac{3}{4} = 45\frac{21}{28}$$

$$-\ 16\frac{1}{28} = 16\frac{1}{28}$$

$$29\ \frac{20}{28} = 29\frac{20 \div 4}{28 \div 4} = 29\ \frac{5}{7}$$

1.193

$$LCD = 6$$

$$15\frac{2}{3} = 15\frac{4}{6}$$

$$-\ 7\frac{1}{6} = \ 7\frac{1}{6}$$

$$8\frac{3}{6} = 8\frac{3 \div 3}{6 \div 3} = 8\frac{1}{2}$$

1.194 LCD = 40

$$23\frac{7}{8} = 23\frac{35}{40}$$

$$-14\frac{19}{40} = 14\frac{19}{40}$$

$$9\frac{16}{40} = 9\frac{16 \div 8}{40 \div 8} = 9\frac{2}{5}$$

1.195 LCD = 10

$$9\frac{4}{5} = 9\frac{8}{10}$$

$$-2\frac{3}{10} = 2\frac{3}{10}$$

$$7\frac{5}{10} = 7\frac{5 \div 5}{10 \div 5} = 7\frac{1}{2}$$

1.196 LCD = 36

$$25\frac{1}{4} = 25\frac{9}{36}$$

$$-8\frac{5}{36} = 8\frac{5}{36}$$

$$17\frac{4}{36} = 17\frac{4 \div 4}{36 \div 4} = 17\frac{1}{9}$$

1.197 LCD = 4

$$11\frac{3}{4} = 11\frac{3}{4}$$

$$+2\frac{1}{2} = 2\frac{2}{4}$$

$$13\frac{5}{4} = 13 + 1\frac{1}{4} = 14\frac{1}{4}$$

1.198 LCD = 30

$$27\frac{3}{5} = 27\frac{18}{30}$$

$$+32\frac{5}{6} = 32\frac{25}{30}$$

$$59\frac{43}{30} = 59 + 1\frac{13}{30} = 60\frac{13}{30}$$

1.199 LCD = 9

$$8\frac{4}{9} = 8\frac{4}{9}$$

$$+2\frac{2}{3} = 2\frac{6}{9}$$

$$10\frac{10}{9} = 10 + 1\frac{1}{9} = 11\frac{1}{9}$$

1.200 LCD = 10

$$4\frac{4}{5} = 4\frac{8}{10}$$

$$+6\frac{7}{10} = 6\frac{7}{10}$$

$$10\frac{15}{10} = 10\frac{15 \div 5}{10 \div 5} = 10\frac{3}{2} =$$

$$10 + 1\frac{1}{2} = 11\frac{1}{2}$$

1.201 LCD = 24

$$68\frac{5}{8} = 68\frac{15}{24}$$

$$+14\frac{7}{12} = 14\frac{14}{24}$$

$$82\frac{29}{24} = 82 + 1\frac{5}{24} = 83\frac{5}{24}$$

1.202 LCD = 36

$$10\frac{5}{9} = 10\frac{20}{36}$$

$$4\frac{3}{4} = 4\frac{27}{36}$$

$$+15\frac{5}{6} = 15\frac{30}{36}$$

$$29\frac{77}{36} = 29 + 2\frac{5}{36} = 31\frac{5}{36}$$

1.203 LCD = 12

$$7\frac{1}{3} = 7\frac{4}{12} = 6 + 1\frac{4}{12} = 6\frac{16}{12}$$

$$-1\frac{3}{4} = 1\frac{9}{12} = \qquad 1\frac{9}{12}$$

$$5\frac{7}{12}$$

1.204 LCD = 18

$$15\frac{4}{9} = 15\frac{8}{18} = 14 + 1\frac{8}{18} = 14\frac{26}{18}$$

$$-11\frac{1}{2} = 11\frac{9}{18} = \qquad 11\frac{9}{18}$$

$$3\frac{17}{18}$$

1.205

$$25 = 24\frac{15}{15}$$

$$-7\frac{4}{15} = 7\frac{4}{15}$$

$$17\frac{11}{15}$$

1.206 LCD = 14

$$35\frac{2}{14} = 35\frac{2}{14} = 34 + 1\frac{2}{14} = 34\frac{16}{14}$$
$$-12\frac{3}{7} = 12\frac{6}{14} = \qquad\qquad 12\frac{6}{14}$$
$$\rule{4cm}{0.4pt}$$
$$22\frac{10}{14}$$
$$= 22\frac{10 \div 2}{14 \div 2} = 22\frac{5}{7}$$

1.207 LCD = 10

$$9\frac{1}{10} = 9\frac{1}{10} = 8 + 1\frac{1}{10} = 8\frac{11}{10}$$
$$-3\frac{4}{5} = 3\frac{8}{10} = \qquad\qquad 3\frac{8}{10}$$
$$\rule{4cm}{0.4pt}$$
$$5\frac{3}{10}$$

1.208 LCD = 12

$$17\frac{7}{12} = 17\frac{7}{12} = 16 + 1\frac{7}{12} = 16\frac{19}{12}$$
$$-16\frac{5}{6} = 16\frac{10}{12} = \qquad\qquad 16\frac{10}{12}$$
$$\rule{4cm}{0.4pt}$$
$$\frac{9}{12}$$
$$= \frac{9 \div 3}{12 \div 3} = \frac{3}{4}$$

1.209 LCD = 8

$$1\frac{1}{2} = 1\frac{4}{8}$$
$$2\frac{7}{8} = 2\frac{7}{8}$$
$$+2\frac{7}{8} = 2\frac{7}{8}$$
$$\rule{4cm}{0.4pt}$$
$$5\frac{18}{8} = 5\frac{18 \div 2}{8 \div 2} = 5\frac{9}{4} = 5 + 2\frac{1}{4}$$
$$= 7\frac{1}{4} \text{ minutes}$$

1.210 LCD = 6

$$15\frac{1}{6} = 15\frac{1}{6}$$
$$+24\frac{2}{3} = 24\frac{4}{6}$$
$$\rule{4cm}{0.4pt}$$
$$39\frac{5}{6} \text{ minutes}$$

1.211 LCD = 24

$$\frac{7}{8} = \frac{21}{24}$$
$$\frac{3}{4} = \frac{18}{24}$$
$$+\frac{2}{3} = \frac{16}{24}$$
$$\rule{4cm}{0.4pt}$$
$$\frac{55}{24} = 2\frac{7}{24} \text{ miles}$$

1.212 LCD = 18

$$3\frac{4}{9} = 3\frac{8}{18} = 2 + 1\frac{8}{18} = 2\frac{26}{18}$$
$$-1\frac{1}{2} = 1\frac{9}{18} = \qquad\qquad 1\frac{9}{18}$$
$$\rule{4cm}{0.4pt}$$
$$1\frac{17}{18} \text{ hours}$$

1.213 LCD = 12

$$\frac{11}{12} = \frac{11}{12}$$
$$+\frac{1}{4} = \frac{3}{12}$$
$$\rule{4cm}{0.4pt}$$
$$\frac{14}{12} = \frac{14 \div 2}{12 \div 2} = \frac{7}{6} = 1\frac{1}{6} \text{ minutes}$$

1.214 LCD = 27

$$7\frac{2}{3} = 7\frac{18}{27}$$
$$-4\frac{15}{27} = 4\frac{15}{27}$$
$$\rule{4cm}{0.4pt}$$
$$3\frac{3}{27} = 3\frac{3 \div 3}{27 \div 3} = 3\frac{1}{9} \text{ gallons}$$

1.215 ACROSS

1. LCD = 6

$$5\frac{7}{6} = 5\frac{7}{6}$$
$$-3\frac{2}{3} = 3\frac{4}{6}$$
$$\rule{4cm}{0.4pt}$$
$$2\frac{3}{6} = 2\frac{3 \div 3}{6 \div 3} = 2\frac{1}{2}$$

3. LCD = 10

$$6\frac{2}{5} = 6\frac{4}{10}$$
$$+7\frac{1}{10} = 7\frac{1}{10}$$
$$\rule{4cm}{0.4pt}$$
$$13\frac{5}{10} = 13\frac{5 \div 5}{10 \div 5} = 13\frac{1}{2}$$

7. LCD = 16

$$59\frac{1}{4} = 59\frac{4}{16}$$

$$+\ 24\frac{5}{16} = 24\frac{5}{16}$$

$$83\frac{9}{16}$$

10. LCD = 12

$$14\frac{5}{6} = 14\frac{10}{12}$$

$$-\ 4\frac{1}{12} = 4\frac{1}{12}$$

$$10\frac{9}{12} = 10\frac{9 \div 3}{12 \div 3} = 10\frac{3}{4}$$

14. LCD = 14

$$9\frac{5}{7} = 9\frac{10}{14}$$

$$+\ 8\frac{1}{2} = 8\frac{7}{14}$$

$$17\frac{17}{14} = 17 + 1\frac{3}{14} = 18\frac{3}{14}$$

17. LCD = 18

$$15\frac{7}{9} = 15\frac{14}{18}$$

$$-\ 13\frac{1}{6} = 13\frac{3}{18}$$

$$2\frac{11}{18}$$

DOWN

1. LCD = 6

$$12\frac{4}{6} = 12\frac{4}{6}$$

$$+\ 9\frac{1}{3} = 9\frac{2}{6}$$

$$21\frac{6}{6} = 21 + 1 = 22$$

3. LCD = 4

$$27\frac{1}{2} = 27\frac{2}{4} = 26 + 1\frac{2}{4} = 26\frac{6}{4}$$

$$-\ 13\frac{3}{4} = \qquad\qquad\qquad 13\frac{3}{4}$$

$$13\frac{3}{4}$$

4. LCD = 16

$$3\frac{1}{2} = 3\frac{8}{16}$$

$$+\ \frac{1}{16} = \frac{1}{16}$$

$$3\frac{9}{16}$$

7. LCD = 14

$$91\frac{9}{14} = 91\frac{9}{14}$$

$$-\ 11\frac{3}{7} = 11\frac{6}{14}$$

$$80\frac{3}{14}$$

10. LCD = 16

$$10\frac{4}{16} = 10\frac{4}{16}$$

$$+\ 7\frac{6}{8} = 7\frac{12}{16}$$

$$17\frac{16}{16} = 17 + 1 = 18$$

13. LCD = 18

$$2\frac{2}{9} = 2\frac{4}{18}$$

$$4\frac{1}{18} = 4\frac{1}{18}$$

$$+\ 1\frac{1}{3} = 1\frac{6}{18}$$

$$7\frac{11}{18}$$

II.   SECTION TWO

2.1   8.3; eight and three tenths

2.2   79.6; seventy-nine and six tenths

2.3   81.05; eighty-one and five hundreths

2.4   1,923.17; one thousand nine hundred twenty-three and seventeen-hundredths

2.5   31.0618; thirty-one and six hundred eighteen ten-thousandths

2.6   300.2; three hundred and two-tenths

2.7   9.009; nine and nine-thousandths

2.8   465; four hundred sixty-five

2.9   0.410; four hundred ten-thousandths

2.10   54.054; fifty-four and fifty-four thousandths

2.11   0.4921; four thousand nine hundred twenty-one ten-thousandths

2.12   0.42

2.13   1.07

2.14   0.15

2.15   0.34

2.16   1.01

2.17   2.12

2.18   0.54

2.19   0.15

2.20   0.82
        0.70
        + 0.25
        1.77

2.21   0.13
        0.08
        + 0.46
        0.67

2.22   0.64
        − 0.38
        0.26

2.23   0.16
        0.90
        + 0.03
        1.09

2.24   0.57
        − 0.40
        0.17

2.25   0.60
        − 0.23
        0.37

2.26 $17.28

2.27   12.88

2.28 $5.55

2.29   23.53

2.30   39.66

2.31   94.51

2.32   2.58

2.33   20.63

2.34 $87.16

2.35   70.12

2.36   3.9

2.37   70.93

2.38 $35.72
        + 2.41
        $38.13

2.39   26.20
        0.07
        + 21.00
        47.27

2.40 $65.21
        − 41.03
        $24.18

2.41
```
    5.46
   21.30
 + 0.02
   26.78
```

2.42
```
   86.9
 − 7.0
   79.9
```

2.43
```
   9.11
 − 0.06
   9.05
```

2.44
```
   32.70
    5.11
 + 0.25
   38.06
```

2.45
```
   0.58
 − 0.21
   0.37
```

2.46
```
   83.74
 − 61.97
   21.77
```

2.47   34.3406

2.48   3.70821

2.49   67.3946

2.50   268.744

2.51   398.411

2.52   101.77651

2.53   27.7109

2.54   25.62896

2.55
```
     0.351
     7.890
 + 431.210
   439.451
```

2.56
```
   582.8900
     3.7682
 + 11.4290
   598.0872
```

2.57
```
 $36,247.52
 − 2,581.87
 $33,665.65
```

2.58
```
   3.6000
 − 0.0075
   3.5925
```

2.59
```
   3,000.000
     608.049
 +     0.020
   3,608.069
```

2.60
```
 $11.00
 − 0.35
 $10.65
```

2.61
```
 $57.31
  47.98
 + 5.00
 $110.29
```

2.62   a.   0.642 carats

   b.
```
    0.642
 − 0.599
   0.043 carats
```

2.63   a.
```
   14.960
   18.034
    9.020
 + 18.700
   60.714
```

   b.
```
   2.100
   2.000
   4.300
 + 4.782
  13.182
```

2.64
```
   72.25
 − 51.50
   20.75 feet
```

2.65
```
   3.12500
 + 0.90632
   4.03132 grains
```

2.66   teacher check

2.67   58.415

219

2.68   205.572

2.69   1.959

2.70   66.984

2.71   6.918

2.72   643.57

2.73   47.53

2.74   707.383

2.75   0.53645

2.76   48.648

2.77   0.0

2.78   7.7345; shell

2.79   5.604; hogs

2.80   563.8; begs

2.81   35.006; goose

2.82   53188.04; hobbies

2.83   46.1375; sleigh

2.84   7,105; soil

2.85   teacher check

2.86 $\dfrac{9}{10}$

2.87 $\dfrac{6}{10} = \dfrac{6 \div 2}{10 \div 2} = \dfrac{3}{5}$

2.88 $3\dfrac{7}{10}$

2.89 $658\dfrac{72}{100} = 658\dfrac{72 \div 4}{100 \div 4} = 658\dfrac{18}{25}$

2.90 $\dfrac{81}{100}$

2.91 $\dfrac{75}{100} = \dfrac{75 \div 25}{100 \div 25} = \dfrac{3}{4}$

2.92 $98\dfrac{24}{100} = 98\dfrac{24 \div 4}{100 \div 4} = 98\dfrac{6}{25}$

2.93 $1\dfrac{2}{10,000} = 1\dfrac{2 \div 2}{10,000 \div 2} = 1\dfrac{1}{5,000}$

2.94 $\dfrac{73}{1,000}$

2.95 $\dfrac{225}{1,000} = \dfrac{225 \div 25}{1,000 \div 25} = \dfrac{9}{40}$

2.96 $16\dfrac{5}{100} = 16\dfrac{5 \div 5}{100 \div 5} = 16\dfrac{1}{20}$

2.97 $48\dfrac{2}{10} = 48\dfrac{2 \div 2}{10 \div 2} = 48\dfrac{1}{5}$

2.98 $\dfrac{1}{2} = 2\overline{)1.0}$

$$\begin{array}{r} 0.5 \\ 2\overline{)1.0} \\ \underline{1\ 0} \\ 0 \end{array}$$

2.99 $\dfrac{9}{16} = 16\overline{)9.0000}$

$$\begin{array}{r} 0.5625 \\ 16\overline{)9.0000} \\ \underline{80} \\ 100 \\ \underline{96} \\ 40 \\ \underline{32} \\ 80 \\ \underline{80} \\ 0 \end{array}$$

$25\dfrac{9}{16} = 25.5625$

2.100 $\dfrac{9}{12} = 12\overline{)9.00}$

$$\begin{array}{r} 0.75 \\ 12\overline{)9.00} \\ \underline{84} \\ 60 \\ \underline{60} \\ 0 \end{array}$$

$31\dfrac{9}{12} = 31.75$

2.101 $\dfrac{2}{5} = 5\overline{)2.0}$

$$\begin{array}{r} 0.4 \\ 5\overline{)2.0} \\ \underline{20} \\ 0 \end{array}$$

2.102 $\dfrac{1}{25} = 25\overline{)1.00}$

$$\begin{array}{r} 0.04 \\ 25\overline{)1.00} \\ \underline{100} \\ 0 \end{array}$$

$7\dfrac{1}{25} = 7.04$

2.103

$$\frac{3}{8} = 8\overline{)3.000}$$
$$\frac{0.375}{24}$$
$$60$$
$$\underline{56}$$
$$40$$
$$\underline{40}$$
$$0$$

$$2\frac{3}{8} = 2.375$$

2.104

$$\frac{7}{8} = 8\overline{)7.000}$$
$$\frac{0.875}{64}$$
$$60$$
$$\underline{56}$$
$$40$$
$$\underline{40}$$
$$0$$

$$3\frac{7}{8} = 3.875$$

2.105

$$\frac{7}{10} = 10\overline{)7.0}$$
$$\frac{0.7}{70}$$
$$0$$

2.106

$$\frac{12}{25} = 25\overline{)12.00}$$
$$\frac{0.48}{100}$$
$$200$$
$$\underline{200}$$
$$0$$

2.107

$$\frac{3}{7} = 7\overline{)3.000}$$
$$\frac{0.428}{28}$$
$$20$$
$$\underline{14}$$
$$60$$
$$\underline{56}$$
$$4$$

$$\frac{3}{7} = 0.43$$

2.108

$$\frac{1}{6} = 6\overline{)1.000}$$
$$\frac{0.166}{6}$$
$$40$$
$$\underline{36}$$
$$40$$
$$\underline{36}$$
$$4$$

$$852\frac{1}{6} = 852.17$$

2.109

$$\frac{2}{3} = 3\overline{)2.000}$$
$$\frac{0.666}{18}$$
$$20$$
$$\underline{18}$$
$$20$$
$$\underline{18}$$
$$2$$

$$\frac{2}{3} = 0.67$$

2.110

$$\frac{2}{9} = 9\overline{)2.000}$$
$$\frac{0.222}{18}$$
$$20$$
$$\underline{18}$$
$$20$$
$$\underline{18}$$
$$2$$

$$\frac{2}{9} = 0.22$$

2.111

$$\frac{7}{21} = 21\overline{)7.000}$$
$$\frac{0.333}{63}$$
$$70$$
$$\underline{63}$$
$$70$$
$$\underline{63}$$
$$7$$

$$19\frac{7}{21} = 19.33$$

2.112 $\quad \dfrac{20}{23} = 23\overline{)20.000}$

$$\begin{array}{r} 0.869 \\ \hline \underline{184} \\ 160 \\ \underline{138} \\ 220 \\ \underline{207} \\ 13 \end{array}$$

$\dfrac{20}{23} = 0.87$

2.113 a. $\quad \dfrac{3 \div 3}{12 \div 3} = \dfrac{1}{4}$

b. $\quad 12\overline{)3.00}$

$$\begin{array}{r} 0.25 \\ \hline \underline{24} \\ 60 \\ \underline{60} \\ 0 \end{array}$$

2.114 a. $\quad 0.65 = \dfrac{65}{100} = \dfrac{65 \div 5}{100 \div 5} = \dfrac{13}{20}$

b. Example: $\dfrac{65}{100}$

2.115 a. $\quad \dfrac{6}{1,000} = \dfrac{6 \div 2}{1,000 \div 2} = \dfrac{3}{500}$

b. Example: $\dfrac{6}{1,000}$

2.116 a. $\quad \dfrac{10 \div 2}{12 \div 2} = \dfrac{5}{6}$

b. $\quad 12\overline{)10.0000} = 0.833$

$$\begin{array}{r} 0.8333 \\ \hline \underline{96} \\ 40 \\ \underline{36} \\ 40 \\ \underline{36} \\ 40 \end{array}$$

2.117 a. Example: $\dfrac{1 \times 2}{4 \times 2} = \dfrac{2}{8}$

b. $\quad 4\overline{)1.00}$

$$\begin{array}{r} 0.25 \\ \hline \underline{8} \\ 20 \\ \underline{20} \\ 0 \end{array}$$

2.118 a. Example: $\dfrac{3 \times 2}{5 \times 2} = \dfrac{6}{10}$

b. $\quad 5\overline{)3.0}$

$$\begin{array}{r} 0.6 \\ \hline \underline{3\,0} \\ 0 \end{array}$$

2.119 a. $\quad \dfrac{3 \div 3}{27 \div 3} = \dfrac{1}{9}$

b. $\quad 27\overline{)3.0000} = 0.111$

$$\begin{array}{r} 0.1111 \\ \hline \underline{27} \\ 30 \\ \underline{27} \\ 30 \\ \underline{27} \\ 30 \\ \underline{27} \\ 3 \end{array}$$

2.120 a. $\quad 56\dfrac{38}{100} = 56\dfrac{38 \div 2}{100 \div 2} = 56\dfrac{19}{50}$

b. Example: $56\dfrac{38}{100}$

2.121

$$78\frac{1}{6} = \qquad 78\frac{4}{24} \qquad \text{or} \qquad 78\frac{1}{6} = 78.17$$

$$+ 18.25 = 18\frac{1}{4} = 18\frac{6}{24} \qquad\qquad + 18.25 = 18.25$$

$$\overline{\qquad\qquad\qquad 96\frac{10}{24} = \qquad} \qquad \overline{\qquad\qquad 96.42 \qquad}$$

$$96\,\frac{10 \div 2}{24 \div 2} = 96\frac{5}{12} \text{ yards}$$

2.122

$$8.7 = 8.70 \qquad\qquad\qquad 8.7 = 8\frac{7}{10} = 8\frac{14}{20}$$

$$+ \frac{3}{4} = 0.75 \qquad \text{or} \qquad + \frac{3}{4} = \quad \frac{3}{4} = \quad \frac{15}{20}$$

$$\overline{\qquad 9.45 \text{ miles} \qquad} \qquad \overline{\qquad\qquad\qquad 8\frac{29}{20} = 8 + 1\frac{9}{20} = 9\,\frac{9}{20} \text{ miles}}$$

2.123

$$4\frac{1}{5} = 4.20 \qquad \text{or} \qquad 4\frac{1}{5} = 4\frac{20}{100} = 3 + 1\frac{20}{100} = 3\frac{120}{100}$$

$$- 3.77 = 3.77 \qquad\qquad - 3.77 = 3\frac{77}{100} = \qquad\qquad 3\frac{77}{100}$$

$$\overline{\qquad 0.43 \text{ minutes} \qquad} \qquad \overline{\qquad\qquad\qquad\qquad \frac{43}{100} \text{ minutes}}$$

2.124

$$2.1 = 2.1 \qquad\qquad\qquad\qquad 2.1 = 2\frac{1}{10}$$

$$+ 1\frac{4}{5} = 1.8 \qquad \text{or} \qquad + 1\frac{4}{5} = 1\frac{8}{10}$$

$$\overline{\qquad 3.9 \text{ hours} \qquad} \qquad\qquad \overline{\qquad 3\frac{9}{10} \text{ hours} \qquad}$$

2.125 Jim's trout:

$$2.60 = 2.6 = \quad 2\frac{6}{10} \qquad\qquad 2.60 = 2.60$$

$$4\frac{2}{10} = \qquad\quad 4\frac{2}{10} \qquad \text{or} \qquad 4\frac{2}{10} = 4.2$$

$$+ 4\,\frac{4}{8} = 4\frac{1}{2} = \ 4\frac{5}{10} \qquad\qquad + 4\frac{4}{8} = 4.5$$

$$\overline{\qquad\qquad\qquad 10\frac{13}{10} = \qquad} \qquad \overline{\qquad 11.3 \text{ pounds} \qquad}$$

$$10 + 1\frac{3}{10} = 11\frac{3}{10} \text{ pounds}$$

2.125 (cont.)

$$11\frac{3}{10} = 11\frac{21}{70}$$

$$-8\frac{2}{7} = 8\frac{20}{70} \quad \text{or}$$

$$3\frac{1}{70} \text{ pounds}$$

$$11.3 = 11.30$$

$$-8\frac{2}{7} = 8.29$$

$$3.01 \text{ pounds}$$

2.126

$$40\frac{2}{25} = 40.0800$$

$$-22.9898 = 22.9898 \quad \text{or}$$

$$17.0902$$

$$40\frac{2}{25} = 40\frac{800}{10,000} = 39\frac{10,800}{10,000}$$

$$-12.9898 = 22\frac{9,898}{10,000} = -22\frac{9,898}{10,000}$$

$$17\frac{902}{10,000} = 17\frac{451}{5,000}$$

2.127

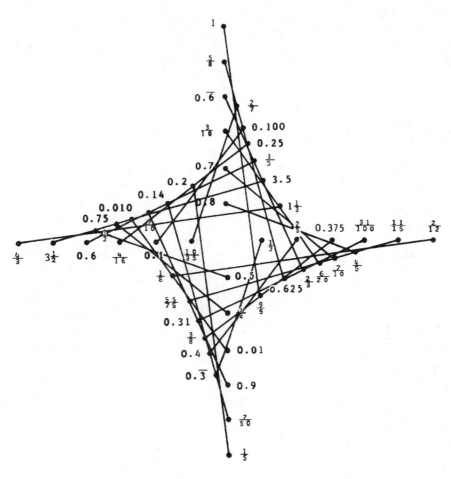

Counterclockwise, the equivalent values are:

$1 = \dfrac{9}{9}$

$\dfrac{5}{8} = 8\overline{)5.000}$ $\quad \dfrac{0.625}{}$

$$\begin{array}{r} 0.625 \\ 8\overline{)5.000} \\ \underline{48} \\ 20 \\ \underline{16} \\ 40 \\ \underline{40} \\ 0 \end{array}$$

$0.\overline{6} = \dfrac{2}{3}$

$\dfrac{3}{10} = \dfrac{3 \times 2}{10 \times 2} = \dfrac{6}{20}$

$0.7 = \dfrac{7}{10}$

$0.8 = \dfrac{8}{10} = \dfrac{8 \div 2}{10 \div 2} = \dfrac{4}{5}$

$0.2 = \dfrac{2}{10} = \dfrac{2 \div 2}{10 \div 2} = \dfrac{1}{5}$

$0.14 = \dfrac{14}{100} = \dfrac{14 \div 2}{100 \div 2} = \dfrac{7}{50}$

$\dfrac{9}{10} = 0.9$

$0.010 = 0.01$

$0.75 = \dfrac{75}{100} = \dfrac{75 \div 25}{100 \div 25} = \dfrac{3}{4}$

$\dfrac{1}{2} = 2\overline{)1.0}$

$$\begin{array}{r} 0.5 \\ 2\overline{)1.0} \\ \underline{1\,0} \\ 0 \end{array}$$

2.127 cont.

$$\frac{4}{3} = 1\frac{1}{3}$$

$$3\frac{1}{2} = 3.5$$

$$0.6 = \frac{6}{10} = \frac{6 \div 2}{10 \div 2} = \frac{3}{5}$$

$$\frac{4}{16} = 16\overline{)4.00}$$

$$\begin{array}{r} 0.25 \\ 16\overline{)4.00} \\ \underline{32} \\ 80 \\ \underline{80} \\ 0 \end{array}$$

$$0.1 = 0.100$$

$$\frac{10}{35} = \frac{10 \div 5}{35 \div 5} = \frac{2}{7}$$

$$\frac{1}{6} = \frac{1 \times 2}{6 \times 2} = \frac{2}{12}$$

$$\frac{55}{75} = \frac{55 \div 5}{75 \div 5} = \frac{11}{15}$$

$$0.31 = \frac{31}{100}$$

$$\frac{3}{8} = 8\overline{)3.000}$$

$$\begin{array}{r} 0.375 \\ 8\overline{)3.000} \\ \underline{24} \\ 60 \\ \underline{56} \\ 40 \\ \underline{40} \\ 0 \end{array}$$

$$0.4 = \frac{4}{10} = \frac{4 \div 2}{10 \div 2} = \frac{2}{5}$$

$$0.\overline{3} = \frac{1}{3}$$

I. SECTION ONE

1.1 $\dfrac{1 \times 2}{3 \times 5} = \dfrac{2}{15}$

1.2 $\dfrac{3 \times 1}{4 \times 2} = \dfrac{3}{8}$

1.3 $\dfrac{5 \times 1}{6 \times 3} = \dfrac{5}{18}$

1.4 $\dfrac{3 \times 3}{4 \times 4} = \dfrac{9}{16}$

1.5 $\dfrac{2 \times 2}{7 \times 3} = \dfrac{4}{21}$

1.6 $\dfrac{3 \times 3}{5 \times 4} = \dfrac{9}{20}$

1.7 $\dfrac{7 \times 1}{8 \times 3} = \dfrac{7}{24}$

1.8 $\dfrac{1 \times 3}{2 \times 4} = \dfrac{3}{8}$

1.9 $\dfrac{1 \times 2}{6 \times 5} = \dfrac{2}{30} = \dfrac{1}{15}$

1.10 $\dfrac{3 \times 1}{7 \times 4} = \dfrac{3}{28}$

1.11 $\dfrac{15 \div 5}{20 \div 5} = \dfrac{3}{4}$

1.12 $\dfrac{14 \div 7}{21 \div 7} = \dfrac{2}{3}$

1.13 $\dfrac{3 \div 3}{9 \div 3} = \dfrac{1}{3}$

1.14 $\dfrac{10 \div 2}{16 \div 2} = \dfrac{5}{8}$

1.15 $\dfrac{105 \div 105}{315 \div 105} = \dfrac{1}{3}$

1.16 $\dfrac{60 \div 30}{90 \div 30} = \dfrac{2}{3}$

1.17 $\dfrac{252 \div 84}{336 \div 84} = \dfrac{3}{4}$

1.18 $\dfrac{504 \div 504}{1,512 \div 504} = \dfrac{1}{3}$

1.19 $\dfrac{140 \div 14}{182 \div 14} = \dfrac{10}{13}$

1.20 $\dfrac{256 \div 32}{288 \div 32} = \dfrac{8}{9}$

1.21 $\dfrac{500 \div 100}{600 \div 100} = \dfrac{5}{6}$

1.22 $\dfrac{324 \div 324}{648 \div 324} = \dfrac{1}{2}$

1.23 $\dfrac{6}{15} = \dfrac{6 \div 3}{15 \div 3} = \dfrac{2}{5}$

1.24 $\dfrac{15}{24} = \dfrac{15 \div 3}{24 \div 3} = \dfrac{5}{8}$

1.25 $\dfrac{6}{21} = \dfrac{6 \div 3}{21 \div 3} = \dfrac{2}{7}$

1.26 $\dfrac{60}{80} = \dfrac{60 \div 20}{80 \div 20} = \dfrac{3}{4}$

1.27 $\dfrac{\overset{1}{2} \times \overset{1}{3}}{\underset{1}{3} \times \underset{2}{4}} = \dfrac{1}{2}$

1.28 $\dfrac{\overset{1}{4} \times \overset{3}{15}}{\underset{1}{5} \times \underset{4}{16}} = \dfrac{3}{4}$

1.29 $\dfrac{\overset{1}{12} \times \overset{3}{39}}{\underset{1}{13} \times \underset{4}{48}} = \dfrac{3}{4}$

1.30 $\dfrac{\overset{1}{6} \times \overset{5}{35}}{\underset{1}{7} \times \underset{8}{48}} = \dfrac{5}{8}$

1.31 $\dfrac{\overset{3}{9} \times \overset{1}{5}}{\underset{2}{10} \times \underset{2}{6}} = \dfrac{3}{4}$

1.32 $\dfrac{7 \times \overset{3}{6}}{8 \times \underset{5}{10}} = \dfrac{21}{40}$

1.33 $\dfrac{\overset{1}{\cancel{12}} \times \overset{\overset{1}{\cancel{3}}}{\cancel{21}}}{\underset{2}{\cancel{14}} \times \underset{\underset{1}{\cancel{3}}}{\cancel{36}}} = \dfrac{1}{2}$

1.34 $\dfrac{\overset{1}{\cancel{3}} \times \overset{3}{\cancel{12}}}{\underset{2}{\cancel{8}} \times \underset{7}{\cancel{21}}} = \dfrac{3}{14}$

1.35 $\dfrac{\overset{1}{\cancel{3}} \times \overset{4}{\cancel{16}}}{\underset{1}{\cancel{4}} \times \underset{7}{\cancel{21}}} = \dfrac{4}{7}$

1.36 improper

1.37 proper

1.38 improper

1.39 proper

1.40 improper

1.41 improper

1.42 $\dfrac{3 \times 8 + 3}{8} = \dfrac{24 + 3}{8} = \dfrac{27}{8}$

1.43 $\dfrac{14 \times 2 + 1}{2} = \dfrac{28 + 1}{2} = \dfrac{29}{2}$

1.44 $\dfrac{6 \times 4 + 3}{4} = \dfrac{24 + 3}{4} = \dfrac{27}{4}$

1.45 $\dfrac{4 \times 5 + 3}{5} = \dfrac{20 + 3}{5} = \dfrac{23}{5}$

1.46 $\dfrac{7 \times 3 + 2}{3} = \dfrac{21 + 2}{3} = \dfrac{23}{3}$

1.47 $\dfrac{9 \times 6 + 1}{6} = \dfrac{54 + 1}{6} = \dfrac{55}{6}$

1.48 $\dfrac{13}{4} \times \dfrac{4}{3} = \dfrac{13 \times \overset{1}{\cancel{4}}}{\underset{1}{\cancel{4}} \times 3} = \dfrac{13}{3}$

1.49 $\dfrac{8}{7} \times \dfrac{14}{5} = \dfrac{8 \times \overset{2}{\cancel{14}}}{\underset{1}{\cancel{7}} \times 5} = \dfrac{16}{5}$

1.50 $\dfrac{16}{5} \times \dfrac{15}{7} = \dfrac{16 \times \overset{3}{\cancel{15}}}{\underset{1}{\cancel{5}} \times 7} = \dfrac{48}{7}$

1.51 $\dfrac{9}{2} \times \dfrac{10}{3} = \dfrac{\overset{3}{\cancel{9}} \times \overset{5}{\cancel{10}}}{\underset{1}{\cancel{2}} \times \underset{1}{\cancel{3}}} = \dfrac{15}{1}$

1.52 $\dfrac{17}{3} \times \dfrac{18}{17} = \dfrac{\overset{1}{\cancel{17}} \times \overset{6}{\cancel{18}}}{\underset{1}{\cancel{3}} \times \underset{1}{\cancel{17}}} = \dfrac{6}{1}$

1.53 $\dfrac{4}{3} = $ $3\overline{)4}$ $= 1\dfrac{1}{3}$
$\dfrac{3}{}$
1

1.54 $\dfrac{18}{7} = $ $7\overline{)18}$ $= 2\dfrac{4}{7}$
$\dfrac{14}{}$
4

1.55 $\dfrac{46}{11} = $ $11\overline{)46}$ $= 4\dfrac{2}{11}$
$\dfrac{44}{}$
2

1.56 $\dfrac{19}{2} = $ $2\overline{)19}$ $= 9\dfrac{1}{2}$
$\dfrac{18}{}$
1

1.57 $\dfrac{23}{4} = $ $4\overline{)23}$ $= 5\dfrac{3}{4}$
$\dfrac{20}{}$
3

1.58 $\dfrac{22}{9} = $ $9\overline{)22}$ $= 2\dfrac{4}{9}$
$\dfrac{18}{}$
4

1.59 $\dfrac{51}{8} = $ $8\overline{)51}$ $= 6\dfrac{3}{8}$
$\dfrac{48}{}$
3

1.60 $\dfrac{67}{5} = $ $5\overline{)67}$ $= 13\dfrac{2}{5}$
$\dfrac{5}{}$
17
$\dfrac{15}{}$
2

1.61 $\dfrac{35}{6} = $ $6\overline{)35}$ $= 5\dfrac{5}{6}$
$\dfrac{30}{}$
5

1.62 $\frac{7}{3} = $ $3\overline{)7}$ gives 2, $\frac{6}{1}$ $= 2\frac{1}{3}$

1.63 $\frac{147}{11} = $ $11\overline{)147}$ gives 13
$\underline{11}$
37
$\underline{33}$
4
$= 13\frac{4}{11}$

1.64 $\frac{9{,}642}{13} = $ $13\overline{)9{,}642}$ gives 741
$\underline{91}$
54
$\underline{52}$
22
$\underline{13}$
9
$= 741\frac{9}{13}$

1.65 $\frac{279}{23} = $ $23\overline{)279}$ gives 12
$\underline{23}$
49
$\underline{46}$
3
$= 12\frac{3}{23}$

1.66 $\frac{1{,}111}{6} = $ $6\overline{)1{,}111}$ gives 185
$\underline{6}$
51
$\underline{48}$
31
$\underline{30}$
1
$= 185\frac{1}{6}$

1.67 $\frac{25}{8} \times \frac{1}{7} = \frac{25}{56}$

1.68 $\frac{4}{7} \times \frac{55}{9} = \frac{220}{63} = 63\overline{)220}$ gives 3
$\underline{189}$
31
$= 3\frac{31}{63}$

1.69 $\frac{4}{{}_1 \cancel{3}} \times \frac{\cancel{81}^{27}}{7} = \frac{108}{7} = 7\overline{)108}$ gives 15
$\underline{7}$
38
$\underline{35}$
3
$= 15\frac{3}{7}$

1.70 $\frac{\cancel{248}^{124}}{11} \times \frac{3}{\cancel{2}_1} = \frac{372}{11} = 11\overline{)372}$ gives 33
$\underline{33}$
42
$\underline{33}$
9
$= 33\frac{9}{11}$

1.71 $\frac{11}{5} \times \frac{27}{4} = \frac{297}{20} = 20\overline{)297}$ gives 14
$\underline{20}$
97
$\underline{80}$
17
$= 14\frac{17}{20}$

1.72 $\frac{1}{{}_1\cancel{10}} \times \frac{\cancel{10}^1}{3} = \frac{1}{3}$

1.73 $\frac{7}{1} \times \frac{25}{8} = \frac{175}{8} = 8\overline{)175}$ gives 21
$\underline{16}$
15
$\underline{8}$
7
$= 21\frac{7}{8}$

1.74 $\frac{\cancel{4}^1}{{}_1\cancel{3}} \times \frac{\cancel{15}^5}{\cancel{4}_1} = \frac{5}{1} = 5$

1.75 $\frac{13}{{}_1\cancel{2}} \times \frac{\cancel{4}^2}{1} = \frac{26}{1} = 26$

1.76 $\frac{\cancel{25}^5}{{}_1\cancel{8}} \times \frac{\cancel{24}^3}{\cancel{5}_1} = \frac{15}{1} = 15$

1.77 $\frac{3}{{}_1\cancel{2}} \times \frac{\cancel{4}^2}{1} = \frac{6}{1} = 6$

1.78 $\frac{\cancel{9}^3}{4} \times \frac{5}{\cancel{3}_1} = \frac{15}{4} = 4\overline{)15}$ gives 3
$\underline{12}$
3
$= 3\frac{3}{4}$

1.79 $\frac{3}{8} \times \frac{7}{1} = \frac{21}{8} = 8\overline{)21}$ gives 2
$\underline{16}$
5
$= 2\frac{5}{8}$

1.80 $\frac{6}{11} \times \frac{4}{1} = \frac{24}{11} = \begin{array}{r} 2 \\ 11\overline{)24} \\ \underline{22} \\ 2 \end{array} = 2\frac{2}{11}$

1.81 $\frac{2}{3} \times \frac{5}{1} = \frac{10}{3} = \begin{array}{r} 3 \\ 3\overline{)10} \\ \underline{9} \\ 1 \end{array} = 3\frac{1}{3}$

1.82 $\frac{3}{4} \times \frac{3}{1} = \frac{9}{4} = \begin{array}{r} 2 \\ 4\overline{)9} \\ \underline{8} \\ 1 \end{array} = 2\frac{1}{4}$

1.83 $\frac{1}{5} \times \frac{4}{3} = \frac{4}{15}$

1.84 $\frac{2}{5} \times \frac{6}{1} = \frac{12}{5} = \begin{array}{r} 2 \\ 5\overline{)12} \\ \underline{10} \\ 2 \end{array} = 2\frac{2}{5}$

1.85 $\frac{19}{9} \div \frac{10}{3} = \frac{19}{{}_3\cancel{9}} \times \frac{\cancel{3}^1}{10} = \frac{19}{30}$

1.86 $\frac{{}^3\cancel{6}}{1} \times \frac{5}{\cancel{4}_2} = \frac{15}{2} = \begin{array}{r} 7 \\ 2\overline{)15} \\ \underline{14} \\ 1 \end{array} = 7\frac{1}{2}$

1.87 $\frac{{}^7\cancel{14}}{3} \times \frac{1}{\cancel{4}_2} = \frac{7}{6} = \begin{array}{r} 1 \\ 6\overline{)7} \\ \underline{6} \\ 1 \end{array} = 1\frac{1}{6}$

1.88 $\frac{3}{7} \div \frac{7}{6} = \frac{3}{7} \times \frac{6}{7} = \frac{18}{49}$

1.89 $\frac{19}{20} \div \frac{19}{5} = \frac{{}^1\cancel{19}}{{}_4\cancel{20}} \times \frac{\cancel{5}^1}{\cancel{19}_1} = \frac{1}{4}$

1.90 $\frac{17}{3} \div \frac{17}{8} = \frac{{}^1\cancel{17}}{3} \times \frac{8}{\cancel{17}_1} = \frac{8}{3} = \begin{array}{r} 2 \\ 3\overline{)8} \\ \underline{6} \\ 2 \end{array} = 2\frac{2}{3}$

1.91 $\frac{3}{8} \times \frac{1}{2} = \frac{3}{16}$

1.92 $\frac{3}{4} \div \frac{6}{7} = \frac{{}^1\cancel{3}}{4} \times \frac{7}{\cancel{6}_2} = \frac{7}{8}$

1.93 $\frac{4}{{}_1\cancel{9}} \times \frac{\cancel{27}^3}{1} = \frac{12}{1} = 12$

1.94 $37 \div \frac{1}{5} = \frac{37}{1} \times \frac{5}{1} = \frac{185}{1} = 185$

1.95 Note "*of*" — multiply.

$\frac{{}^1\cancel{2}}{{}_1\cancel{3}} \times \frac{\cancel{3}^1}{\cancel{4}_2} = \frac{1}{2}$ cup

1.96 Note "*into*" —divide.

$\frac{5}{6} \div 12 = \frac{5}{6} \times \frac{1}{12} = \frac{5}{72}$ of an hour

1.97 $\frac{1}{2} \div 6 = \frac{1}{2} \times \frac{1}{6} = \frac{1}{12}$ ton

1.98 $14\frac{1}{4} \div \frac{3}{4} = \frac{{}^{19}\cancel{57}}{{}_1\cancel{4}} \times \frac{\cancel{4}^1}{\cancel{3}_1} = 19$ pieces

1.99 $\frac{1}{4} \times 2\frac{1}{2} = \frac{1}{4} \times \frac{5}{2} = \frac{5}{8}$ cup

1.100 $36 \div 4\frac{1}{2} = 36 \div \frac{9}{2} = \frac{{}^4\cancel{36}}{1} \times \frac{2}{\cancel{9}_1} = 8$ tiles

1.101 $7\frac{1}{2} \div 9 = 7\frac{1}{2} \times \frac{1}{9} = \frac{{}^5\cancel{15}}{2} \times \frac{1}{\cancel{9}_3} = $

$\frac{5}{6}$ ounce

1.102 $\frac{3}{\underset{10}{50}} \times \frac{{}^{249}\cancel{1{,}245}}{1} = \frac{747}{10} = \begin{array}{r} 74 \\ 10\overline{)747} \\ \underline{70} \\ 47 \\ \underline{40} \\ 7 \end{array} = 74\frac{7}{10}$ lbs.

1.103 $35\frac{3}{4} \div 1\frac{5}{8} = \frac{{}^{11}\cancel{143}}{{}_1\cancel{4}} \times \frac{\cancel{8}^2}{\cancel{13}_1} = 22$ orbits

1.104 $255 \div 2\frac{1}{2} = 255 \div \frac{5}{2} = \frac{{}^{51}\cancel{255}}{1} \times \frac{2}{\cancel{5}_1}$

$= 102$ packages

1.105 $7\frac{1}{2} \div \frac{1}{4} = \frac{15}{2} \div \frac{1}{4} = \frac{15}{2} \times \frac{4}{1} = \frac{30}{1}$

= 30 houses

1.106 $6\frac{3}{4} \div 50 = \frac{27}{4} \div 50 = \frac{27}{4} \times \frac{1}{50}$

$= \frac{27}{200}$ second

1.107 $93\frac{1}{2} \div 13 = \frac{187}{2} \div 13 = \frac{187}{2} \times \frac{1}{13} =$

$\frac{187}{26} = 26\overline{)187} = 7\frac{5}{26}$ yards

$\underline{182}$

5

1.108 $2\frac{3}{4} \times 5 = \frac{11}{4} \times \frac{5}{1} = \frac{55}{4} = 4\overline{)55} =$

13 $\frac{3}{4}$ feet

13

$\underline{4}$

15

$\underline{12}$

3

1.109 $6\frac{3}{4} \times 5\frac{1}{2} = \frac{27}{4} \times \frac{11}{2} = \frac{297}{8} = 8\overline{)297}$

37

$\underline{24}$

57

$\underline{56}$

1

$= 37\frac{1}{8}$ hours

1.110 $13\frac{1}{2} \div 2\frac{1}{4} = \frac{27}{2} \times \frac{4}{9} = 6$ degrees

II. SECTION TWO

2.1 $\frac{2}{5} = 5\overline{)2.0}$

0.4

$\underline{2\ 0}$

0

$2\frac{2}{5} = 2.4$

2.2 $\frac{5}{8} = 8\overline{)5.000}$

0.625

$\underline{48}$

20

$\underline{16}$

40

$4\frac{5}{8} = 4.625$

2.3 $\frac{1}{4} = 4\overline{)1.00}$

0.25

$\underline{8}$

20

$\underline{20}$

0

$1\frac{1}{4} = 1.25$

2.4 $\frac{1}{8} = 8\overline{)1.000}$

0.125

$\underline{8}$

20

$\underline{16}$

40

$\underline{40}$

0

2.5 $\frac{3}{4} = 4\overline{)3.00}$

0.75

$\underline{28}$

20

$\underline{20}$

0

2.6 $\frac{3}{8} = 8\overline{)3.000}$

0.375

$\underline{24}$

60

$\underline{56}$

40

$\underline{40}$

0

2.7 $\frac{1}{5} = 5\overline{)1.0}$

0.2

$\underline{10}$

0

$6\frac{1}{5} = 6.2$

2.8 $\frac{1}{2} = 2\overline{)1.0}$

0.5

$\underline{10}$

0

$3\frac{1}{2} = 3.5$

2.9 $5.125 = 5\frac{125}{1,000} = 5\frac{125 \div 125}{1,000 \div 125} = 5\frac{1}{8}$

2.10 $0.4 = \frac{4}{10} = \frac{4 \div 2}{10 \div 2} = \frac{2}{5}$

2.11 $2.6 = 2\frac{6}{10} = 2\frac{6 \div 2}{10 \div 2} = 2\frac{3}{5}$

2.12 $\quad 4.75 = 4\frac{75}{100} = 4\frac{75 \div 25}{100 \div 25} = 4\frac{3}{4}$

2.13 $\quad 0.5 = \frac{5}{10} = \frac{5 \div 5}{10 \div 5} = \frac{1}{2}$

2.14 $\quad 1.8 = 1\frac{8}{10} = 1\frac{8 \div 2}{10 \div 2} = 1\frac{4}{5}$

2.15 $\quad 3\frac{9}{10}$

2.16 $\quad 8.375 = 8\frac{375}{1,000} = 8\frac{375 \div 125}{1,000 \div 125} = 8\frac{3}{8}$

2.17
$$
\begin{array}{r}
47.3 \\
\times\ 0.128 \\
\hline
3784 \\
946 \\
473 \\
\hline
6.0544
\end{array}
$$

2.18
$$
\begin{array}{r}
0.0047 \\
\times\ \ \ 634 \\
\hline
188 \\
141 \\
282 \\
\hline
0.29798
\end{array}
$$

2.19
$$
\begin{array}{r}
82 \\
\times\ 0.436 \\
\hline
492 \\
246 \\
328 \\
\hline
35.752
\end{array}
$$

2.20
$$
\begin{array}{r}
0.16 \\
\times\ 7.5 \\
\hline
80 \\
112 \\
\hline
1.200 \text{ or } 1.2
\end{array}
$$

2.21 $\quad \frac{3}{4} \times \frac{5}{16} = \frac{15}{64}$

$\qquad \frac{3}{4} = \quad 0.75$

$\times \quad \frac{5}{16} = \underline{0.3125}$
$$
\begin{array}{r}
375 \\
150 \\
75 \\
225 \\
\hline
0.234375
\end{array}
$$

$\frac{15}{64} = 0.234375$

2.22 $\quad \frac{1}{3} \times \frac{6}{11} = \frac{2}{11}$

$\qquad \frac{1}{3} = \quad 0.33333$

$\times \quad \frac{6}{11} = \underline{0.54545}$
$$
\begin{array}{r}
166665 \\
133332 \\
166665 \\
133332 \\
\underline{166665} \\
0.1818148485
\end{array}
$$

$\frac{2}{11} = 0.1818182$

Note small difference in 7th decimal place caused by rounding.

2.23 $\quad \frac{3}{8} \times \frac{1}{5} = \frac{3}{40}$

$\qquad \frac{3}{8} = \quad 0.375$

$\times \quad \frac{1}{5} = \underline{\quad 0.2}$

$\qquad\qquad 0.0750$

$\frac{3}{40} = \quad 0.075$

2.24 $\quad \frac{4}{1} \times \frac{6}{7} = \frac{24}{7} = 3\frac{3}{7}$

$\qquad \frac{6}{7} = 0.857142857$

$\times \quad 4 = \underline{\qquad\qquad 4}$

$\qquad\qquad 3.428571428$

$3\frac{3}{7} = 3.42857143$

Note small difference in 8th decimal place caused by rounding.

2.25 $\quad 9\frac{1}{16} \times 1\frac{1}{2} = \frac{145}{16} \times \frac{3}{2} = \frac{435}{32} = 13\frac{19}{32}$

$\qquad \frac{145}{16} = \quad 9.0625$

$\times \quad \frac{3}{2} \quad = \underline{\quad\quad 1.5}$
$$
\begin{array}{r}
453125 \\
\underline{90625} \\
13.59375
\end{array}
$$

$13\frac{19}{32} = 13.59375$

2.26
```
       6.4218
  x     0.47
     449526
     256872
    3.018246
```

2.27
```
       0.0381
  x    10.73
       1143
       2667
       3810
     0.408813
```

2.28
```
        41.76
   x     7.4
       16704
       29232
      309.024
```

2.29 64,210

2.30 838.4

2.31 4.83

2.32 157,300

2.33
```
              4.322 → 4.32
    1.476)6380.000
           5904
           4760
           4428
           3320
           2952
           3680
           2952
            728
```

2.34
```
            12.808 → 12.81
    14.6)187.0.000
          146
          410
          292
         1180
         1168
         1200
         1168
           32
```

2.35
```
              85.282 → 85.28
    0.4104)35.0000.000
            32832
            21680
            20520
            11600
             8208
            33920
            32832
            10880
             8208
             2672
```

2.36 0.358

2.37 5.213

2.38 0.064

2.39 0.0495

2.40
```
            19.4152 → 19.415
    0.46)8.93.1000
          46
          433
          414
          191
          184
           70
           46
          240
          230
          100
           92
            8
```

2.41
```
             2 54
    0.25)63.50
          50
          135
          125
          100
          100
            0
```

2.42

$$224.7549 \rightarrow 224.755$$

$0.0816\overline{)18.3400.0000}$

```
      1632
      2020
      1632
      3880
      3264
      6160
      5712
      4480
      4080
      4000
      3264
      7360
      7344
        16
```

2.43

$$14.6428 \rightarrow 14.643$$

$0.028\overline{)0.410.0000}$

```
     28
    130
    112
    180
    168
    120
    112
     80
     56
    240
    224
     16
```

2.44

$$176.3157 \rightarrow 176.316$$

$0.0038\overline{)0.6700.0000}$

```
     38
    290
    266
    240
    228
    120
    114
     60
     38
    220
    190
    300
    266
     34
```

2.45

$$0.0105 \rightarrow 0.011$$

$3.8\overline{)0.0.4000}$

```
    38
   200
   190
    10
```

2.46

$$14.614 \qquad 14.61 \text{ miles per gallon}$$

$14.8\overline{)216.3.000}$

```
    148
    683
    592
    910
    888
    220
    148
    720
    592
    128
```

2.47

```
     94.8
   x  5.6
     5688
     4740
   530.88  miles
```

2.48

```
      120
   x 0.022
      240
      240
    2.640 → 2.64  pounds
```

2.49

```
        $1.95
   250)487.50
        250
        2375
        2250
        1250
        1250
           0
```

2.50

```
   16,420
   x 0.003
   49.260 → 49.26 inches
```

2.51

```
    $2.375
   x   7.5
    11875
    16625
    $17.8125 → $17.81
```

2.52
```
      6.68
    x 15.5
     3340
     3340
      668
   103.540 → 103.54 pounds
```

2.53
```
          $0.150 → $0.15
      52)7.850
         52
         265
         260
          50
```

2.54
```
          2.4 grams
       8)19.2
         16
         32
         32
          0
```

2.55
```
           0.2692 → 0.269
      78)21.0000
         156
         540
         468
         720
         702
         180
         156
          24
```

2.56
```
      26.3
    x 1.50
    13150
     263
    39.450 = 39.45
```

2.57
```
      0.0001
    x   150
     00050
     0001
    00.0150 → 0.015 ounces
```

2.58
```
      18.09
    x  8.33
     5427
     5427
    14472
    150.6897 → 150.69 pounds
```

2.59
```
          0.7875 yards
       8)63000
         56
         70
         64
         60
         56
         40
         40
          0
```

2.60
```
      14.5
    x 0.72
     290
    1015
    10.440 = $10.44
```

III. SECTION THREE

3.1   0.47

3.2   0.163

3.3   1.84

3.4   0.0005

3.5   59%

3.6   780%

3.7   0.3%

3.8   160%

3.9   a. 0.375

     b. $0.375 = \frac{375}{1,000} = \frac{375 \div 125}{1,000 \div 125} = \frac{3}{8}$

3.10 a. 80%

     b. $0.8 = \frac{8}{10} = \frac{8 \div 2}{10 \div 2} = \frac{4}{5}$

3.11 a. $\frac{3}{4} = 3 \div 4 = 0.75 = 75\%$

     b. 0.75

3.12 a. $= \frac{1}{9} = 1 \div 9 = 0.\overline{1}$ or 0.111...

                 = 11.1%

     b. = 0.111

3.13 a. 41.7%

      b. $0.417 = \frac{417}{1,000}$

3.14 a. 1.2

      b. $1.2 = 1\frac{2}{10} = 1\frac{2 \div 2}{10 \div 2} = 1\frac{1}{5}$

3.15 a. 450

      b. 19%

      c. 85.5

316   a. 34.95

      b. 103%

      c. 36

3.17 a. 47.93

      b. 0.3%

      c. 0.144

3.18 a. 28

      b. 25%

      c. 7

3.19 percentage

3.20 base

3.21 rate

3.22 rate

3.23 base

3.24 The base is 147.
The rate is 97% = 0.97.
The percentage is 147 x 0.97 =

$$\begin{array}{r} 147 \\ \times\ 0.97 \\ \hline 1029 \\ \underline{1323} \\ 142.59 \end{array}$$

3.25 The base is 375.
The rate is 13% = 0.13.
The percentage is 375 x 0.13 =

$$\begin{array}{r} 375 \\ \times\ 0.13 \\ \hline 1125 \\ \underline{375} \\ 48.75 \end{array}$$

3.26 The base is 600.
The rate is 84% = 0.84.
The percentage is 600 x 0.84 =

$$\begin{array}{r} 600 \\ \times\ 0.84 \\ \hline 2400 \\ \underline{4800} \\ 504.00 = 504 \end{array}$$

3.27

$$\begin{array}{r} 8,600 \\ \times\ \ \ 0.1 \\ \hline 860.0 = \$860 \end{array}$$

3.28

$$\begin{array}{r} 268 \\ \times\ 0.42 \\ \hline 536 \\ \underline{1072} \\ 112.56 = \text{about 113 votes} \end{array}$$

3.29

$$\begin{array}{r} 625 \\ \times\ 0.61 \\ \hline 625 \\ \underline{3750} \\ 381.25 \rightarrow \text{381 dentists} \end{array}$$

3.30 a.

$$\begin{array}{r} 25 \\ \times\ 0.3 \\ \hline 7.5 = \$7.50 \end{array}$$

      b.

$$\begin{array}{r} \$25.00 \\ -\ \ \ 7.50 \\ \hline \$17.50 \end{array}$$

3.31

$$\begin{array}{r} 2,565 \\ \times\ 0.07 \\ \hline \$179.55 \end{array}$$

3.32 The base is 73. The percentage
is 27. The rate is 27 ÷ 73 =

$$\begin{array}{r} 0.369 \rightarrow 0.37 = 37\% \\ 73\overline{)27.00} \\ \underline{219} \\ 510 \\ \underline{438} \\ 720 \\ \underline{657} \\ 63 \end{array}$$

3.33   The base is 300.
        The percentage is 162.
        The rate is 162 ÷ 300 =

$$\begin{array}{r} 0.54 = 54\% \\ 300\overline{)162.00} \\ \underline{1500} \\ 1200 \\ \underline{1200} \\ 0 \end{array}$$

3.34   The base is 230.
        The percentage is 347.
        The rate is 347 ÷ 230 =

$$\begin{array}{r} 1.508 \rightarrow 1.51 = 151\% \\ 230\overline{)347.000} \\ \underline{230} \\ 1170 \\ \underline{1150} \\ 2000 \\ \underline{1840} \\ 160 \end{array}$$

3.35
$$\begin{array}{r} 0.48 = 48\% \\ 114,000\overline{)54,720.00} \\ \underline{456000} \\ 912000 \\ \underline{912000} \\ 0 \end{array}$$

3.36
$$\begin{array}{r} 0.64 = 64\% \\ 114\overline{)73.00} \\ \underline{684} \\ 460 \\ \underline{456} \\ 4 \end{array}$$

3.37
$$\begin{array}{r} 0.8 = 80\% \\ 25\overline{)20.0} \\ \underline{200} \\ 0 \end{array}$$

3.38
$$\begin{array}{r} 0.6 \quad 60\% \\ 35\overline{)21.0} \\ \underline{210} \\ 0 \end{array}$$

3.39
$$\begin{array}{r} 0.734 \rightarrow 0.73 = 73\% \\ 64\overline{)47000} \\ \underline{448} \\ 220 \\ \underline{192} \\ 280 \\ \underline{256} \\ 24 \end{array}$$

3.40   The rate is 73% = 0.73.
        The percentage is 121.
        The base is 121 ÷ 0.73 =

$$\begin{array}{r} 165.753 = 165.75 \\ 0.73\overline{)121.00,000} \\ \underline{73} \\ 480 \\ \underline{438} \\ 420 \\ \underline{365} \\ 550 \\ \underline{511} \\ 390 \\ \underline{365} \\ 250 \\ \underline{219} \\ 31 \end{array}$$

3.41   The rate is 16% = 0.16.
        The percentage is 332.
        The base is 332 ÷ 0.16 =

$$\begin{array}{r} 2,012.5 \\ 0.16\overline{)322.00.0} \\ \underline{32} \\ 020 \\ \underline{16} \\ 40 \\ \underline{32} \\ 80 \\ \underline{80} \\ 0 \end{array}$$

3.42   The rate is 140% = 1.4.
        The percentage is 78.
        The base is 78 ÷ 1.4 =

$$\begin{array}{r} 55.714 \rightarrow 55.71 \\ 1.4\overline{)78.0.000} \\ \underline{70} \\ 80 \\ \underline{70} \\ 100 \\ \underline{98} \\ 20 \\ \underline{14} \\ 60 \\ \underline{56} \\ 4 \end{array}$$

3.43

```
         1750
0.08)140.00.
         8
         60
         56
          40
          40
           0
```

3.44

```
        130
0.3)39.0.
        3
        09
         9
         0
```

3.45

```
           1956.5→ 1,957
0.23)450.00.0
      23
      220
      207
        130
        115
         150
         138
          120
          115
            5
```

3.46

```
      $3,000
0.8)2400.0.
      24
       0
```

3.47

```
      $12,000
0.05)600.00.
       5
       10
       10
        0
```

3.48 The base is $4,800.
The percentage is $3,840.
The rate is $3,840 ÷ $4,800 =
```
            0.8 = 80%
4,800)3,840.0
      38400
          0
```

3.49 The base is $10,000.
The rate is 3% = 0.03.
The percentage is 10,000 x 0.03 =

```
10,000
x  0.03
300.00 = 300
```

3.50 The base is $1,000.
The percentage is $60.
The rate is $60 ÷ $1,000 =
```
          0.06  = 6%
1,000)60.00
       6000
          0
```

3.51 The rate is 75% = 0.75.
The percentage is $48.
The base is 48 ÷ 0.75 =

```
        $64
0.75)48.00.
      450
      300
      300
        0
```

3.52 The rate is 15% = 0.15.
The percentage is $44.50.
The base is $44.50 ÷ 0.15 =
```
          296.666 → $296.67
0.15)44.50.000
      30
      145
      135
       100
        90
       100
        90
        100
         90
         100
          90
          10
```

3.53 The base is 425.
The percentage is 136.
The rate is 136 ÷ 425 =
```
          0.32  = 32%
425)136.00
     1275
      850
      850
        0
```

238

I. SECTION ONE

1.1 $A = 47 \times 12$

$$
\begin{array}{r}
47 \\
\times\,12 \\
\hline
94 \\
47 \\
\hline
\end{array}
$$

$A = 564$

1.2 $A = 5\frac{1}{2} \times 2\frac{1}{2}$

$\quad = \frac{11}{2} \times \frac{5}{2}$

$\quad = \frac{55}{4}$

$A = 13\frac{3}{4}$

1.3 $A = 19.3 \times 0.57$

$$
\begin{array}{r}
19.3 \\
\times\,0.57 \\
\hline
1351 \\
965 \\
\hline
\end{array}
$$

$A = 11.001$

1.4 $A = 13 \times 12$

$$
\begin{array}{r}
13 \\
\times\,12 \\
\hline
26 \\
13 \\
\hline
\end{array}
$$

$A = 156$

1.5 $100 = 20 \times W$
          $W = 5$

1.6 $50 = L \times 2$
          $L = 25$

1.7 $6 = 2 \times W$
          $W = 3$

1.8 $28 = L \times 4$
          $L = 7$

1.9 $d = r \times t$

$\quad = $

$$
\begin{array}{r}
0.7 \\
\times\,2.3 \\
\hline
21 \\
14 \\
\hline
\end{array}
$$

$\quad$ 1.61 miles

1.10 $d = r \times t$

$\quad = 20$

$\quad = \frac{\overset{10}{20}}{1} \times \frac{3}{\underset{1}{2}}$

$\quad = \frac{30}{1}$

$\quad = 30$ miles

1.11 $d = r \times t$

$\quad = 3\frac{1}{5} \times \frac{1}{4}$

$\quad = \frac{\overset{4}{16}}{5} \times \frac{1}{\underset{1}{4}} = \frac{4}{5}$ miles

1.12 $d = r \times t$

$\quad = 55 \times 9$

$\quad = 495$ miles

1.13 $d = r \times t$

$\quad 12 = r \times 3$

$\quad r = 4$

1.14 $d = r \times t$

$\quad 6 = r \times 1$

$\quad r = 6$

1.15 $d = r \times t$

$\quad 16 = 8 \times t$

$\quad t = 2$

1.16 $d = r \times t$

$\quad 200 = 50 \times t$

$\quad t = 4$

1.17 $S = C + P$

$\quad S = \$67.50 + \$32.18$

$\quad S = \$99.68$

1.18 $S = C + P$

$\quad \$20 = \$13 + P$

$\quad P = \$7$

1.19 $S = C + P$

$\quad \$4.50 = C + \$1.50$

$\quad C = \$3.00$

1.20 $S = C + P$

$\quad S = \$89.95 + \$45.15$

$\quad S = \$135.10$

1.21 $S = C + P$
$18 = C + 9
$C = 9

1.22 $S = C + P$
$1,000 = $600 + P$
$P = 400

1.23 $I = R \times P$
$I = 0.06 \times 950$
$\begin{array}{r} 0.06 \\ \times\ 950 \\ \hline 300 \\ 54 \\ \hline \end{array}$
$I = 57.00

1.24 $I = R \times P$
$I = 0.07 \times 400$
$\begin{array}{r} 0.07 \\ \times\ 400 \\ \hline \end{array}$
$I = 28.00

1.25 $I = R \times P$
$I = 0.12 \times 450$
$\begin{array}{r} 0.12 \\ \times\ 450 \\ \hline 600 \\ 48 \\ \hline \end{array}$
$I = 54.00

1.26 $I = R \times P$
$I = 0.1 \times $1,649$
$I = 164.90

1.27 $I = R \times P$

1.28 $d = r \times t$

1.29 $A = L \times W$

1.30 $S = C + P$

1.31   Example:
$P = W \times H$

1.32   Teacher check:
$P = W \times H$ is a good formula because P reminds you of *pay*, W reminds you of *wages*, and H reminds you of *hours*. Any student answer showing thought about the formula is acceptable.

1.33 $P = W \times H$
$P = 6.50×40
$\begin{array}{r} $6.50 \\ \times\ 40 \\ \hline \end{array}$
$P = 260.00

1.34 $P = W \times H$
$200 = W \times 40$
$W = 5

1.35 $P = W \times H$
$75 = $1.50 \times H$
$H = 50$

1.36   Example:
$P = 4 \times S$

1.37 $P = 4 \times S$
$8 = 4 \times S$
$S = 2$

1.38 $P = 4 \times S$
$P = 4 \times 4$
$P = 16$

1.39 $P = 4 \times S$
$P = 4 \times 11.361$
$P = 45.444$

1.40   Examples:
$N = L - D$ or $P = L - D$ or
$N = P - D$

1.41 $N = L - D$
$N = $49.95 - 10.00
$N = 39.95

1.42 $N = L - D$
$12.00 = L - 2.00
$L = 14.00

1.43 $N = L - D$
$600 = $750 - D$
$D = 150

1.44 $mpg = m \div g$

1.45 $mpg = m \div g$
$20 = 200 \div g$
$g = 10$

1.46 $mpg = m \div g$
$15 = m \div 10$
$m = 150$

1.47 $mpg = m \div g$
       $mpg = 672 \div 32$

$$32\overline{)672} \quad \begin{array}{r} 21 \\ \hline \end{array}$$

           $\begin{array}{r} 21 \\ 32\overline{)672} \\ \underline{64} \\ 32 \\ \underline{32} \\ 0 \end{array}$

       $mpg = 21$

1.48 $mpg = m \div g$
       $16 = m \div 29$
       $m = 464$

1.49 $P = W \times H$
       $P = \$3.75 \times 38$

           $\begin{array}{r} \$3.75 \\ \times\ \ 38 \\ \hline 3000 \\ \underline{1125} \\ \$14250 \end{array}$

       $P = \$142.50$

1.50 $A = L \times W$
       $A = 35 \times 25$

           $\begin{array}{r} 35 \\ \times 25 \\ \hline 175 \\ \underline{70} \\ \end{array}$

       $A = 875$

1.51 $S = C + P$
       $S = \$149.50 + \$100.00$
       $S = \$249.50$

1.52 $I = R \times P$
       $I = 0.15 \times 300$

           $\begin{array}{r} 0.15 \\ \times\ \ 300 \\ \hline \end{array}$

       $I = \$45.00$

1.53 $15 + 3 = 18$

1.54 $3 + 3 = 6$

1.55 $7 - 4 = 3$

1.56 $14 - 4 = 10$

1.57 $4 - 4 = 0$

1.58 $\begin{array}{r} 1.4 \\ \times\ 3.1 \\ \hline 14 \\ \underline{42} \\ 4.34 \end{array}$

1.59 $\begin{array}{r} 0.03 \\ \times\ \ 3.1 \\ \hline 003 \\ \underline{009} \\ 0.093 \end{array}$

1.60 $\begin{array}{r} 0.16 \\ \times\ \ 3.1 \\ \hline 016 \\ \underline{048} \\ 0.496 \end{array}$

1.61 $\begin{array}{r} 7.1 \\ \times\ 3.1 \\ \hline 71 \\ \underline{213} \\ 22.01 \end{array}$

1.62 $56 \div 7 = 8$

1.63 $14 \div 7 = 2$

1.64 $7 \div 7 = 1$

1.65 $\div 3$

1.66 $+ 3$

1.67 $\times 0$

1.68 $0 \times 2 = 0$
       $0 + 3 = 3$
       $3 \div 1.5 = 2$

1.69 $15 \times 2 = 30$
       $30 + 3 = 33$
       $33 \div 1.5 = 22$

1.70 $6 \div 3 = 2$
       $2 + 1 = 3$
       $3 \times 2 = 6$

1.71 $15 \div 3 = 5$
       $5 + 1 = 6$
       $6 \times 2 = 12$

1.72 $3 \div 3 = 1$
       $1 + 1 = 2$
       $2 \times 2 = 4$

1.73 $\frac{3}{8} + \frac{1}{2} = \frac{3}{8} + \frac{4}{8} = \frac{7}{8}$

       $\frac{7}{8} \div \frac{1}{3} = \frac{7}{8} \times \frac{3}{1} = \frac{21}{8} = 2\frac{5}{8}$

1.74 $\frac{3}{4} + \frac{1}{2} = \frac{3}{4} + \frac{2}{4} = \frac{5}{4}$

  $\frac{5}{4} \div \frac{1}{3} = \frac{5}{4} \times \frac{3}{1} = \frac{15}{4} = 3\frac{3}{4}$

1.75 $\frac{1}{6} + \frac{1}{2} = \frac{1}{6} + \frac{3}{6} = \frac{4}{6} = \frac{2}{3}$

  $\frac{2}{3} \div \frac{1}{3} = \frac{2}{\underset{1}{3}} \times \frac{\overset{1}{3}}{1} = \frac{2}{1} = 2$

1.76 $7 \times 8 = 56$

1.77 $\frac{1}{2} \times \frac{3}{4} = \frac{3}{8}$

1.78 $6 \div 3 = 2$

1.79 $3 \div 6 = \frac{1}{2}$ or 0.5

1.80 $3.81 + 6.3 = 10.11$

1.81 $1\frac{3}{4} + \frac{2}{3} = 1\frac{9}{12} + \frac{8}{12} =$

  $1\frac{17}{12} = 1 + 1\frac{5}{12} = 2\frac{5}{12}$

1.82 $1{,}473 + 0.04 = 1{,}473.04$

1.83 $14 \div 7 = 2$

1.84 $3.46 \div 0.2 = 17.3$

1.85 $\frac{1}{4} \div \frac{1}{7} =$

  $\frac{1}{4} \times \frac{7}{1} = \frac{7}{4} = 1\frac{3}{4}$

1.86 $3\frac{1}{8} \div \frac{3}{5} =$

  $\frac{25}{8} \times \frac{5}{3} = \frac{125}{24} = 5\frac{5}{24}$

1.87 $37.4 - 3.51 = 33.89$

1.88 $\frac{3}{5} - \frac{4}{9} =$

  $\frac{27}{45} - \frac{20}{45} = \frac{7}{45}$

1.89 $1\frac{1}{8} - \frac{4}{5} =$

  $1\frac{5}{40} - \frac{32}{40} =$

  $\frac{45}{40} - \frac{32}{40} = \frac{13}{40}$

1.90
$$\begin{array}{r} 8.41 \\ \times\ 1.3 \\ \hline 2523 \\ 841\ \ \\ \hline 10.933 \end{array}$$

1.91 $6\frac{2}{3} \times \frac{1}{8} =$

  $\frac{\overset{5}{20}}{3} \times \frac{1}{\underset{2}{8}} = \frac{5}{6}$

1.92 $\frac{147}{1} \times \frac{3}{4} = \frac{441}{4} = 110\frac{1}{4}$

II. SECTION TWO

2.1 true

2.2 false

2.3 true; $3 = 3$

2.4 open

2.5 open

2.6 open

2.7 open

2.8 $R = 84$

2.9 $x = 19$

2.10 $16 = K$

2.11 $x = 141$

2.12 $\frac{4}{8} + \frac{5}{8} = x$

  $\frac{9}{8} = x$

  $1\frac{1}{8} = x$

2.13 $x = 8$

2.14 $K = 18$

2.15 $x = 4$

2.16 $R = 120$

2.17 $x = 30 - 19$
   $x = 11$

2.18 $57 - 47 = x$
   $10 = x$

2.19 $200 - 143 = x$
   $57 = x$

2.20 $x = 166 - 133$
   $x = 33$

2.21 $x = 4 + 19$
   $x = 23$

2.22 $16 + 23 = R$
   $39 = R$

2.23 $130 + 65 = x$
   $195 = x$

2.24 $T = 78 + 3$
   $T = 81$

2.25 $x = 108 \div 9$
   $x = 12$

2.26 $77 \div 7 = W$
   $11 = W$

2.27 $x = 286 \div 143$
   $x = 2$

2.28 $x = 1,209 \div 3$
   $x = 403$

2.29 $R = 68.5 \cdot 1.3$
   68.5
   $\underline{\times\ 1.3}$
   2055
   $\underline{685\ }$
   $R = 89.05$

2.30 $100 \cdot 4 = x$
   $400 = x$

2.31 $6 \cdot 2 = T$
   $12 = T$

2.32 $x = 6 \cdot 14$
   $x = 84$

2.33 $R = 2 \cdot 7$
   $R = 14$

2.34 $x = 18 \div 3$
   $x = 6$

2.35 $x = 4 + 19$
   $x = 23$

2.36 $64 - 1 = x$
   $63 = x$

2.37 $18 + 23 = R$
   $41 = R$

2.38 $42 \div 6 = x$
   $7 = x$

2.39 $A = L \times W$
   $41.5 = L \times 5$
   $41.5 \div 5 = L$
   $8.3 = L$

2.40 $d = r \times t$
   $400 = r \times 8$
   $400 \div 8 = r$
   $50 = r$

2.41 $I = R \times P$
   $\$18 = 0.12 \times P$
   $\$18 \div 0.12 = P$

$$\begin{array}{r} 150 \\ 12\overline{)1800} \\ \underline{12\ } \\ 60 \\ \underline{60} \\ 0 \end{array}$$

   $\$150 = P$

2.42 $P = W \times H$
   $\$150 = \$4.20 \times H$
   $\$150 \div \$4.20 = H$

$$\begin{array}{r} 35.71 \\ 420\overline{)15000.00} \\ \underline{1260} \\ 2400 \\ \underline{2100} \\ 3000 \\ \underline{2940} \\ 600 \\ \underline{420} \\ 180 \end{array}$$

   $35.71 = H$

2.43 $P = 4 \times S$
   $18.5 = 4 \times S$
   $18.5 \div 4 = S$

2.43 cont.

$$4\overline{)18.500} = 4.625$$

$$
\begin{array}{r}
4.625 \\
4\overline{)18.500} \\
\underline{16} \\
25 \\
\underline{24} \\
10 \\
\underline{8} \\
20 \\
\underline{20} \\
0
\end{array}
$$

$$4.625 = S$$

2.44

$$N = L - D$$
$$\$7.95 = L - \$1.04$$
$$\$7.95 + \$1.04 = L$$
$$\$8.99 = L$$

2.45

$$mpg = m \div g$$
$$7.2 = m \div 14.5$$
$$7.2 \times 14.5 = m$$

$$
\begin{array}{r}
14.5 \\
\times\ 7.2 \\
\hline
290 \\
1015\ \\
\hline
10440
\end{array}
$$

$$104.4 = m$$

III. SECTION THREE

3.1   3:7

3.2   7:3

3.3   9:11

3.4   4:1

3.5   6:30 = 1:5

3.6   30:6 = 5:1

3.7   75:50 = 3:2

3.8   12:21 = 4:7

3.9   12:33 = 4:11

3.10   21:33 = 7:11

3.11   21:12 = 7:4

3.12   300:700 = 3:7

3.13   400:300 = 4:3

3.14   300:400 = 3:4

3.15   400:700 = 4:7

3.16
a.   9 x 16 = 144
b.   4 x 36 = 144
c.   yes

3.17
a.   1 x 18 = 18
b.   6 x 4 = 24
c.   no

3.18
a.   10 x 19 = 190
b.   5 x 37 = 185
c.   no

3.19
a.   3 x 8 = 24
b.   12 x 2 = 24
c.   yes

3.20

$$4 \cdot x = 3 \cdot 20$$
$$4 \cdot x = 60$$
$$x = 60 \div 4$$
$$x = 15$$

3.21

$$3.4 \cdot 4 = 1.7 \cdot x$$
$$13.6 = 1.7 \cdot x$$
$$13.6 \div 1.7 = x$$

$$
\begin{array}{r}
8 \\
17\overline{)136} \\
\underline{136} \\
0
\end{array}
$$

$$8 = x$$

3.22

$$28 \cdot 12 = 21 \cdot R$$
$$336 = 21 \cdot R$$
$$336 \div 21 = R$$

$$
\begin{array}{r}
16 \\
21\overline{)336} \\
\underline{21} \\
126 \\
\underline{126} \\
0
\end{array}
$$

$$16 = R$$

3.23

$$3 \cdot 9 = x \cdot 27$$
$$27 = x \cdot 27$$
$$27 \div 27 = x$$
$$1 = x$$

3.24

$$x \cdot 2 = 100 \cdot 14$$
$$x \cdot 2 = 1,400$$
$$x = 1,400 \div 2$$
$$x = 700$$

3.25 $45 \cdot 39 = 15 \cdot Q$
$1{,}755 = 15 \cdot Q$
$1{,}755 \div 15 = Q$

$$
\begin{array}{r}
117 \\
15{\overline{)1{,}755}} \\
\underline{15} \\
25 \\
\underline{15} \\
105 \\
\underline{105} \\
0
\end{array}
$$

$117 = Q$

3.26 $3{:}5 = 12{:}x$
$5 \cdot 12 = 3 \cdot x$
$60 = 3 \cdot x$
$60 \div 3 = x$
$20 = x$
$x = 20$

3.27 $14{:}21 = x{:}24$
$21 \cdot x = 14 \cdot 24$
$21 \cdot x = 336$
$x = 336 \div 21$

$$
\begin{array}{r}
16 \\
21{\overline{)336}} \\
\underline{21} \\
126 \\
\underline{126} \\
0
\end{array}
$$

$x = 16$

3.28 $3{:}5 = x{:}40$
$5 \cdot x = 3 \cdot 40$
$5 \cdot x = 120$
$x = 120 \div 5$
$x = 24$ inches

245

I. SECTION ONE

1.1 887, 704, 744, 767, 217

1.2 88, 770; 47, 447; 67, 217;
     63, 350; 25, 839
     (The next two are 21, 206
     and 76,630—we go to the
     beginning of the next
     row.)

1.3 4, 0, 3, 3, 2, 0, 3, 8, 2, 6

1.4 40, 33, 20, 38, 26, 13, 89,
     51, 03, 74

1.5 true

1.6 false

1.7 false

1.8 true

1.9 false

1.10 c

1.11 a

1.12 d

1.13 b

1.14 f

1.15 79

1.16 70

1.17 3

1.18 77

1.19 Tally: 90| 95||
               91||| 96|||
               92|| 97||||
               93|||| 98|||
               94||| 99||||

     most frequent _97_

1.20 Tally: 10|| 60||||
               20|||| 70|
               30|| 80|||
               40||| 90||
               50||

     most frequent _60_

1.21 Tally: 110| 115||
               111|||| 116|||
               112||| 117|
               113| 118||
               114|| 119||||

     most frequent _111 and 119_

1.22

Number	Frequency
90	1
91	3
92	2
93	4
94	3
95	2
96	3
97	5
98	3
99	4
	30

1.22 cont.

<u>Relative Frequency</u>

$\frac{1}{30}$ $\qquad = 0.033 = 3.3\%$

$\frac{3}{30} = \frac{1}{10}$ $\quad = 0.1 \quad = 10.0\%$

$\frac{2}{30} = \frac{1}{15}$ $\quad = 0.067 = 6.7\%$

$\frac{4}{30} = \frac{2}{15}$ $\quad = 0.133 = 13.3\%$

$\frac{3}{30} = \frac{1}{10}$ $\quad = 0.1 \quad = 10.0\%$

$\frac{2}{30} = \frac{1}{15}$ $\quad = 0.067 = 6.7\%$

$\frac{3}{30} = \frac{1}{10}$ $\quad = 0.1 \quad = 10.0\%$

$\frac{5}{30} = \frac{1}{6}$ $\quad = 0.167 = 16.7\%$

$\frac{3}{30} = \frac{1}{10}$ $\quad = 0.1 \quad = 10.0\%$

$\frac{4}{30} = \frac{2}{15}$ $\quad = 0.133 = 13.3\%$

$\qquad\qquad\qquad\qquad 100.0\%$

1.23

Number	Frequency
10	2
20	4
30	2
40	3
50	2
60	5
70	1
80	3
90	2
	24

<u>Relative Frequency</u>

$\frac{2}{24} = \frac{1}{12}$ $\quad = 0.083 = 8.3\%$

$\frac{4}{24} = \frac{1}{6}$ $\quad = 0.167 = 16.7\%$

$\frac{2}{24} = \frac{1}{12}$ $\quad = 0.083 = 8.3\%$

$\frac{3}{24} = \frac{1}{8}$ $\quad = 0.125 = 12.5\%$

$\frac{2}{24} = \frac{1}{12}$ $\quad = 0.083 = 8.3\%$

$\frac{5}{24} =$ $\qquad 0.208 = 20.8\%$

$\frac{1}{24} =$ $\qquad 0.042 = 4.2\%$

1.23 cont.

$\frac{3}{24} = \frac{1}{8}$ $\quad = 0.125 = 12.5\%$

$\frac{2}{24} = \frac{1}{12}$ $\quad = 0.083 = \frac{8.3\%}{99.9\%}$

1.24

Number	Frequency
110	1
111	4
112	3
113	1
114	2
115	2
116	3
117	1
118	2
119	4
	23

<u>Relative Frequency</u>

$\frac{1}{23} = 0.0435 = 4.3\%$

$\frac{4}{23} = 0.1739 = 17.4\%$

$\frac{3}{23} = 0.1304 = 13.0\%$

$\frac{1}{23} = 0.0435 = 4.3\%$

$\frac{2}{23} = 0.0870 = 8.7\%$

$\frac{2}{23} = 0.0870 = 8.7\%$

$\frac{3}{23} = 0.1304 = 13.0\%$

$\frac{1}{23} = 0.0435 = 4.3\%$

$\frac{2}{23} = 0.0870 = 8.7\%$

$\frac{4}{23} = 0.1739 = \underline{17.4\%}$

$\qquad\qquad\qquad\quad 99.8\%$

II. SECTION TWO

2.1 28

2.2 120

2.3 $8,000.00

2.4 none

2.5 1,500 and 1,550

2.6   140 and 170

2.7   3
4
17
23
29
median <u>17</u>

2.8   median <u>130</u>

2.9
3,000	3,016
3,000	3,017
3,015	3,033
3,016	3,098
3,016	

median <u>3,016</u>

2.10 $15 \div 2 = 7$ remainder 1;
$7 + 1 = $ 8th number

2.11 $100 \div 2 = 50$;
50th and 51st numbers

2.12 $12 \div 2 = 6$;
6th and 7th numbers

2.13 $37 \div 2 = 18$ remainder 1;
$18 + 1 = $ 19th number

2.14   5, 6, 12, 12, 12, 13, 14,
17, 18, 18, 18, 20, 23,
34, 46, 65
median $= \frac{17 + 18}{2} = \frac{35}{2} = 17\frac{1}{2}$
or 17.5

2.15 $2.27, $2.50, $3.19, $3.25,
$3.30, $4.50, $4.50, $13.75
median $= \frac{\$3.25 + \$3.30}{2} =$

$\frac{\$6.55}{2} = \$3.275$

2.16   median $\frac{7+8}{2} = \frac{15}{2} = 7\frac{1}{2}$ or 7.5

2.17   The set contains 22 numbers.
median $= 22 \div 2 = 11$;
the 11th and 12th numbers are
22.

2.18   The set contains 48 numbers.
median $= 48 \div 2 = 24$ the 24th
and 25th numbers are 300 and
400.
median $= \frac{300 + 400}{2} = \frac{700}{2} = 350$.

2.19   The set contains 57 numbers.
median $= 57 \div 2 = 28$ remainder 1;
$28 + 1 = $ 29th number.
median $= 6,030$.

2.20 $121 + 347 + 45 + 21 + 300 +$
$614 + 312 + 333 + 421 = 2,514$
$2,514 \div 9 = 279.3$

2.21 $64 + 65 + 63 + 66 + 61$
$65 + 66 + 68 + 69 + 66 +$
$63 = 716$
$716 \div 11 = 65.1$

2.22 $3.57 + $7.50 + $2.21 +
$4.57 + $6.60 + $3.25
$7.67 + $5.50 = $40.87
$40.87 \div 8 = 05.11$

2.23

Number	Frequency	Number x Frequency
1	3	1 x 3 = 3
2	5	2 x 5 = 10
3	7	3 x 7 = 21
4	9	4 x 9 = 36
5	7	5 x 7 = 35
6	5	6 x 5 = 30
7	<u>3</u>	7 x 3 = <u>21</u>
	Total 39	156

Mean $= 156 \div 39 = 4$

2.24

Number	Frequency	Number x Frequency
364	2	728
365	3	1,095
366	4	1,464
367	<u>10</u>	<u>3,670</u>
	Total 19	6,957

Mean $= 6,957 \div 19 = 366.2$

2.25

Number	Frequency	Number x Frequency
$ 6,000	7	$42,000
$ 7,000	7	$49,000
$ 8,000	7	$56,000
$ 9,000	7	$63,000
$10,000	7	$70,000
	35	$280,000

Mean = $280,000 ÷ 35 = $8,000

2.26 through 2.28 Any order

2.26 mean

2.27 median

2.28 mode

2.29 mean—sum of data points ÷ count

2.30 median—data point in middle

2.31 mode—most frequent element

2.32 a. Mean: 14 + 15 + 12 + 18 + 17 + 13 + 12 + 12 + 14 + 15 + 19 + 12 = 173
173 ÷ 12 = 14. 4

b. 12, 12, 12, 12, 13, 14, 14, 15, 15, 17, 18, 19
The set contains 12 numbers.
Median: 12 ÷ 2 = 6;
the 6th and 7th numbers are 14. Median = 14.

c. Mode 12

2.33 a.

Number	Frequency
$6.50	1
$7.00	5
$7.50	3
$8.00	8
$8.50	3
	20

Number x Frequency
$6.50 x 1 = $ 6.50
$7.00 x 5 =   35.00
$7.50 x 3 =   22.50
$8.00 x 8 =   64.00
$8.50 x 3 =   25.50
                $153.50

2.33 cont.

Mean = $153.50 ÷ 20 = $7.68

b. The set contains 20 numbers.
Median: 20 ÷ 2 = 10;
10th and 11th numbers are $8.00. Median = $8.00.

c. Mode = $8.00

2.34 a.

Number	Frequency
100	3
110	7
120	3
130	5
140	7
150	3
	28

Number x Frequency
100 x 3 =   300
110 x 7 =   770
120 x 3 =   360
130 x 5 =   650
140 x 7 =   980
150 x 3 =   450
              3, 510
Mean = 3,510 ÷ 28 = 125.4

b. The set contains 28 numbers. Median: 28 ÷ 2 = 14; the 14th and 15th numbers are both 130. Median = 130.

c. Mode = 110 and 140

2.35 Example:
The three measures are not always equal, but they can be. For instance, in the set 1, 2, 3, 4, 5, 4, 3, 2, 3, the mean, median, and mode are all 3.

2.36 24 − 2 = 22

2.37 179 − 100 = 79

2.38 $120.00 − $3.00 = $117.00

2.39 Range = 24 − 2 = 22.
Average deviation =
22 ÷ 12 = 1.8.

2.40 Range = 179 − 100 = 79.
Average deviation =
79 ÷ 8 = 9.9.

2.41 Range = $120.00 − $3.00 =
$117.00.
Average deviation =
$117.00 ÷ 7 = $16.71.

2.42

Number	a.	b.
90	|	1
91	|||	3
92	||	2
93	||||	4
94	|||	3
95	||	2
96	|||	3
97	|||||	5
98	|||	3
99	||||	4
		30

c.

$\frac{1}{30}$ =   0.033 = 3.3%

$\frac{3}{30}$ = $\frac{1}{10}$ =   0.1   = 10.0%

$\frac{2}{30}$ = $\frac{1}{15}$ =   0.067 = 6.7%

$\frac{4}{30}$ = $\frac{2}{15}$ =   0.133 = 13.3%

$\frac{3}{30}$ = $\frac{1}{10}$ =   0.1   = 10.0%

$\frac{2}{30}$ = $\frac{1}{15}$ =   0.067 = 6.7%

$\frac{3}{30}$ = $\frac{1}{10}$ =   0.1   = 10.0%

$\frac{5}{30}$ = $\frac{1}{6}$ =   0.167 = 16.7%

$\frac{3}{30}$ = $\frac{1}{10}$ =   0.1   = 10.0%

$\frac{4}{30}$ = $\frac{2}{15}$ =   0.133 = 13.3%

                              100.0%

d. Mode = 97

2.42 cont.

e. The set contains 30 numbers. Median:
30 ÷ 2 = 15; 15th and 16th numbers are 95 and 96.
Median = $\frac{95 + 96}{2}$ = $\frac{191}{2}$ = 95.5.

f.

Number	Frequency
90	1
91	3
92	2
93	4
94	3
95	2
96	3
97	5
98	3
99	4
	30

Number x
Frequency

90 x 1 =   90
91 x 3 =   273
92 x 2 =   184
93 x 4 =   372
94 x 3 =   282
95 x 2 =   190
96 x 3 =   288
97 x 5 =   485
98 x 3 =   294
99 x 4 =   396
          2,854

Mean = 2,854 ÷ 30 = 95.1

g. 99 − 90 = 9

h. Range = 9.
Average deviation =
9 ÷ 30 = 0.3.

2.43

Number	a.	b.
10	||	2
20	||||	4
30	||	2
40	|||	3
50	||	2
60	|||||	5
70	|	1
80	|||	3

Math 709 Answer Key

2.43 cont.

90 || $\frac{2}{24}$

c.

$\frac{2}{24}$ = $\frac{1}{12}$	= 0.083 =	8.3%
$\frac{4}{24}$ = $\frac{1}{6}$	= 0.167 =	16.7%
$\frac{2}{24}$ = $\frac{1}{12}$	= 0.083 =	8.3%
$\frac{3}{24}$ = $\frac{1}{8}$	= 0.125 =	12.5%
$\frac{2}{24}$ = $\frac{1}{12}$	= 0.083 =	8.3%
$\frac{5}{24}$ =	0.208 =	20.8%
$\frac{1}{24}$ =	0.042 =	4.2%
$\frac{3}{24}$ = $\frac{1}{8}$	= 0.125 =	12.5%
$\frac{2}{24}$ = $\frac{1}{12}$	= 0.083 =	8.3%

99.9%

d. Mode = 60

e. The set contains 24 numbers. Median = 24 ÷ 2 = 12; the 12th and 13th numbers are both 50. Median = 50.

f.
Number	Frequency
10	2
20	4
30	2
40	3
50	2
60	5
70	1
80	3
90	2
	24

Number x Frequency
10 x 2 = 20
20 x 4 = 80
30 x 2 = 60
40 x 3 = 120
50 x 2 = 100
60 x 5 = 300
70 x 1 = 70
80 x 3 = 240
90 x 2 = 180
1,170

2.43 cont.

Mean = 1,170 ÷ 24 = 48.8

g. 90 − 10 = 80
Average deviation = 80 ÷ 24 = 3.3.

2.44

Number	a.	b.
110	I	1
111	IIII	4
112	III	3
113	I	1
114	II	2
115	II	2
116	III	3
117	I	1
118	II	2
119	IIII	4
		23

c.

$\frac{1}{23}$ = 0.0434 = 4.3%
$\frac{4}{23}$ = 0.1739 = 17.4%
$\frac{3}{23}$ = 0.1304 = 13.0%
$\frac{1}{23}$ = 0.0434 = 4.3%
$\frac{2}{23}$ = 0.0870 = 8.7%
$\frac{2}{23}$ = 0.0870 = 8.7%
$\frac{3}{23}$ = 0.1304 = 13.0%
$\frac{1}{23}$ = 0.0434 = 4.3%
$\frac{2}{23}$ = 0.0870 = 8.7%
$\frac{4}{23}$ = 0.1739 = 17.4%

99.8%

2.44 cont.

d. Mode = 111 and 119

e. The set contains 23 numbers. Median: 23 ÷ 2 = 11 remainder 1; 11 + 1 = 12th number. Median = 115.

f.
Number	Frequency
110	1
111	4
112	3
113	1
114	2
115	2
116	3
117	1
118	2
119	4
	23

Number x Frequency
110 x 1 = 110
111 x 4 = 444
112 x 3 = 336
113 x 1 = 113
114 x 2 = 228
115 x 2 = 230
116 x 3 = 348
117 x 1 = 117
118 x 2 = 236
119 x 4 = 476
          2,638

Mean = 2,638 ÷ 23 = 114.7

g. 119 - 110 = 9

h. Range = 9
Average deviation = 9 ÷ 23 = 0.4

III. SECTION THREE

3.1

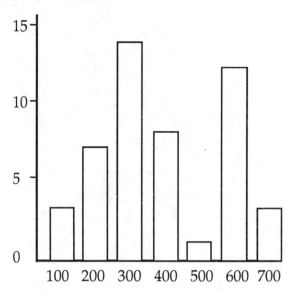

3.2

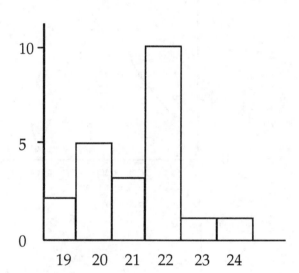

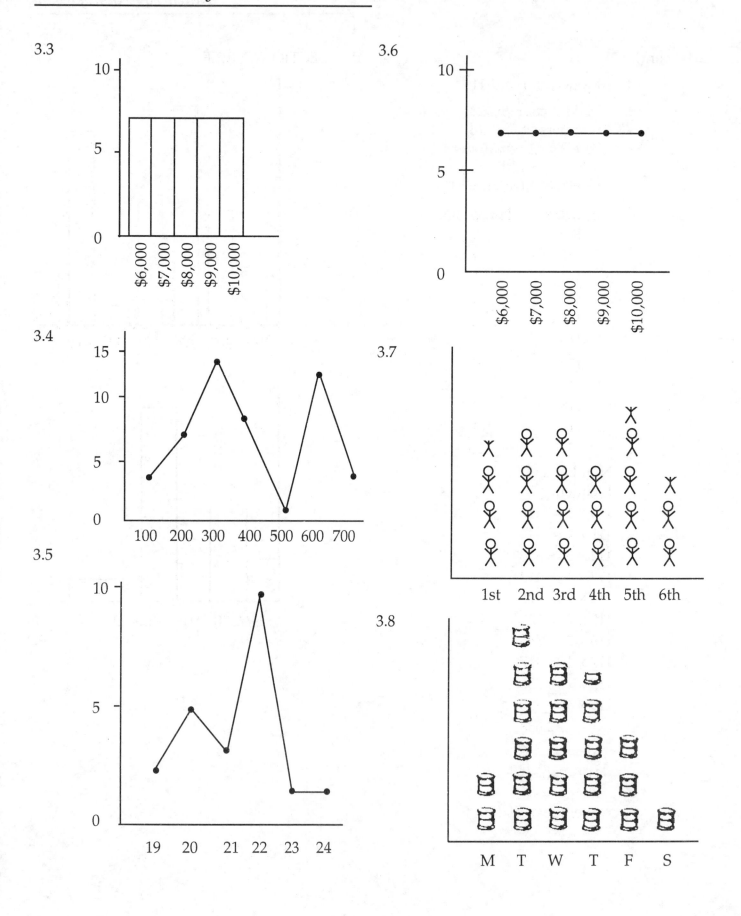

3.9

3.10

3.11 through 3.16

3.17

3.18

3.19

3.20

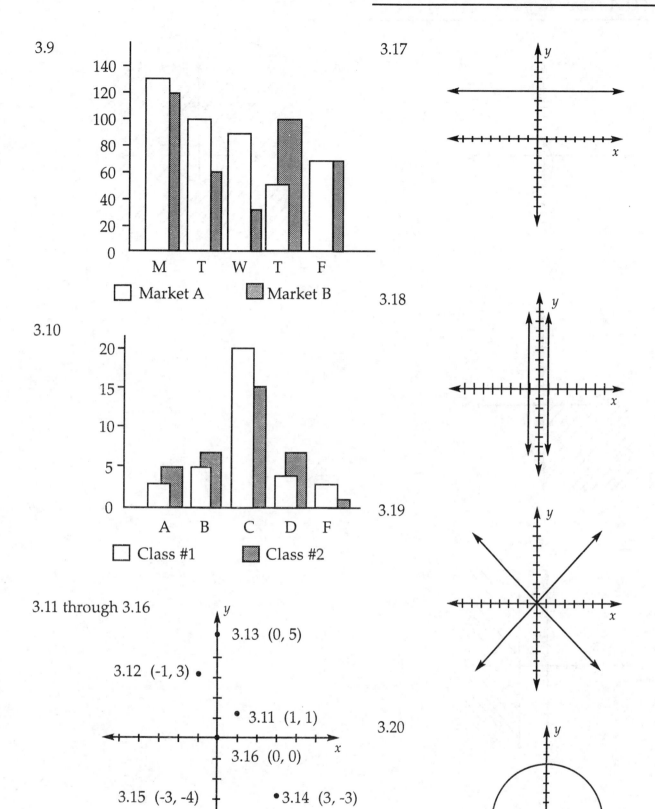

3.21

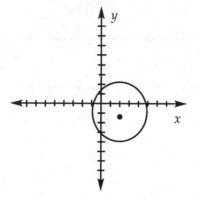

3.24

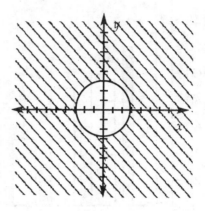

3.22

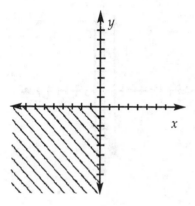

3.23

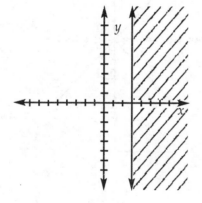

I.   SECTION ONE

1.1   6,000 + 500 + 10 + 3

1.2   100,000 + 60,000 + 7,000 + 400 + 10 + 9

1.3   300 + 2

1.4   8,000 + 80 + 1

1.5   30,000 + 4,000 + 700 + 90 + 8

1.6   700 + 7 = 707

1.7   82,000 + 300 + 40 = 82,340

1.8   600,000 + 9 = 600,009

1.9   a.   2,000 + 800 + 70 + 3
       b.   500 + 2
       c.   100 + 30 + 6
       d.   700 + 10 + 4

1.10   300,000 + 50,000 + 9,000 + 400 + 80 + 9

1.11   10,000 + 2,000 + 700 + 20 + 1

1.12   13 > 10

1.13   84 < 112

1.14   false

1.15   130

1.16   1,300

1.17   70

1.18   a.   7,000
       b.   6,800
       c.   6,760

1.19   a.   4,000
       b.   4,400
       c.   4,350

1.20   118

1.21   897

1.22   211

1.23   1,192,207

1.24   a. 183
       b. 16
       c. 13
       d. 68
       e. 36
       f. 40
       g. 25
       h. 9
       i. 17
       j. 85

1.25   a. $\underline{FTM} = 7 \times 1 = 7$
            $\underline{FGM} = 7 \times 2 = \underline{14}$
            21
       b. $\underline{FTM} = 1 \times 1 = 1$
            $\underline{FGM} = 4 \times 2 = \underline{8}$
            9
       c. $\underline{FTM} = 3 \times 1 = 3$
            $\underline{FGM} = 7 \times 2 = \underline{14}$
            17
       d. $\underline{FTM} = 0 \times 1 = 0$
            $\underline{FGM} = 11 \times 2 = \underline{22}$
            22
       e. $\underline{FTM} = 5 \times 1 = 5$
            $\underline{FGM} = 7 \times 2 = \underline{14}$
            19
       f.   162
       g.   27
       h.   16
       i.   57
       j.   36
       k.   32
       l.   17
       m. 7
       n.   10
       o.   88

1.26   a.   3 + 4 = 7
       b.   7 + 10 + 1 + 7 = 25
       c.   25 + 1 + 7 = 33
       d.   33 + 10 + 6 = 49
       e.   49 + 10 + 3 = 62
       f.   62 + 3 + 6 = 71
       g.   71 + 10 + 4 = 85
       h.   85 + 4 + 3 = 92
       i.   92 + 9 = 101
       j.   101 + 10 + 3 + 6 = 120

1.27   a. $3 + 4 = 7$
        b. $7 + 10 + 10 = 27$
        c. $27 + 10 + 7 = 44$
        d. $44 + 10 + 6 = 60$
        e. $60 + 10 + 10 = 80$
        f. $80 + 10 + 10 = 100$
        g. $100 + 10 + 4 = 114$
        h. $114 + 4 + 3 = 121$
        i. $121 + 9 = 130$
        j. $130 + 10 + 10 + 6 = 156$

1.28   300

1.29   The answer is 7,105; SOIL.

1.30   The answer is 376,608; BOGGLE.

1.31   112

1.32   4,988

1.33   6, 182

1.34   1,351

1.35   7

1.36   124

1.37   103

1.38   21

1.39   12

1.40   60

1.41   11

1.42   The answer is 608; BOG

1.43   255, 450

1.44   12,084

1.45   45,167

1.46   41,168

1.47 $18 \times 24 = 432$

1.48 $24 \times 17 = 408$

1.49 $15 \times 16 = 240$

1.50 $432 + 408 + 240 = 1,080$

1.51 $52 \times 6 = 312$ sq. ft.

1.52 $360 \times 225 = 81,000$ sq. ft.

1.53 $300 \times 165 = 49,500$ sq. ft.

1.54 $81,000 - 49,500 = 31,500$ sq. ft.

1.55   a. $40 \times 36 = 1,440$ minutes
        b. $1,440 \div 60 = 24$

1.56   The answer is 37,818; BIBLE.

1.57   The answer is 55,378; BLESS.

1.58

$$
\begin{array}{r}
6,531 \\
7{,}204\overline{)47{,}049{,}324} \\
\underline{43224} \\
38253 \\
\underline{36020} \\
22332 \\
\underline{21612} \\
7204 \\
\underline{7204} \\
0
\end{array}
$$

1.59

$$
\begin{array}{r}
31,405 \\
641\overline{)20{,}130{,}605} \\
\underline{1923} \\
900 \\
\underline{641} \\
2596 \\
\underline{2564} \\
3205 \\
\underline{3205} \\
0
\end{array}
$$

1.60

$$
\begin{array}{r}
3,284 \\
96\overline{)315{,}264} \\
\underline{288} \\
272 \\
\underline{192} \\
806 \\
\underline{768} \\
384 \\
\underline{384} \\
0
\end{array}
$$

1.61
$$
\begin{array}{r}
56 \\
5{,}413\overline{)303{,}128} \\
\underline{27065} \\
32478 \\
\underline{32478} \\
0
\end{array}
$$

1.62 $98 \div 7 = 14$ yd.

1.63 $5{,}280 \div 11 = 480$ ft.

1.64 a. $78 + 71 = 149$
         $149 \div 12 = 12\frac{5}{12}$
      b. yes: 5

1.65 $324 \div 18 = 18$ points

1.66 $2{,}808 \div 36 = 78$ ft.

1.67 $16 \times 3 = 48$ games.
       $5{,}904 \div 48 = 123$

1.68 The answer is 938; BEG.

1.69 The answer is 710; OIL.

II. SECTION TWO

2.1   Answers will vary but should be close to 1 inch.

2.2   Answers should be close to $3\frac{1}{4}$ inches.

2.3 $3\frac{1}{4}$ inches.

2.4   a. $\approx 2\frac{1}{4}''$
      b. $2\frac{1}{4}''$

2.5   a. $\approx 5''$
      b. $5''$

2.6   a. $\approx 3\frac{1}{2}''$
      b. $3\frac{1}{2}''$

2.7   a. $\approx 6''$
      b. $6''$

2.8   Answers will vary.

2.9 through 2.11
      Answers for a. will vary but should be fairly close to actual measures for b.

2.12 Answers will vary.

2.13 through 2.15
      Answers for a. will vary but should be fairly close to actual measures for b.

2.16 Answers will vary.

2.17 through 2.19
      Answers for a. will vary but should be fairly close to actual measures for b.

2.20 Answers will vary.

2.21 through 2.23
      Answers will vary.

2.24 Answers will vary.

2.25 through 2.27
      Answers for a. will vary but should be fairly close to actual measures for b.

2.28 Answers will vary.

2.29 through 2.31
      Answers will vary.

2.32 This angle is usually about 30°.

2.33 This angle is usually about 45°.

2.34 This angle is usually about 60°.

2.35 a. $\approx 19°$
      b. 19°

2.36 a. $\approx 50°$
      b. 50°

2.37 a. $\approx 74°$
      b. 74°

2.38   a. $\approx 125°$
        b. $125°$

(2.39–2.44   Note: Answers may vary slightly)

2.39   6 inches; 14.9 cm; 90°;
        25°; 65°

2.40 $6\frac{1}{4}$ inches; 15.5 cm; 140°;
        25°; 15°

2.41 $5\frac{3}{4}$ inches; 14.7 cm; 60°;
        60°; 60°

2.42 $7\frac{3}{4}$ inches; 19.8 cm; 111°;
        69°; 111°; 69°

2.43 $7\frac{1}{4}$ inches; 18.4 cm; 131°;
        49°; 69°; 111°

2.44 $7\frac{1}{4}$ inches; 18.4 cm; 118°;
        62°; 118°; 62°

2.45   b

2.46   e

2.47   a

2.48   c

2.49   6 + 6 + 52 + 52 = 12 + 104 =
        116 feet

2.50   20 + 20 + 44 + 44 = 40 + 88 =
        128 inches

2.51   60 + 60 + 60 + 60 = 240 feet.

2.52   a.   49 + 10 + 48 + 22 + 72 +
           48 + 72 + 10 + 49 + 22 =
           402
        b. $402 \div 12 = 33\frac{1}{2} = 33$ feet,
           6 inches

2.53 $C = \pi \times d$
        $C = 3.14 \times 6$
        $C = 18.84$ feet

2.54 $A = \pi \times r^2$
        $r = 6 \div 2 = 3$
        $A = 3.14 \times 3^2$
        $A = 3.14 \times 9$
        28.26 sq. feet

2.55 $(8 \times 12) + 2\frac{1}{2} = 98\frac{1}{2}$

2.56 $C = \pi \times d$
        $C = 3.14 \times 98\frac{1}{2}$
        $C = 3.14 \times \frac{197}{2}$
        $C = 1.57 \times 197$
        $C = 309.29$ inches

2.57 $C = \pi \times d$
        $C = 3.14 \times 7$
        $C = 21.98$ feet

2.58 $A = \pi \times r^2$
        $r = 7 \div 2 = 3.5$
        $A = 3.14 \times 3.5^2$
        $A = 3.14 \times 12.25$
        $A = 38.465$ sq. ft.

2.59 $\notin$

2.60 $\notin$

2.61 $\in$

2.62 $\notin$

2.63 $\in$

2.64 $\in$

2.65   {Kevin, Greg, Mike}

2.66 ϕ

2.67   {David, Sam, Jerry}

2.68   {David, Pat, Sam, Sue,
        Jerry, Cheryl, Nancy,
        Vicki, Dawn}

2.69   {Willie, John, Kevin,
        Greg, Jeff, Fred, Richard,
        Mike, Sam, David, Pat,
        Sue, Jerry}

2.70

2.71

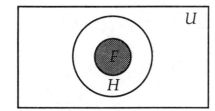

2.72

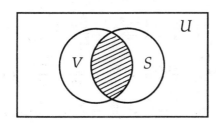

2.73   a. 1, 5

         b. 101_2

2.74   a. 3, 3

         b. 1001_2

2.75   a. 2, 3

         b. 110_2

2.76   a. 2, 5

         b. 1010_2

2.77   2, 2, 2, 2, 3

2.78 $14 = 2 \cdot 7$

         $24 = 2 \cdot 2 \cdot 2 \cdot 3$

         $LCM = 2 \cdot 7 \cdot 2 \cdot 2 \cdot 3 = 168$

2.79 $51 = ③ \cdot 17$

         $69 = ③ \cdot 23$

         $GCF = 3$

2.80 $18 = 2 \cdot 3 \cdot 3$

         $36 = 2 \cdot 2 \cdot 3 \cdot 3$

         $GCF = 2 \cdot 3 \cdot 3 = 18$

2.81 $12 = 2 \cdot 2 \cdot 3$

         $18 = 2 \cdot 3 \cdot 3$

         $30 = 2 \cdot 3 \cdot 5$

         $LCM = 2 \cdot 2 \cdot 3 \cdot 3 \cdot 5 = 180$

2.82   a. Yes; the last digit, 2, is evenly divisible by 2.

         b. No; the sum of the digits = 3 + 4 + 7 + 1 + 2 = 17 and 17 is not divisible by 3.

         c. No; the last digit is not 0 or 5.

2.83   a. No; the last digit is not evenly divisible by 2.

         b. Yes; the sum of the digits = 6 + 1 + 2 + 3 + 8 + 1 = 21 and 21 is divisible by 3.

         c. No; the last digit is not 0 or 5.

2.84   a. No; the last digit is not evenly divisible by 2.

         b. No; the sum of the digits = 2 + 4 + 2 + 6 + 5 = 19 and 19 is not divisible by 3.

         c. Yes; the last digit is 5.

III. SECTION THREE

3.1   a. $\frac{1}{2} + \frac{1}{9} = \frac{9}{18} + \frac{2}{18} = \frac{11}{18}$

       b. $\frac{1}{2} - \frac{1}{9} = \frac{9}{18} - \frac{2}{18} = \frac{7}{18}$

       c. $\frac{1}{2} \times \frac{1}{9} = \frac{1}{18}$

       d. $\frac{1}{2} \div \frac{1}{9} = \frac{1}{2} \times \frac{9}{1} = \frac{9}{2}$ or $4\frac{1}{2}$

3.2 $3\frac{1}{8} = \frac{(3 \times 8) + 1}{8} = \frac{25}{8}$

       a. $\frac{1}{2} + \frac{25}{8} = \frac{4}{8} + \frac{25}{8} = \frac{29}{8}$ or $3\frac{5}{8}$

       b. $\frac{25}{8} - \frac{1}{2} = \frac{25}{8} - \frac{4}{8} = \frac{21}{8}$ or $2\frac{5}{8}$

       c. $\frac{1}{2} \times \frac{25}{8} = \frac{25}{16}$ or $1\frac{9}{16}$

       d. $\frac{1}{2} \div \frac{25}{8} = \frac{1}{\cancel{2}_1} \times \frac{\cancel{8}^4}{25} = \frac{4}{25}$

3.3   a. $\frac{1}{2} + 6 = 6\frac{1}{2}$

       b. $6 - \frac{1}{2} = \frac{12}{2} - \frac{1}{2} = \frac{11}{2}$ or $5\frac{1}{2}$

       c. $6 \times \frac{1}{2} = \frac{\cancel{6}^3}{1} \times \frac{1}{\cancel{2}_1} = \frac{3}{1} = 3$

       d. $6 \div \frac{1}{2} = \frac{6}{1} \times \frac{2}{1} = \frac{12}{1} = 12$

3.4 a. $\frac{1}{2} + \frac{2}{3} = \frac{3}{6} + \frac{4}{6} = \frac{7}{6}$ or $1\frac{1}{6}$

b. $\frac{2}{3} - \frac{1}{2} = \frac{4}{6} - \frac{3}{6} = \frac{1}{6}$

c. $\frac{1}{\cancel{2}_1} \times \frac{\cancel{2}^1}{3} = \frac{1}{3}$

d. $\frac{1}{2} \div \frac{2}{3} = \frac{1}{2} \times \frac{3}{2} = \frac{3}{4}$

3.5 a. $\frac{1}{2} + \frac{4}{5} = \frac{5}{10} + \frac{8}{10} = \frac{13}{10}$ or $1\frac{3}{10}$

b. $\frac{4}{5} - \frac{1}{2} = \frac{8}{10} - \frac{5}{10} = \frac{3}{10}$

c. $\frac{1}{\cancel{2}_1} \times \frac{\cancel{4}^2}{5} = \frac{2}{5}$

d. $\frac{1}{2} \div \frac{4}{5} = \frac{1}{2} \times \frac{5}{4} = \frac{5}{8}$

3.6 $2\frac{2}{3} = \frac{(2 \times 3) + 2}{3} = \frac{8}{3}$

a. $\frac{8}{3} + \frac{1}{9} = \frac{24}{9} + \frac{1}{9} = \frac{25}{9}$ or $2\frac{7}{9}$

b. $\frac{8}{3} - \frac{1}{9} = \frac{24}{9} - \frac{1}{9} = \frac{23}{9}$ or $2\frac{5}{9}$

c. $\frac{8}{3} \times \frac{1}{9} = \frac{8}{27}$

d. $\frac{8}{3} \div \frac{1}{9} = \frac{8}{\cancel{3}_1} \times \frac{\cancel{9}^3}{1} = \frac{24}{1} = 24$

3.7 $3\frac{1}{8} = \frac{(3 \times 8) + 1}{8} = \frac{25}{8}$

a. $\frac{8}{3} + \frac{25}{8} = \frac{64}{24} + \frac{75}{24} = \frac{139}{24}$ or $5\frac{19}{24}$

b. $\frac{25}{8} - \frac{8}{3} = \frac{75}{24} - \frac{64}{24} = \frac{11}{24}$

c. $\frac{\cancel{8}^1}{3} \times \frac{25}{\cancel{8}_1} = \frac{25}{3}$ or $8\frac{1}{3}$

d. $\frac{8}{3} \div \frac{25}{8} = \frac{8}{3} \times \frac{8}{25} = \frac{64}{75}$

3.8 a. $\frac{8}{3} + 6 = \frac{8}{3} + \frac{18}{3} = \frac{26}{3}$ or $8\frac{2}{3}$

b. $6 - \frac{8}{3} = \frac{18}{3} - \frac{8}{3} = \frac{10}{3}$ or $3\frac{1}{3}$

c. $\frac{8}{3} \times 6 = \frac{8}{\cancel{3}_1} \times \frac{\cancel{6}^2}{1} = \frac{16}{1} = 16$

d. $\frac{8}{3} \div 6 = \frac{\cancel{8}^4}{3} \times \frac{1}{\cancel{6}_3} = \frac{4}{9}$

3.9 a. $\frac{8}{3} + \frac{2}{3} = \frac{10}{3}$ or $3\frac{1}{3}$

b. $\frac{8}{3} - \frac{2}{3} = \frac{6}{3} = 2$

c. $\frac{8}{3} \times \frac{2}{3} = \frac{16}{9}$ or $1\frac{7}{9}$

d. $\frac{8}{3} \div \frac{2}{3} = \frac{\cancel{8}^4}{\cancel{3}_1} \times \frac{\cancel{3}^1}{\cancel{2}_1} = \frac{4}{1} = 4$

3.10 a. $\frac{8}{3} + \frac{4}{5} = \frac{40}{15} + \frac{12}{15} = \frac{52}{15}$ or $3\frac{7}{15}$

b. $\frac{8}{3} - \frac{4}{5} = \frac{40}{15} - \frac{12}{15} = \frac{28}{15}$ or $1\frac{13}{15}$

c. $\frac{8}{3} \times \frac{4}{5} = \frac{32}{15}$ or $2\frac{2}{15}$

d. $\frac{8}{3} \div \frac{4}{5} = \frac{\cancel{8}^2}{3} \times \frac{5}{\cancel{4}_1} = \frac{10}{3}$ or $3\frac{1}{3}$

3.11 a. $7 + \frac{1}{9} = 7\frac{1}{9}$

b. $7 - \frac{1}{9} = \frac{63}{9} - \frac{1}{9} = \frac{62}{9}$ or $6\frac{8}{9}$

c. $7 \times \frac{1}{9} = \frac{7}{1} \times \frac{1}{9} = \frac{7}{9}$

d. $7 \div \frac{1}{9} = \frac{7}{1} \times \frac{9}{1} = \frac{63}{1} = 63$

3.12 $3\frac{1}{8} = \frac{(3 \times 8) + 1}{8} = \frac{25}{8}$

a. $7 + \frac{25}{8} = 7 + 3\frac{1}{8} = 10\frac{1}{8}$

b. $7 - \frac{25}{8} = \frac{56}{8} - \frac{25}{8} = \frac{31}{8}$ or $3\frac{7}{8}$

c. $7 \times \frac{25}{8} = \frac{7}{1} \times \frac{25}{8} = \frac{175}{8}$ or $21\frac{7}{8}$

d. $7 \div \frac{25}{8} = \frac{7}{1} \times \frac{8}{25} = \frac{56}{25}$ or $2\frac{6}{25}$

3.13 a. $7 + 6 = 13$

b. $7 - 6 = 1$

c. $7 \times 6 = 42$

d. $7 \div 6 = \frac{7}{6}$ or $1\frac{1}{6}$

3.14 a. $7 + \frac{2}{3} = 7\frac{2}{3}$

b. $7 - \frac{2}{3} = \frac{21}{3} - \frac{2}{3} = \frac{19}{3}$ or $6\frac{1}{3}$

c. $7 \times \frac{2}{3} = \frac{7}{1} \times \frac{2}{3} = \frac{14}{3}$ or $4\frac{2}{3}$

d. $7 \div \frac{2}{3} = \frac{7}{1} \times \frac{3}{2} = \frac{21}{2}$ or $10\frac{1}{2}$

3.15 a. $7 + \frac{4}{5} = 7\frac{4}{5}$

b. $7 - \frac{4}{5} = \frac{35}{5} - \frac{4}{5} = \frac{31}{5}$ or $6\frac{1}{5}$

c. $7 \times \frac{4}{5} = \frac{7}{1} \times \frac{4}{5} = \frac{28}{5}$ or $5\frac{3}{5}$

d. $7 \div \frac{4}{5} = \frac{7}{1} \times \frac{5}{4} = \frac{35}{4}$ or $8\frac{3}{4}$

3.16 a. $\frac{3}{10} + \frac{1}{9} = \frac{27}{90} + \frac{10}{90} = \frac{37}{90}$

b. $\frac{3}{10} - \frac{1}{9} = \frac{27}{90} - \frac{10}{90} = \frac{17}{90}$

c. $\frac{\overset{1}{3}}{10} \times \frac{1}{\underset{3}{9}} = \frac{1}{30}$

d. $\frac{3}{10} \div \frac{1}{9} = \frac{3}{10} \times \frac{9}{1} = \frac{27}{10}$ or $2\frac{7}{10}$

3.17 $3\frac{1}{8} = \frac{(3 \times 8) + 1}{8} = \frac{25}{8}$

a. $\frac{3}{10} + \frac{25}{8} = \frac{12}{40} + \frac{125}{40} = \frac{137}{40}$ or $3\frac{17}{40}$

b. $\frac{25}{8} - \frac{3}{10} = \frac{125}{40} - \frac{12}{40} = \frac{113}{40}$ or $2\frac{33}{40}$

c. $\frac{3}{\underset{2}{10}} \times \frac{\overset{5}{25}}{8} = \frac{15}{16}$

d. $\frac{3}{10} \div \frac{25}{8} = \frac{3}{\underset{5}{10}} \times \frac{\overset{4}{8}}{25} = \frac{12}{125}$

3.18 a. $\frac{3}{10} + 6 = 6\frac{3}{10}$

b. $6 - \frac{3}{10} = \frac{60}{10} - \frac{3}{10} = \frac{57}{10}$ or $5\frac{7}{10}$

c. $\frac{3}{10} \times 6 = \frac{3}{\underset{5}{10}} \times \frac{\overset{3}{6}}{1} = \frac{9}{5}$ or $1\frac{4}{5}$

d. $\frac{3}{10} \div 6 = \frac{3}{10} \times \frac{1}{\underset{2}{6}} = \frac{1}{20}$

3.19 a. $\frac{3}{10} + \frac{2}{3} = \frac{9}{30} + \frac{20}{30} = \frac{29}{30}$

b. $\frac{2}{3} - \frac{3}{10} = \frac{20}{30} - \frac{9}{30} = \frac{11}{30}$

c. $\frac{\overset{1}{3}}{\underset{5}{10}} \times \frac{\overset{1}{2}}{\underset{1}{3}} = \frac{1}{5}$

d. $\frac{3}{10} \div \frac{2}{3} = \frac{3}{10} \times \frac{3}{2} = \frac{9}{20}$

3.20 a. $\frac{3}{10} + \frac{4}{5} = \frac{3}{10} + \frac{8}{10} = \frac{11}{10}$ or $1\frac{1}{10}$

b. $\frac{4}{5} - \frac{3}{10} = \frac{8}{10} - \frac{3}{10} = \frac{5}{10} = \frac{1}{2}$

c. $\frac{3}{\underset{5}{10}} \times \frac{\overset{2}{4}}{5} = \frac{6}{25}$

d. $\frac{3}{10} \div \frac{4}{5} = \frac{3}{\underset{2}{10}} \times \frac{\overset{1}{5}}{4} = \frac{3}{8}$

3.21 a. $\frac{4}{5} + \frac{1}{9} = \frac{36}{45} + \frac{5}{45} = \frac{41}{45}$

b. $\frac{4}{5} - \frac{1}{9} = \frac{36}{45} - \frac{5}{45} = \frac{31}{45}$

c. $\frac{4}{5} \times \frac{1}{9} = \frac{4}{45}$

d. $\frac{4}{5} \div \frac{1}{9} = \frac{4}{5} \times \frac{9}{1} = \frac{36}{5}$ or $7\frac{1}{5}$

3.22 $3\frac{1}{8} = \frac{(3 \times 8) + 1}{8} = \frac{25}{8}$

a. $\frac{4}{5} + \frac{25}{8} = \frac{32}{40} + \frac{125}{40} = \frac{157}{40}$ or $3\frac{37}{40}$

b. $\frac{25}{8} - \frac{4}{5} = \frac{125}{40} - \frac{32}{40} = \frac{93}{40}$ or $2\frac{13}{40}$

c. $\frac{\overset{1}{4}}{5} \times \frac{\overset{5}{25}}{\underset{2}{8}} = \frac{5}{2}$ or $2\frac{1}{2}$

d. $\frac{4}{5} \div \frac{25}{8} = \frac{4}{5} \times \frac{8}{25} = \frac{32}{125}$

3.23 a. $\frac{4}{5} + 6 = 6\frac{4}{5}$

b. $6 - \frac{4}{5} = \frac{30}{5} - \frac{4}{5} = \frac{26}{5}$ or $5\frac{1}{5}$

c. $\frac{4}{5} \times 6 = \frac{4}{5} \times \frac{6}{1} = \frac{24}{5}$ or $4\frac{4}{5}$

d. $\frac{4}{5} \div 6 = \frac{\overset{2}{4}}{5} \times \frac{1}{\underset{3}{6}} = \frac{2}{15}$

3.24 a. $\frac{4}{5} + \frac{2}{3} = \frac{12}{15} + \frac{10}{15} = \frac{22}{15}$ or $1\frac{7}{15}$

b. $\frac{4}{5} - \frac{2}{3} = \frac{12}{15} - \frac{10}{15} = \frac{2}{15}$

c. $\frac{4}{5} \times \frac{2}{3} = \frac{8}{15}$

d. $\frac{4}{5} \div \frac{2}{3} = \frac{\overset{2}{4}}{5} \times \frac{3}{\underset{1}{2}} = \frac{6}{5}$ or $1\frac{1}{5}$

3.25 a. $\frac{4}{5} + \frac{4}{5} = \frac{8}{5}$ or $1\frac{3}{5}$

b. $\frac{4}{5} - \frac{4}{5} = 0$

c. $\frac{4}{5} \times \frac{4}{5} = \frac{16}{25}$

d. $\frac{4}{5} \div \frac{4}{5} = \frac{\overset{1}{4}}{5} \times \frac{\overset{1}{5}}{\underset{1}{4}} = 1$

3.26

```
        0.6831
606)414.0000
    3636
     5040
     4848
      1920
      1818
       1020
        606
        414
```

$0.6831 = 0.683$ rounded to three decimal places. $0.683 \times 1{,}000 = 683$

3.27

```
       0.3156
453)143.0000
    1359
     710
     453
    2570
    2265
    3050
    2718
     332
```

0.3156 = 0.316 rounded to
three decimal places.
0.316 x 1,000 = 316

3.28

```
       0.4113
158)65.0000
    632
    180
    158
    220
    158
    620
    474
    146
```

0.4113 = 0.411 rounded to
three decimal places.
0.411 x 1,000 = 411

3.29

```
        0.2837
652)185.0000
    1304
    5460
    5216
    2440
    1956
    4840
    4564
     276
```

0.2837 = 0.284 rounded to
three decimal places.
0.284 x 1,000 = 284

3.30

```
        0.240
125)30.000
    250
    500
    500
      0
```

0.240 x 1,000 = 240

3.31

```
      0.4426
61)27.0000
   244
   260
   244
   160
   122
   380
   366
    14
```

0.4426 = 0.443 rounded to
three decimal places.
0.443 x 1,000 = 443

3.32

```
      0.4354
62)27.0000
   248
   220
   186
   340
   310
   300
   248
    52
```

0.4354 = 0.435 rounded
to three decimal places.
0.435 x 1,000 = 435

3.33

```
      0.4444
63)28.0000
   252
   280
   252
   280
   252
   280
   252
    28
```

0.4444 = 0.444 rounded
to three decimal places.
0.444 x 1,000 = 444

3.34

a. 63 + 1 = 64
b. 28 + 0 = 28

```
        .4375
64)28.0000
   256
   240
   192
   480
   448
   320
   320
     0
```

3.34 (cont.)

0.4375 = 0.438 rounded to three decimal places.
0.438 x 1,000 = 438

3.35
a. 64 + 1 = 65
b. 28 + 0 = 28
c.
```
      0.4307
65)28.0000
   260
    200
    195
      500
      455
       45
```

0.4307 = 0.431 rounded to three decimal places.
0.431 x 1,000 = 431

3.36 23%

3.37 42%

3.38 780%

3.39 1.5%

3.40 0.49

3.41 0.06

3.42 0.002

3.43 1.5

3.44
a. 58
b. 20%
c. 11.6

3.45
a. 85
b. 67%
c. 57

3.46
a. 105
b. 95%
c. 99.75

3.47
a. 72
b. 50%
c. 36

3.48
percentage = 43
rate = 82% = 0.82
base = 43 ÷ 0.82 = 52.4

```
        52.43          = 52.4 rounded
0.82)43.00.00            to the nearest
     410                 tenth
      200
      164
       360
       328
       320
       246
        74
```

3.49
percentage = 480
rate = 39% = 0.39
base = 480 ÷ 0.39 - 1,230.8

```
        1230.76    =   1,230.8
0.39)480.00.00          rounded to
     39                 the nearest
      90                tenth
      78
      120
      117
       300
       273
       270
       234
        36
```

3.50
percentage = 89
rate = 5% = 0.05
base 89 ÷ 0.05 = 1,780

```
        1780.0   =    1,780
0.05)89.00.0
     5
     39
     35
      40
      40
       0
```

3.51   percentage = 16
        rate = 18% = 0.18
        base = 16 ÷ 0.18 = 88.9

```
            88. 88  =   88.9 rounded
      0.18)16.00.00      to the near-
            144         est tenth
            160
            144
            160
            144
            160
            144
             16
```

3.52   percentage = 19
        base = 45
        rate = 19 ÷ 45 = 0.422 = 42.2%

```
         0.4222    =   0.422 rounded to
      45)19.0000       three decimal
         180          places
         100
          90
         100
          90
         100
          90
          10
```

3.53   percentage = 12
        base = 112
        rate = 12 ÷ 112 = 0.107 =
                          10.7%

```
          0.1071   =   0.107 rounded to
      112)12.0000      three decimal
          112         places
          800
          784
          160
          112
           48
```

3.54   percentage = 76
        base = 70
        rate 76 ÷ 70 = 1.086 = 108.6%

```
           1.0857   =   1.086 rounded to
      70)76.0000        three decimal
         70            places
         600
         560
         400
         350
         500
         490
          10
```

3.55   percentage = 9
        base = 236
        rate = 9 ÷ 236 = 0.038 = 3.8%

```
          0.0381   =   0.038 rounded to
      236)9.0000       three decimal
          708         places
         1920
         1888
          320
          236
           84
```

3.56   base = 39
        rate = 29% = 0.29
        percentage = 39 x 0.29 = 11.31

```
         0.29
       x 39
        261
         87
      11.31
```

3.57   base = 248
        rate = 88% = 0.88
        percentage = 248 x 0.88 = 218.24

```
         248
       x 0.88
        1984
       1984
      218.24
```

3.58   base = 9
        rate = 2% = 0.02
        percentage = 9 x 0.02 = 0.18

```
       0.02
       x 9
      0.18
```

3.59   base = 99
rate = 10% = 0.1
percentage = 99 x 0.1 = 9.9
```
  99
x 0.1
 9.9
```

3.60   a.
```
  0.8333
6)5.0000
  48
  20
  18
  20
  18
  20
  18
   2
```
0.8333 = 0.833
rounded to three
decimal places.
0.833 = 83.3%

b.
```
   0.5789
19)11.0000
   95
   150
   133
   170
   152
   180
   171
     9
```
0.5789 = 0.579
rounded to three
decimal places.
0.579 = 57.9%

c. _FTM_ = 5 x 1 =   5
FGM = 11 x 2 = 22
                 27

3.61   a. 3 ÷ 3 = 1
1 = 100%
b.
```
  0.4444
9)4.0000
  36
  40
  36
  40
  36
  40
  36
   4
```

3.61 (cont.)
0.4444 = 0.444
rounded to three
decimal places.
0.444 = 44.4%
c. _FTM_ = 3 x 1 = 3
FGM = 4 x 2 = 8
                11

3.62   a.
```
  0.7
10)7.0
   70
    0
```
0.7 = 70%
b.
```
   0.4736
19)9.0000
   76
   140
   133
    70
    57
   130
   114
    16
```
0.4736 = 0.474
rounded to three decimal
places.
0.474 = 47.4%
c. _FTM_ = 7 x 1 =   7
FGM = 9 x 2 = 18
                 25

3.63   a.
```
  0.75
4)3.00
  28
  20
  20
   0
```
0.75 = 75%
b.
```
  0.3333
6)2.0000
  18
  20
  18
  20
  18
  20
  18
   2
```
0.3333 = 0.333
rounded to three
decimal places.

267

3.63 (cont.)

$0.333 = 33.3\%$

c. $\underline{FTM} = 3 \times 1 = 3$
$\underline{FGM} = 2 \times 2 = \underline{4}$
$ 7$

3.64

a. $3 \div 3 = 1$
$1 = 100\%$

b.
```
    0.4375
16)7.0000
    64
    60
    48
    120
    112
     80
     80
      0
```
$0.4375 = 0.438$ rounded
to three decimal places.
$0.438 = 43.8\%$

c. $\underline{FTM} = 3 \times 1 = 3$
$\underline{FGM} = 7 \times 2 = \underline{14}$
$ 17$

3.65

a. 0

b.
```
   0.25
4)1.00
   8
   20
   20
    0   0.25 = 25%
```

c. $\underline{FTM} = 0 \times 1 = 0$
$\underline{FGM} = 1 \times 2 = \underline{2}$
$ 2$

3.66

a. $4 \div 4 = 1$
$1 = 100\%$

b.
```
    0.3846
13)5.0000
    39
    110
    104
     60
     52
     80
     78
      2
```
$0.3846 = 0.385$ rounded
to three decimal places.
$0.385 = 38.5\%$

3.66 (cont.)

c. $\underline{FTM} = 4 \times 1 = 4$
$\underline{FGM} = 5 \times 2 = \underline{10}$
$ 14$

3.67

a. 25
b. 30
c. 39
d. 86
e. $27 + 11 + 25 + 7 +$
$17 + 2 + 14 = 103$

3.68
```
    0.8333
30)25.0000
    240
    100
     90
    100
     90
    100
     90
     10
```
$0.8333 = 0.833$ rounded
to three decimal places.
$0.833 = 83.3\%$

3.69
```
    0.4534
86)39.0000
    344
    460
    430
    300
    258
    420
    344
     76
```
$0.4534 = 0.453$ rounded
to three decimal places.
$0.453 = 45.3\%$

3.70

$12 + 14.5 + 3.5 + 27 = 57.0$ yd.
```
   14.25  = average yards per
4)57.00     carry
   4
   17
   16
   10
    8
   20
   20
    0
```

3.71   53 + 9 + 1.5 + 6.5 + 4 +
61 = 135.0 yd.

```
  22.5  = average yards per
6)135.0    carry
  12
  15
  12
   30
   30
    0
```

3.72   93 + 3 + 6 = 102 yd.

```
  34   = average yards per
3)102     carry
  9
  12
  12
   0
```

3.73   4.5 + 6 + 3 + 12 + 19 +
3 + 1 + 6 + 3 = 57.5 yd.

```
  6.38 = 6.4 average yards
9)57.50   per carry (rounded
  54      to the nearest
  35      tenth)
  27
   80
   72
    8
```

3.74   19.8 + 19.9 + 21.3 = 61.0

```
  20.33 = 20.3 rounded to the
3)61.00            nearest tenth
  6
  010
   9
   10
    9
    1
```

3.75   20.1 + 19.3 + 20.0 = 59.4

```
  19.8
3)59.4
  3
  29
  27
   24
   24
    0
```

3.76   19.3 + 19.7 + 19.3 = 58.3

```
  19.43 = 19.4 rounded to the
3)58.30          nearest tenth
  3
  28
  27
   13
   12
   10
    9
    1
```

3.77   18.9 + 19.7 + 20.3 = 58.9

```
  19.63 = 19.6 rounded to the
3)58.90          nearest tenth
  3
  28
  27
   19
   18
   10
    9
    1
```

3.78   21.4 + 21.0 + 20.6 + 63.0

```
  21.0
3)63.0
  6
  03
   3
   0
```

3.79   19.3 + 19.1 + 20.1 = 58.5

```
  19.5
3)58.5
  3
  28
  27
   15
   15
    0
```

3.80   Pat; she ran the dash
in the shortest time
(on the average).

IV.   SECTION FOUR

4.1   e

4.2   a

4.3   d

4.4   g

4.5   b

4.6   c

4.7

$$\begin{array}{r} 7.44 \\ 19.3\overline{)143.6.00} \\ \underline{1351} \\ 856 \\ \underline{772} \\ 840 \\ \underline{772} \\ 68 \end{array}$$

= 7.4 rounded to the nearest tenth

4.8

$mpg = m \div g$

$g = m \div mpg$

$$\begin{array}{r} 11.22 \\ 19.1.\overline{)214.4.00} \\ \underline{191} \\ 234 \\ \underline{191} \\ 430 \\ \underline{382} \\ 480 \\ \underline{382} \\ 98 \end{array}$$

= 11.2 rounded to the nearest tenth

4.9

$mpg = m \div g$

$mpg \times g = m$

$\quad m = mpg \times g$

$$\begin{array}{r} 12.8 \\ \times\ 12.3 \\ \hline 384 \\ 256\ \ \\ 128\ \ \ \\ \hline 157.44 \end{array}$$ = 157.4 rounded to the nearest tenth

4.10

$6\% = 0.06$

$$\begin{array}{r} 8,000 \\ \times\ 0.06 \\ \hline \$480.00 \end{array}$$ or \$480

4.11

$I = R \times P$

$I \div P = R$

$\quad R = I \div P$

$$\begin{array}{r} 0.05 \\ 120\overline{)6.00} \\ \underline{600} \\ 0 \end{array}$$ = 5%

4.12

$I = R \times P$

$I \div R = P$

$P = I \div R$

$10\% = 0.1$

$$\begin{array}{r} 475.00 \\ 0.1.\overline{)47.5.00} \\ \underline{4} \\ 07 \\ \underline{7} \\ 05 \\ \underline{5} \\ 0 \end{array}$$ = \$475

4.13

$7\frac{1}{2} = \frac{(2 \times 7) + 1}{2} = \frac{15}{2}$

$\frac{^2\cancel{4}}{1} \times \frac{15}{\cancel{2}_1} = \frac{30}{1} = 30$

4.14

$P = 4 \times S$

$P \div 4 = S$

$\quad S = P \div 4$

$$\begin{array}{r} 16.25 \\ 4\overline{)65.00} \\ \underline{4} \\ 25 \\ \underline{24} \\ 10 \\ \underline{8} \\ 20 \\ \underline{20} \\ 0 \end{array}$$ or $16\frac{1}{4}$

4.15

$6\frac{1}{3} \times 2\frac{1}{8} =$

$\frac{19}{3} \times \frac{17}{8} = \frac{323}{24}$ or $13\frac{11}{24}$

4.16

$A = L \times W$

$A \div W = L$

$\quad L = A \div W$

$$\begin{array}{r} 25.83 \\ 24\overline{)620.00} \\ \underline{48} \\ 140 \\ \underline{120} \\ 200 \\ \underline{192} \\ 80 \\ \underline{72} \\ 8 \end{array}$$ = 25.8 rounded to the nearest tenth

4.17
$$\underline{A} = L \times W$$
$$\underline{A} \div L = W$$
$$W = A \div L$$

$$\begin{array}{r} 20 \\ 72\overline{)1,440} \\ \underline{144} \\ 00 \end{array}$$

4.18
$$\begin{array}{r} 55 \\ \times\,3.5 \\ \hline 275 \\ \underline{165} \\ 192.5 \end{array}$$

4.19
$$D = R \times T$$
$$D \div T = R$$
$$R = D \div T$$

$$\begin{array}{r} 46.41 \\ 9.2\overline{)427.0.00} \\ \underline{368} \\ 590 \\ \underline{552} \\ 380 \\ \underline{368} \\ 120 \\ \underline{92} \\ 28 \end{array}$$
= 46.4 rounded to the nearest tenth

4.20
$$\underline{D} = R \times T$$
$$\underline{D} \div R = T$$
$$T = D \div R$$

$$\begin{array}{r} 0.875 \\ 40\overline{)35.000} \\ \underline{320} \\ 300 \\ \underline{280} \\ 200 \\ \underline{200} \\ 0 \end{array}$$

4.21
$$\begin{array}{r} 3.50 \\ \times\,\,40 \\ \hline 140.00 \end{array} = \$140$$

4.22
$$\underline{P} = W \times H$$
$$\underline{P} \div H = W$$
$$W = P \div H$$

$$\begin{array}{r} 3.65 \\ 40\overline{)146.00} \\ \underline{120} \\ 260 \\ \underline{240} \\ 200 \\ \underline{200} \\ 0 \end{array}$$

4.23
$$\underline{P} = W \times H$$
$$\underline{P} \div W = H$$
$$H = P \div W$$
$$40 = 40$$

$$\begin{array}{r} 6.5\overline{)260.0} \\ \underline{260} \\ 00 \end{array}$$

4.24 $\$1.93 + 1.01 = \$2.94$

4.25
$$\underline{S} = C + P$$
$$\underline{S} - P = C$$
$$C = S - P$$
$$\$64.95 - \$19.95 = \$45.00 \text{ or } \$45$$

4.26
$$\underline{S} = C + P$$
$$\underline{S} - C = P$$
$$P = S - C$$
$$\$603 - \$419 = \$184$$

4.27
$$\underline{A} = L \times W$$
$$\underline{A} = 85 \times 47$$

$$\begin{array}{r} 85 \\ \times\,47 \\ \hline 595 \\ \underline{340} \\ \end{array}$$
$$\underline{A} = 3,995 \text{ sq. ft}$$

4.28
$$\underline{D} = R \times T$$
$$\underline{D} = 1,320 \times 5$$

$$\begin{array}{r} 1,320 \\ \times\,5 \\ \hline \end{array}$$
$$\underline{D} = 6,600 \text{ ft.}$$

4.29
$$\underline{S} = C + P$$
$$\$49.50 = \$21.14 + P$$
$$\$49.50 - \$21.14 = P$$
$$\underline{P} = \$28.36$$

4.30 $P = W \times H$
       $P = \$4.15 \times 36$

$$\begin{array}{r} \$4.15 \\ \times\ 36 \\ \hline 2490 \\ 1245\ \ \\ \hline \end{array}$$
       $P = \$149.40$

4.31 $P = 4 \times S$
       $88 = 4 \times S$
       $88 \div 4 = S$
       $S = 22$ ft.

4.32 $I = R \times P$
       $I = 7\frac{1}{2}\% \times 4{,}750$
       $I = 7.5\% \times 4{,}750$
       $I = 0.075 \times 4{,}750$

$$\begin{array}{r} 4{,}750 \\ \times\ 0.075 \\ \hline 23750 \\ 33250\ \ \ \\ \hline \end{array}$$
       $I = 356.250 = \$356.25$

4.33   false; $10 \neq 14$

4.34   open

4.35   open

4.36   open

4.37   true; $8.937 = 8.937$

4.38   open

4.39   false; $15 \neq 17$

4.40   open

4.41   false; $0.14 \neq 16$

4.42   true; $\frac{33}{8} \times \frac{3}{4} = \frac{99}{32} = 3\frac{3}{32}$

4.43   open

4.44   open

4.45   a. $7 \times \underline{\ \ \ } = 28$
       b. 4

4.46   a. $9 \div 3 = \underline{\ \ \ }$
       b. 3

4.47   a. $n - 10 = 4$
       b. 14

4.48   a. $6\frac{1}{2} + 19\frac{3}{8} =$
       b. $6\frac{4}{8} + 19\frac{3}{8} = 25\frac{7}{8}$

4.49   a. $8 \div \underline{\ \ \ } = 1.25$
       b. 6.4

4.50   a. $n \times 7.6 = 53.2$
       b. 7

4.51   a. $4.13 + \underline{\ \ \ } = 18.341$
       b. 14.211

4.52   For (a) and (b) the total number of students = $19 + 7 = 26$.
       a. 19:26
       b. 7: 26
       c. 19:7
       d. 7:19

4.53   For (a), (b), and (c), the number of losses = $18 - 14 = 4$.
       a. 14:18 or 7:9
       b. 4:18 or 2:9
       c. 14:4 or 7:2
       d. 4:14 or 2:7

4.54   a. 14:24 or 7:12
       b. 8:24 or 1:3
       c. 2:24 or 1:12
       d. 14:2 or 7:1
       e. 8:2 or 4:1

4.55 $7{:}8 = 14{:} \underline{\ \ \ }$
       $8 \times 14 = 7 \times \underline{\ \ \ }$
       $112 = 7 \times \underline{\ \ \ }$
       $112 \div 7 = \underline{16}$

4.56 $144{:}12 = \underline{\ \ \ }{:}3$
       $12 \times \underline{\ \ \ } = 144 \times 3$
       $12 \times \underline{\ \ \ } = 432$
       $\underline{36} = 432 \div 12$

4.57 $63{:}3 = \underline{\ \ \ }{:}2$
       $3 \times \underline{\ \ \ } = 63 \times 2$
       $3 \times \underline{\ \ \ } = 126$
       $\underline{42} = 126 \div 3$

4.58 $\underline{\ \ \ } : 25 = 4{:}5$

4.58 (cont.)

$$25 \times 4 = \underline{\quad} \times 5$$
$$100 = \underline{\quad} \times 5$$
$$100 \div 5 = \underline{20}$$

4.59

$$129 : \underline{\quad} = 3:1$$
$$\underline{\quad} \times 3 = 129:1$$
$$\underline{\quad} \times 3 = 129$$
$$\underline{43} = 129 \div 3$$

4.60 24

4.61

Score	Tally
1	|
2	
3	||
4	|
5	|
6	|
7	||
8	|
9	||
10	++++ |
11	|
12	||
13	||
14	||

4.62

Score	Frequency
1	1
2	0
3	2
4	1
5	1
6	1
7	2
8	1
9	2
10	6
11	1
12	2
13	2
14	2
	24

Relative Frequency

$\frac{1}{24}$	= 0.042	=	4.2%
$\frac{0}{24}$	= 0.00	=	0.0%
$\frac{2}{24}$	= $\frac{1}{12}$ = 0.083	=	8.3%
$\frac{1}{24}$	= 0.042	=	4.2%
$\frac{1}{24}$	= 0.042	=	4.2%
$\frac{1}{24}$	= 0.042	=	4.2%
$\frac{2}{24}$	= $\frac{1}{12}$ = 0.083	=	8.3%
$\frac{1}{24}$	= 0.042	=	4.2%
$\frac{2}{24}$	= $\frac{1}{12}$ = 0.083	=	8.3%
$\frac{6}{24}$	= $\frac{1}{4}$ = 0.25	=	25.0%
$\frac{1}{24}$	= 0.042	=	4.2%
$\frac{2}{24}$	= $\frac{1}{12}$ = 0.083	=	8.3%
$\frac{2}{24}$	= $\frac{1}{12}$ = 0.083	=	8.3%
$\frac{2}{24}$	= $\frac{1}{12}$ = 0.083	=	8.3%
			100.0%

4.63

a. The set contains 24 numbers. Median: $24 \div 2 = 12$; the 12th and 13th numbers are both 10. Median = 10.

b. Mode = 10

c.

Score	Frequency	Score x Frequency
1	1	1 x 1 = 1
2	0	2 x 0 = 0
3	2	3 x 2 = 6
4	1	4 x 1 = 4
5	1	5 x 1 = 5
6	1	6 x 1 = 6
7	2	7 x 2 = 14
8	1	8 x 1 = 8
9	2	9 x 2 = 18
10	6	10 x 6 = 60
11	1	11 x 1 = 11
12	2	12 x 2 = 24
13	2	13 x 2 = 26
14	2	14 x 2 = 28
	24	211

4.63 (cont.)

       $211 \div 24 = 8.79$ or 8.8 or round-
         to the nearest tenth.

4.64   a. $14 - 1 = 13$

          b. Range = 13

             Average deviation =
                    $13 \div 24 = 0.54$

4.65

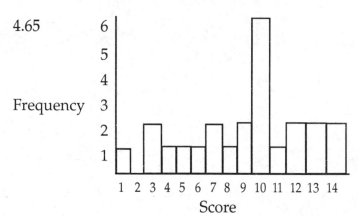

SELF TEST 1

1.01   583 = 500 + 80 + 3

1.02   723,521 = 700,000 + 20,000
        + 3,000 + 500 + 20 + 1

1.03   1,502,060 = 1,000,000 + 500,000
        + 2,000 + 60

1.04   64,722 = 60,000 + 4,000
        + 700 + 20 + 2

1.05   9,854 = 9,000 + 800 + 50 + 4

1.06   Four thousand, forty-five
        = 4,000 + 45 = 4,045

1.07   Three million, five thousand,
        sixty-two = 3,000,000 + 5,000
        + 62 = 3,005,062

1.08   One thousand, six hundred
        fifty-eight = 1,000 + 600
        + 58 = 1,658

1.09   Sixty thousand, seventy-eight
        = 60,000 + 78 = 60,078

1.010   Two thousand, nine = 2,000
        + 9 = 2,009

1.011   2(100) + 3(10) + 5
        = 200 + 30 + 5 = 235

1.012   6(1,000) + 4(10) + 2
        = 6,000 + 40 + 2 = 6,042

1.013   2(100,000) + 3(1,000) + 5
        = 200,000 + 3,000 + 5
        = 203,005

1.014   7(10,000) + 7(100) + 7
        = 70,000 + 700 + 7 = 70,707

1.015   3(1,000,000) + 6(10,000)
        + 9(10) = 3,000,000 + 60,000
        + 90 = 3,060,090

1.016   16,024; 16,025

1.017   9,999; 10,000; 10,001

1.018   true

1.019   true

1.020   false; 60 < 61 < 62

1.021   a.   780
        b.   800
        c.   1,000

1.022   a.   5,290
        b.   5,300
        c.   5,000

1.023   a.   602,530
        b.   602,500
        c.   603,000

SELF TEST 2

2.01   7,940,308 = 7,000,000
        + 900,000 + 40,000 + 300 + 8

2.02   589,985 = 500,000 + 80,000
        + 9,000 + 900 + 80 + 5

2.03   222 = 200 + 20 + 2

2.04   Seven hundred seventy-nine
        = 700 + 79 = 779

2.05   Seven thousand, seventy-nine
        = 7,000 + 79 = 7,079

2.06   Seven million, seventy-nine
        = 7,000,000 + 79 = 7,000,079

2.07   4(100) + 3(10) + 2 = 400 + 30 + 2
        = 432

2.08   7(1,000) + 4(100) + 9(10) + 1
        = 7,000 + 400 + 90 + 1 = 7,491

2.09   5(100,000) + 6(100) + 8
        = 500,000 + 600 + 8 = 500,608

2.010   678 =
    a.   **680 rounded to the nearest ten**
    b.   **700 rounded to the nearest hundred**
    c.   **1,000 rounded to the nearest thousand**

2.011   4,166 =
    a.   **4,170 rounded to the nearest ten**
    b.   **4,200 rounded to the nearest hundred**
    c.   **4,000 rounded to the nearest thousand**

2.012   742,381 =
    a.   **742,380 rounded to the nearest ten**
    b.   **742,400 rounded to the nearest hundred**
    c.   **742,000 rounded to the nearest thousand**

2.013
$$\begin{array}{r} 6 \\ 8 \\ \hline 14 \end{array}$$

2.014
$$\begin{array}{r} {}_1 9 \\ 17 \\ \hline 26 \end{array}$$

2.015
$$\begin{array}{r} 1 \\ 29 \\ 8 \\ \hline 37 \end{array}$$

2.016
$$\begin{array}{r} 4 \\ 8 \\ 7 \\ \hline 19 \end{array}$$

2.017
$$\begin{array}{r} 1 \\ 68 \\ 29 \\ \hline 97 \end{array}$$

2.018
$$\begin{array}{r} 1 \\ 73 \\ 89 \\ \hline 162 \end{array}$$

2.019
$$\begin{array}{r} 1 \\ 55 \\ 96 \\ \hline 151 \end{array}$$

2.020
$$\begin{array}{r} 1 \\ 72 \\ 96 \\ 83 \\ \hline 251 \end{array}$$

2.021
$$\begin{array}{r} 1 \\ 692 \\ 956 \\ \hline 1,648 \end{array}$$

2.022
$$\begin{array}{r} 11 \\ 496 \\ 817 \\ \hline 1,313 \end{array}$$

2.023
$$\begin{array}{r} 11 \\ 706 \\ 499 \\ \hline 1,205 \end{array}$$

2.024
$$\begin{array}{r} 21 \\ 755 \\ 496 \\ 852 \\ \hline 2,103 \end{array}$$

2.025
$$\begin{array}{r} 1\,1\,1 \\ 4,962 \\ 8,345 \\ 4,016 \\ \hline 17,323 \end{array}$$

2.026
$$\begin{array}{r} 1\ \ 12 \\ 5,948 \\ 2,006 \\ 3,015 \\ 5,062 \\ \hline 16,031 \end{array}$$

2.027   582 + 496 + 723
= 600 + 500 + 700 = 1,800

2.028   5,123 + 964 + 3,102
= 5,000 + 1,000 + 3,000
= 9,000

2.029   82 + 76 + 105 = 80 + 80 + 110
= 270

2.030
$$10 + N = 22$$
$$10 + N - 10 = 22 - 10$$
$$N = 22 - 10$$
$$N = 12$$

2.031
$$50 + 32 + N = 96$$
$$82 + N = 96$$
$$82 + N - 82 = 96 - 82$$
$$N = 96 - 82$$
$$N = 14$$

2.032
$$42 + N > 66$$
$$42 + N - 42 > 66 - 42$$
$$N > 66 - 42$$
$$N > 24$$
Any number greater than 24 will make the sentence true.

SELF TEST 3

3.01 $233 = 200 + 30 + 3$

3.02 $6,520 = 6,000 + 500 + 20$

3.03 $17,458 = 10,000 + 7,000 + 400 + 50 + 8$

3.04 Two hundred seventy-five
$= 200 + 75 = 275$

3.05 One thousand, sixty-six
$= 1,000 + 66 = 1,066$

3.06 Seventy thousand, seventy
$= 70,000 + 70 = 70,070$

3.07 $6(1,000) + 3(100) + 2(10) + 5$
$= 6,000 + 300 + 20 + 5 = 6,325$

3.08 $2(100) + 3 = 200 + 3 = 203$

3.09 $9(1,000) + 9(10) = 9,000 + 90$
$= 9,090$

3.010 false; $12 < 15$ is true, but
$15 < 14$ is false.

3.011 true

3.012 true

3.013 $602 + 533 + 778$
$= 600 + 500 + 800 = 1,900$

3.014 $6,521 + 1,063 + 901$
$= 7,000 + 1,000 + 1,000$
$= 9,000$

3.015 $785 + 1,293 + 47$
$= 790 + 1,290 + 50$
$= 2,130$

3.016
$$\begin{array}{r} 8 \\ 6 \\ +\ 9 \\ \hline 23 \end{array}$$

3.017
$$\begin{array}{r} 1 \\ 23 \\ 48 \\ +\ 72 \\ \hline 143 \end{array}$$

3.018
$$\begin{array}{r} {}^{6}\!\!\!\!\!{}_1 \\ \cancel{7}6 \\ -\ 48 \\ \hline 28 \end{array}$$

3.019
$$\begin{array}{r} {}^{112}\!{}_1 \\ \cancel{2}\cancel{3}3 \\ -155 \\ \hline 78 \end{array}$$

3.020
$$\begin{array}{r} 73 \\ -41 \\ \hline 32 \end{array}$$

3.021
$$\begin{array}{r} 11 \\ 488 \\ 123 \\ +\ 707 \\ \hline 1,318 \end{array}$$

3.022
$$\begin{array}{r} {}^{8}\!{}_1 \\ 4\cancel{9}5 \\ -\ 327 \\ \hline 168 \end{array}$$

3.023
$$\begin{array}{r} {}^{8\ 10} \\ 6\,2,\,\cancel{9}\,\cancel{1}^{1}1 \\ -\ 4\,1,\,7\,8\,4 \\ \hline 2\,1,\,1\,2\,7 \end{array}$$

3.024
$$\begin{array}{r} 1\ 1 \\ 7,030 \\ 1,945 \\ +\ 2,281 \\ \hline 11,256 \end{array}$$

3.025 $652 + 543 + 38 = 1,233$

3.026 $1,056 - 697 = 359$

3.027 $69,230 + 54,612 + 1,056$
$= 124,898$

3.028 $72,850 - 36,125 = 36,725$

3.029 $412 - 288 = 124$

SELF TEST 4

4.01 $7,070,707 = 7,000,000$
$+ 70,000 + 700 + 7$

4.02 $4,680 = 4,000 + 600 + 80$

4.03   Three million, four hundred
thousand, seven hundred sixty-
five $= 3,000,000 + 400,000$
$+ 700 + 65 = 3,400,765$

4.04   Eight hundred thousand,
ninety-two $= 800,000 + 92$
$= 800,092$

4.05 $9(100,000) + 8(10,000)$
$+ 7(1,000) + 6(100) = 900,000$
$+ 80,000 + 7,000 + 600$
$= 987,600$

4.06 $2(10,000) + 3(1,000) + 4(100)$
$+ 5(10) + 6 = 20,000 + 3,000$
$+ 400 + 50 + 6 = 23,456$

4.07 $490 + 315 + 840 = 500 + 300$
$+ 800 = 1,600$

4.08 $7,519 + 3,041 + 1,102$
$= 8,000 + 3,000 + 1,000$
$= 12,000$

4.09 $687 - 234 = 690 - 230 = 460$

4.010
$$\begin{array}{r} 2 \\ 17 \\ 18 \\ + 19 \\ \hline 54 \end{array}$$

4.011
$$\begin{array}{r} 1 \\ 59 \\ 60 \\ + 61 \\ \hline 180 \end{array}$$

4.012
$$\begin{array}{r} 1 \\ 12 \\ 34 \\ + 56 \\ \hline 102 \end{array}$$

4.013
$$\begin{array}{r} 1\,1 \\ 123 \\ 456 \\ + 789 \\ \hline 1,368 \end{array}$$

4.014
$$\begin{array}{r} 1\,1 \\ 444 \\ 555 \\ + 666 \\ \hline 1,665 \end{array}$$

4.015
$$\begin{array}{r} 1 \\ 19 \\ {}^1191 \\ + {}^11,911 \\ \hline 2,121 \end{array}$$

4.016
$$\begin{array}{r} {}^6{}_1 \\ 777 \\ - 484 \\ \hline 293 \end{array}$$

4.017
$$\begin{array}{r} {}^8{}_1 \\ 591 \\ - 375 \\ \hline 216 \end{array}$$

4.018
$$\begin{array}{r} 0\,{}_1\,8{}_1 \\ 1,092 \\ - \ 759 \\ \hline 333 \end{array}$$

4.019
$$\begin{array}{r} 81112{}_1 \\ 9,234 \\ - 8,645 \\ \hline 589 \end{array}$$

4.020
$$\begin{array}{r} 0{}_1\,615{}_1 \\ 15,762 \\ - 7,283 \\ \hline 8,479 \end{array}$$

4.021
$$\begin{array}{r} 5{}_1 \ \ 5{}_1 \\ 64,646 \\ - 25,252 \\ \hline 39,394 \end{array}$$

4.022 $321 + 654 + 987 = 1,962$

4.023 $48,509 - 35,042 = 13,467$

4.024 $6,077 + 3,465 + 5,111 = 14,653$

4.025 $578 - 389 = 189$

4.026 $3,456 - 2,345 = 1,111$

4.027 $a + 13 = b$
   a. $1 + 13 = b$
        $14 = b$
   b. $6 + 13 = b$
        $19 = b$
   c. $9 + 13 = b$
        $22 = b$
   d. $a + 13 = 15$
        $a = 2$
   e. $a + 13 = 30$
        $a = 17$
   f. $a + 13 = 50$
        $a = 37$

The table for $a + 13 = b$ is

a	1	6	9	d. 2	e. 17	f. 37
b	a. 14	b. 19	c. 22	15	30	50

4.028 $a - b = 15$
   a. $20 - b = 15$
        $b = 5$
   b. $35 - b = 15$
        $b = 20$
   c. $50 - b = 15$
        $b = 35$
   d. $a - 30 = 15$
        $a = 45$
   e. $a - 10 = 15$
        $a = 25$
   f. $a - 0 = 15$
        $a = 15$

The table for $a - b = 15$ is

a	20	35	50	d. 45	e. 25	f. 15
b	a. 5	b. 20	c. 35	30	10	0

4.029 $x + 7 = y$
   a. $3 + 7 = y$
        $10 = y$
The number pair is (3, 10).
   b. $4 + 7 = y$
        $11 = y$
The number pair is (4, 11).
   c. $5 + 7 = y$
        $12 = y$
The number pair is (5, 12).

4.030 $x - y = 5$
   a. $10 - y = 5$
        $y = 5$
The number pair is (10, 5).
   b. $6 - y = 5$
        $y = 1$
The number pair (6, 1).
   c. $8 - y = 5$
        $y = 3$
The number pair is (8, 3).
   d. $25 - y = 5$
        $y = 20$
The number pair is (25, 20).
   e. $58 - y = 5$
        $y = 53$
The number pair is (58, 53).

4.031   {(3, 5), (7, 9), (4, 6),
(5, b.____), (10, c.____)}
   a. Each number pair has a
        difference of 2. Since the
        first number is smaller
        than the second number in
        each number pair, the first
        number plus 2 equals the
        second number. Therefore,
        the sentence is $a + 2 = b$
        (or $b - 2 = a$, or $b - a = 2$)
   b. $5 + 2 = b$
        $7 = b$
The number pair is (5, 7).
   c. $10 + 2 = b$
        $12 = b$
The number pair is (10, 12).

4.032   {10, 6), (11, 7), (12, 8),
         (13, b ___.), (14, c___ .)}
         a.   Each number pair has a
               difference of 4. Since the
               first number is larger than
               the second number in each
               number pair, the first number
               minus 4 equals the second
               number, or the first number
               minus the second number
               equals 4. Therefore, the
               sentence is $a - 4 = b$ (or
               $a - b = 4$, or $b + 4 = a$)
         b. $13 - 4 = b$
                    $9 = b$
         The number pair is (13, 9).
         c. $14 - 4 = b$
                   $10 = b$
         The number pair is (14, 10).

4.033   111
         1965
         + 35
         2000 Sam will be 35 years
                old in the year 2000.

4.034   The words "How much more" are
         a clue to subtract.
           2_1
         \$345
         –165
         \$180 Lisa needs \$180 more.

SELF TEST 1

1.01
```
      4
     47
      7
    ───
    329
```

1.02
```
      2
      0
     24
     52
    ───
     48
    120
    ─────
    1,248
```

1.03
```
      2
      1
     68
     32
    ───
    136
    204
    ─────
    2,176
```

1.04
```
      2
      2
     97
     34
    ───
    388
    ¹291
    ─────
    3,298
```

1.05
```
     00
     41
     10
    452
    193
    ─────
    ¹1¹356
    4068
    452
    ──────
    87,236
```

1.06
```
     23
     00
     34
    768
    405
    ──────
    ¹3840
    000
    3¹072
    ───────
    311,040
```

1.07
```
        1
        2
        2
      931
      687
    ───────
    ¹6517
    ¹7448
    ¹5586
    ────────
    639,597
```

1.08
```
      5 1 2
      2 0 1
      4 1 2
      1 0 0
      2,924
      6,352
    ─────────
    ¹5848
    ¹1²46200
    8772
    1¹7544
    ──────────
    18,573,248
```

1.09
```
         21
         00
         64
         64
      4,086
      3,178
    ───────────
    ²3¹2¹688
    ¹28 602
    40 86
    12258
    ────────────
    12,985,308
```

1.010 25 is the multiplicand (b.)
       x 8 is the multiplier (a.)
       ───
       200 is the product (c.)

1.011 4^3

a. 3 is the exponent
b. 4^3 means 4 multiplied
    by itself 3 times, or
    $4 \times 4 \times 4$.

1.012 The Commutative Principle
       says that the order of
       multiplication makes no
       difference in finding the
       product. $A \times B = B \times A$.
       Therefore, $2 \times 7 = 7 \times 2$.

1.013 $3^4 = 3 \times 3 \times 3 \times 3 = 81$

1.014 $4^3 = 4 \times 4 \times 4 = 64$

1.015 $3^3 = 3 \times 3 \times 3 = 27$

1.016 1, 4, 16, a.____, b.____, c.____
Multiplying each member by
4 results in the member
following it.
$1 \times 4 = 4$
$4 \times 4 = 16$
The next three members of
the pattern are
a. $16 \times 4 = 64$
b. $64 \times 4 = 256$
c. $256 \times 4 = 1,024$

1.017 4, 8, 16, a. ____, b. ____, c. ____
Multiplying each member by
2 results in the member
following it.
$4 \times 2 = 8$
$8 \times 2 = 16$
The next three members of
the pattern are
a. $16 \times 2 = 32$
b. $32 \times 2 = 64$
c. $64 \times 2 = 128$

1.018 1, 3, 9, 27, a. ____, b. ____,
c. ____
Each member is a power of 3.
$3^1 = 3$
$3^2 = 3 \times 3 = 9$
$3^3 = 3 \times 3 \times 3 = 27$
The next three members of the
pattern are
a. $3^4 = 3 \times 3 \times 3 \times 3 = 81$
b. $3^5 = 3 \times 3 \times 3 \times 3 \times 3 = 243$
c. $3^6 = 3 \times 3 \times 3 \times 3 \times 3 \times 3$
   $= 729$

1.019 $8 \times 205 = 8(200 + 5) = 8(200)$
   $+ 8(5) = 1,600 + 40$
   $= 1,640$

1.020 $4 \times \$4.95 = 4(\$5.00 - 0.05)$
   $= 4(\$5.00)$
   $- 4(\$0.05) = \$20.00$
   $- 0.20 = \$19.80$

1.021 $7 \times 197 = 7(200 - 3) = 7(200)$
   $- 7(3) = 1,400 - 21$
   $= 1,379$

1.022 20 gallons x 19 miles per
gallon = 380 miles

1.023 In 1 box are 125 envelopes.
Therefore, for 24 boxes,
125 envelopes x 24 boxes
= 3,000 envelopes.

1.024 Bill reads 230 words in
one minute. A half hour
contains 30 minutes.
Therefore, 230 x 30 = 6,900
words he could read in a
half hour.

Function rule: Multiply by
5, then add 7.

Number	Function
1.025 0	$0 \times 5 = 0$ $0 + 7 = ⑦$
1.026 2	$2 \times 5 = 10$ $10 + 7 = ⑰$
1.027 5	$5 \times 5 = 25$ $25 + 7 = ㉜$
1.028 12	$12 \times 5 = 60$ $60 + 7 = ㊲$
1.029 17	$17 \times 5 = 85$ $85 + 7 = ㊲$
1.030 20	$20 \times 5 = 100$ $100 + 7 = ⑩⑦$

SELF TEST 2

2.01 72 is the multiplicand (a.)
x 7 is the multiplier (b.)
504 is the product (c.)

2.02 $8\overline{)760}$ 95
a. 760 is the dividend.
b. 8 is the divisor.
c. 95 is the quotient.

2.03 $7^3 = 7 \times 7 \times 7 = 343$

2.04 $27 \times 10^6 = 27 \times 10 \times 10 \times 10$
                         $10 \times 10 \times 10$
                     $= 27,000,000$

2.05 $5^4 = 5 \times 5 \times 5 \times 5 = 625$

2.06   5
         48
      $\underline{7}$
        336

2.07   3
          4
         89
      $\underline{45}$
       ¹445
      $\underline{¹356}$
      4,005

2.08   31
         41
         10
        452
      $\underline{793}$
      ¹1¹356
      4068
      $\underline{3164}$
      358,436

2.09   52
         42
         31
        763
      $\underline{975}$
      ¹3815
      ¹5341
      $\underline{¹6867}$
      743,925

2.010   2 1
         0 0
        6 4 2
        0 0 0
        4,753
      $\underline{3,091}$
      ¹4¹753
      4¹2777
      0000
      $\underline{14259}$
      14,691,523

2.011   1, 3, 9, 27, a. ____ ,
        b. ____ , c. ____
        Multiplying each member by
        3 results in the member
        following it.
        $1 \times 3 = 3$
        $3 \times 3 = 9$
        $9 \times 3 = 27$
        The next three members are
        a. $27 \times 3 = 81$
        b. $81 \times 3 = 243$
        c. $243 \times 3 = 729$

2.012   0, 5, 10, 15, a. ____ ,
        b. ____ , c. ____
        Each member is a multiple
        of 5.
        $5 \times 0 = 0$
        $5 \times 1 = 5$
        $5 \times 2 = 10$
        $5 \times 3 = 15$
        The next three members are
        a. $5 \times 4 = 20$
        b. $5 \times 5 = 25$
        c. $5 \times 6 = 30$

2.013   1,024; 256; 64; a. ____ ;
        b ____ ; c. ____
        To find the divisor of the
        pattern, divide any member
        by the one following it.
        $256 \div 64 = 4$
        The divisor is 4. The next
        three members are
        a. $64 \div 4 = 16$
        b. $16 \div 4 = 4$
        c. $4 \div 4 = 1$

2.014   Function rule: Multiply by
        4, then add 3

Number	Function
0	$0 \times 4 = 0$
	$0 + 3 = ⑶$
3	$3 \times 4 = 12$
	$12 + 3 = ⑮$
15	$15 \times 4 = 60$
	$60 + 3 = ⑥③$

2.015   Function rule: Divide by
        5, then add 7

Number	Function
5	$5 \div 5 = 1$
	$1 + 7 = ⑧$
25	$25 \div 5 = 5$
	$5 + 7 = ⑫$
45	$45 \div 5 = 9$
	$9 + 7 = ⑯$

2.016 Function rule: Divide by
3, then subtract 2

Number	Function
6	$6 \div 3 = 2$
	$2 - 2 = \textcircled{0}$
24	$24 \div 3 = 8$
	$8 - 2 = \textcircled{6}$
45	$45 \div 3 = 15$
	$15 - 2 = \textcircled{13}$

2.017

```
      112 R31
47)5,295
    47
    59
    47
   125
    94
    31
```

2.018

```
      178  R 34    Check:
38)6,798            22
   38               66
  299              178
  266               38
  338             1424
  304              534
   34            6,764
                 + 34
                 6,798
```

2.019

```
      119 R36
79)9,437
   79
  153
   79
  747
  711
   36
```

2.020

```
          369 R237    Check:
345)127,542             22
   1035                 23
   2404                 34
   2070                369
   3342                345
   3105              ²1¹845
    237              1476
                     1107
                  127,305
                  +   237
                  127,542
```

2.021

```
       807
756)610,092
    6048
    5292
    5292
       0
```

2.022

```
     1,260  R40    Check:
496)625,000          1 2
   496               2 5
  1290               1 3
   992             1,260
  2980               496
  2976             7560
    40           1¹1340
                   5040
                 62¹4¹960
                  +  40
                 625,000
```

2.023

```
        15,854  R213
407)6,452,791
    407
   2382
   2035
   3477
   3256
   2219
   2035
   1841
   1628
    213
```

2.024

```
        7,997  R766
1,175)9,397,241      Check:
     8225             6 6 4
    11722             4 4 3
    10575             7,997
    11474             1,175
    10575           ³³²9¹985
     8991           ²55 9 79
     8225            79 97
      766           ²799 7
                 9,39¹6¹,4¹75
                  +   766
                  9,397,241
```

2.025

```
          7,498
942)7,063,116
     6594
     4691
     3768
     9231
     8478
      7536
      7536
         0
```

2.026   Add the scores.   2
                         82
                         88
                         86
                         88
                         90
                       + 94
                         528 total

Divide the total by the
number of scores.

```
    88
6)528        Sue's average
   48        grade was 88.
   48
   48
    0
```

2.027   50 mph
      x 9 hrs.
      450 miles

2.028   30 gal.
      15)450
          45
          00
          00
           0

2.029   30 gal.
      x   60¢
          00
      180
      1,800¢ = $18.00

2.030   12
      x 3
      36 chocolates

          2 Each student received
      18)36 2 chocolates.
         36
          0
```
```

SELF TEST 1

1.01 $\overleftrightarrow{RT}$

1.02 $\angle ABC$

1.03 $\angle 3$

1.04 $\overline{NO}$

1.05 $\overleftrightarrow{UP}$

1.06 $1\frac{11}{16}$

1.07   50°

1.08   130°

1.09   4.9

1.010   3.9

1.011
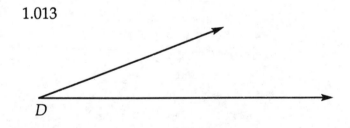

1.012
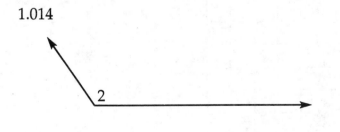

1.013

1.014

1.015
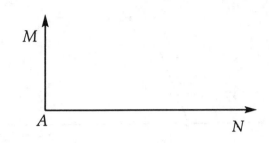

1.016   line
1.017   protractor
1.018   degree
1.019   inch
1.020   a. 0
       b. 180

SELF TEST 2

2.01

2.02

2.03

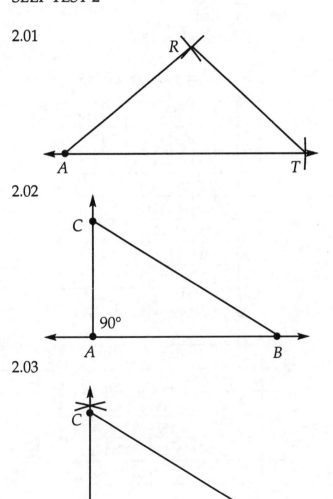

2.04

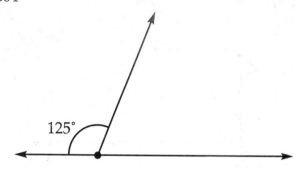

2.05

A   C

2.06 $10\frac{1}{8}$ inches

2.07 $\angle A = 30°$
$\angle B = 90°$
$\angle C = 60°$
Sum of $\angle$'s $= 180°$

2.08   Area is approximately 36

2.09   Area is approximately 52

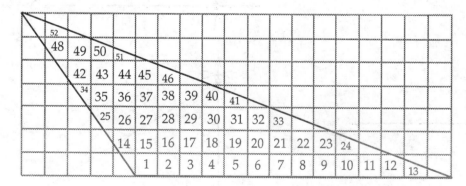

2.010	$\angle ABC = 70°$
2.011	d
2.012	k
2.013	m
2.014	o
2.015	a
2.016	j
2.017	g
2.018	c
2.019	l
2.020	n
2.021	b
2.022	f
2.023	i
2.024	e

SELF TEST 3
NOTE: Due to printing variations, these measurements may vary.

3.01	4
3.02	360°
3.03	4
3.04	90°
3.05	parallel or equal
3.06	1
3.07	$P = 13.5$ cm

3.08 $P = 6\frac{3}{8}$ inches

3.09 $P = 13.1$ cm

3.010 $P = 6\frac{5}{8}$ inches

3.011

3.012

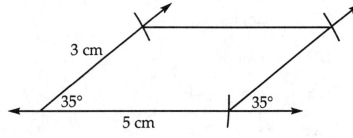

3 cm   35°   5 cm   35°

3.013 $A = 28$ sq. units

3.014

		1	2	3	4	5			
	6	7	8	9	10	11	12		
	13	14	15	16	17	18	19	20	
21	22	23	24	25	26	27	28	29	30

$A \doteq 30$ sq. units

3.015 $A = 90$ sq. units

		1	2	3	4	5	6	7	8	9	10	11	12	13	14	15	16	17	18
	19	20	21	22	23	24	25	26	27	28	29	30	31	32	33	34	35	36	
37	38	39	40	41	42	43	44	45	46	47	48	49	50	51	52	53	54		
55	56	57	58	59	60	61	62	63	64	65	66	67	68	69	70	71	72		
73	74	75	76	77	78	79	80	81	82	83	84	85	86	87	88	89	90		

3.016   e

3.017   h

3.018   d

3.019   a

3.020   c

3.021   b

3.022   f

SELF TEST 4

4.01   6

4.02   angles

4.03   twice

4.04   720

4.05   1

4.06   90

4.07   circumference

4.08   90

4.09   4

4.010   3.14

4.011

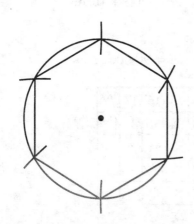

4.012

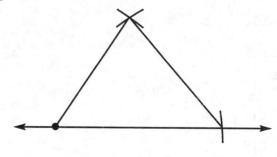

4.013

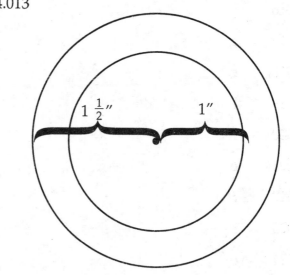

4.014

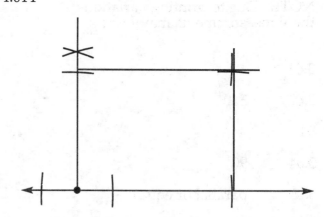

4.015 $A = \pi\, r^2$

$r = \dfrac{d}{2}$

$r = \dfrac{14}{2} = 7$

$A = 3\tfrac{1}{7} \times 7 \times 7$

$A = \dfrac{22}{7} \times 49$

$A = \dfrac{22}{1\,\cancel{7}} \times \dfrac{\cancel{49}^{7}}{1}$

$A = 154$ sq. inches

4.016 $C = 2\pi r$
$C = 2 \times 3.14 \times 7$
$C = 43.96$ inches

4.017   A hexagon has six sides
$P = 6 \times 6 = 36$ inches

4.018

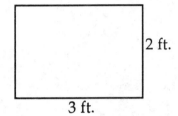

The length of the string =
the perimeter of the
rectangle.
$P = 2 + 3 + 2 + 3$
$P = 10$ ft. = length of the string

4.019 $P = 6 + 8 + 12$
$P = 26$ inches

4.020

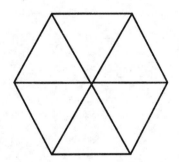

SELF TEST 1

1.01 $\frac{5}{7}$

1.02   seven-ninths

1.03 $\frac{3}{10}$

1.04   Examples:

$$\frac{4}{9} = \frac{4 \times 2}{9 \times 2} = \frac{8}{18}$$

$$\frac{4}{9} = \frac{4 \times 3}{9 \times 3} = \frac{12}{27}$$

1.05 $\frac{3}{6} = \frac{\overset{1}{\cancel{3}} \times 1}{\underset{1}{\cancel{3}} \times 2} = \frac{1}{2}$

1.06 $\frac{15}{35} = \frac{\overset{1}{\cancel{5}} \times 3}{\underset{1}{\cancel{5}} \times 7} = \frac{3}{7}$

1.07 $12 \div 3 = 4$

$$\frac{1}{3} = \frac{1 \times 4}{3 \times 4} = \frac{4}{12}$$

1.08 $16 \div 4 = 4$

$$\frac{4}{5} = \frac{4 \times 4}{5 \times 4} = \frac{16}{20}$$

1.09 $\frac{25}{5} = 25 \div 5 = 5$

1.010 $\frac{27}{7} = 27 \div 7 = 3\frac{6}{7}$

1.011 $\frac{6}{1}$

1.012 $\frac{9}{4} = 9 \div 4 = 2\frac{1}{4}$

1.013

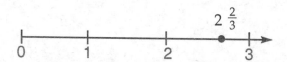

1.014   Multiply $5 \times 1 = 5$ and add 2; $1\frac{2}{5} = \frac{7}{5}$.

1.015 $\frac{5}{6}$ is larger than $\frac{5}{7}$ because sixths are larger than sevenths.

1.016   c

$$\frac{3}{4} = \frac{3}{4}$$

$$\frac{6}{8} = \frac{3 \times 2}{4 \times 2} = \frac{3}{4}$$

$$\frac{9}{12} = \frac{3 \times 3}{4 \times 3} = \frac{3}{4}$$

$$\frac{12}{16} = \frac{3 \times 4}{4 \times 4} = \frac{3}{4}$$

1.017   d

1.018   b

$$\frac{7}{13} = \frac{7 \times 2}{13 \times 2} = \frac{14}{26}$$

$$\frac{7}{13} = \frac{7 \times 3}{13 \times 3} = \frac{21}{39}$$

1.019   a

1.020   c

1.021   c

1.022   d

1.023   g

1.024   a

1.025   i

1.026   h

1.027   b

1.028   f

1.029 Examples:

a. $\frac{3}{8} = \frac{3 \times 2}{8 \times 2} = \frac{6}{16}$

b. $\frac{3}{8} = \frac{3 \times 3}{8 \times 3} = \frac{9}{24}$

c. $\frac{3}{8} = \frac{3 \times 4}{8 \times 4} = \frac{12}{32}$

d. $\frac{3}{8} = \frac{3 \times 5}{8 \times 5} = \frac{15}{40}$

e. $\frac{3}{8} = \frac{3 \times 6}{8 \times 6} = \frac{18}{48}$

f. $\frac{3}{8} = \frac{3 \times 7}{8 \times 7} = \frac{21}{56}$

SELF TEST 2

2.01   seven and three tenths

2.02   56.14

2.03 $\frac{27}{54} = \frac{27 \div 27}{54 \div 27} = \frac{1}{2}$

2.04   0.035

2.05   45 is larger

2.06   Multiply 3 x 4 = 12 and add 2; $4\frac{2}{3} = \frac{14}{3}$.

2.07 $\frac{4}{5} = 0.8$ from the table; or

$\frac{4}{5} = \frac{4 \times 20}{5 \times 20} = \frac{80}{100} = 0.8$

2.08 $0.33 = \frac{33}{100}$

2.09 $1.281 = 1\frac{281}{1,000}$

2.010   Move the decimal point two places to the right and add the % sign.
0.45 = 45%

2.011   Move the decimal point two places to the left and drop the % sign.
8% = 0.08

2.012 $\frac{4}{5}$ is larger than $\frac{4}{7}$ because fifths are larger than sevenths.

2.013 $\frac{47}{5} = 47 \div 5 = 9\frac{2}{5}$

2.014 $\frac{61}{100} = 61\%$

2.015   Move the decimal point two places to the right and add the % sign.
0.344 = 34.4%

2.016 $35 \div 7 = 5$
$\frac{4}{7} = \frac{4 \times 5}{7 \times 5} = \frac{20}{35}$

2.017 $44\% = \frac{44}{100} = \frac{44 \div 4}{100 \div 4} = \frac{11}{25}$

2.018   twenty-three twenty- fifths

2.019   c

2.020   b

2.021   a

2.022   a

2.023   c

2.024   a

2.025   b

2.026   g

2.027   h

2.028   d

2.029   f

2.030   e

2.031   c

SELF TEST 3

3.01 $\frac{23}{45}$ or 23:45

3.02 $\frac{1}{2} = 0.5 = 50\%$

3.03   Convert each amount to cents.
4 dimes = 4 x 10 = 40¢
3 quarters = 3 x 25 = 75¢
$\frac{40}{75} = \frac{40 \div 5}{75 \div 5} = \frac{8}{15} = 8{:}15$

3.04   7 and 8

3.05   2:3 = 6:?
3 x 6 = 2 x ?
18 = 2 x ?
? = 9

3.06 $45\% = \frac{45}{100} = \frac{45 \div 5}{100 \div 5} = \frac{9}{20}$

3.07 $\frac{6}{3} = 6 \div 3 = 2$ or 2.0

3.08 $\frac{32}{5} = 32 \div 5 = 6\frac{2}{5}$

3.09 $\frac{2}{3} = 0.66\frac{2}{3} = 0.666\ldots = 0.\overline{6}$

3.010 $4.\overline{87}$

3.011   Move the decimal point two places to the left and drop the % sign.
0.9% = 0.009

3.012   Move the decimal point two places to the right and add the % sign.
0.00004 = 0.004%

3.013   Move the decimal point two places to the left and drop the % sign.
345% = 3.45

3.014   Move the decimal point two places to the right and add the % sign.
3.87 = 387%

3.015   3.78 is smaller than 37.8

3.016   b

3.017   a
13¢ of $1.00 =
13 out of 100 = 13%

3.018   d

3.019   c

3.020   a

100:1 = 450:?
1 x 450 = 100 x ?
450 = 100 x ?
? = 4.5 hg

3.021   d
$\frac{7}{8} = \frac{7 \times 2}{8 \times 2} = \frac{14}{16}$

3.022   c
Move the decimal point two places to the left and drop the % sign.

3.023   d

3.024   e

3.025   c

3.026   g

3.027   h

3.028   f

3.029   a

3.030   k

3.031   j

3.032   b

SELF TEST 1

1.01	c
1.02	h
1.03	g
1.04	f
1.05	a
1.06	j
1.07	i
1.08	e
1.09	d
1.010	b
1.011	The elements are a, g, d, and t.
1.012	finite
1.013	{The whole number between 0 and 6}
1.014	{a, e , i, o, u}
1.015	proper
1.016	The set X is a subset of set Y.
1.017	{6, 24, 56, 117}
1.018	$x = \{4\}$
1.019	$\overline{AB}$ is the set of points between points A and B.
1.020	Examples: {set of tellers}, {set of desks}
1.021	∅
1.022	because the intersection is empty
1.023	{e, f, g, a, b, c}
1.024	∈
1.025	⊂
1.026	b
1.027	b
1.028	d
1.029	c
1.030	a

SELF TEST 2

2.01	c
2.02	f
2.03	k
2.04	l
2.05	d
2.06	e
2.07	b
2.08	h
2.09	g
2.010	j
2.011	i
2.012	a
2.013	$346 = 300 + 40 + 6 =$ SSS∧∧∧∧∧IIIIII
2.014	$200 + 30 + 4 = 234$
2.015	The elements are red, white, and blue.
2.016	Set A intersects set B.
2.017	$2,653 = 2,000 + 500 + 100 + 50 + 3 =$ XX⌐H⌐///
2.018	$1,000 + 500 + 200 + 50 + 10 + 3 = 1,763$

2.019

2.020   50

2.021   5 is 100 times larger then 0.05

2.022   100 x 0.08 = 8

2.023 10^4

2.024 10^{-5}

2.025   {4}

2.026 $1000_2 = 1 \times 2^3$
          $= 1 \times 8$
          $= 8$

2.027 $22_5 = 2 \times 5^1 + 2 \times 1$
          $= 10 + 2$
          $= 12$

2.028   c

$$\begin{array}{r} 1 \\ 2\overline{)2} \\ 2 \\ 0 \end{array} \qquad \begin{array}{r} 0 \\ 1\overline{)0} \\ 0 \\ 0 \end{array}$$

          $2 = 10_2$

2.029   c
          Move the decimal point two places
          to the left to multiply by $\frac{1}{100}$.

2.030   b

2.031   a
          $12_5 = 1 \times 5^1 + 2 \times 1$
          $= 5 + 2$
          $= 7$

2.032   d

SELF TEST 3

3.01   d

3.02   h

3.03   i

3.04   f

3.05   c

3.06   k

3.07   l

3.08   a

3.09   e

3.010   j

3.011   b

3.012   g

3.013   15 is 100 times greater than 0.15

3.014   {1, 2, 3, 4, 6, 12}

3.015   2, 3, 5, 7, 11

3.016   {red, white, blue}

3.017 $50{,}000 = 5 \times 10{,}000 = 5 \times 10^4$

3.018   4, 6, 8, 9, 10, 12

3.019
$$\begin{array}{r} 1 \\ 4\overline{)6} \\ 4 \\ 2 \end{array} \qquad \begin{array}{r} 1 \\ 2\overline{)2} \\ 2 \\ 0 \end{array} \qquad \begin{array}{r} 0 \\ 1\overline{)0} \\ 0 \\ 0 \end{array}$$

          $6 = 110_2$

3.020   28 is divisible by 2:
              $28 = 2 \times 14$
          14 is divisible by 2:
              $28 = 2 \times 2 \times 7$

3.021 $2^2 \times 3^2 \times 5 = 4 \times 9 \times 5 = 180$

3.022 $25 = 20 + 5 = XXV$

3.023 $12 = 2 \times \boxed{2 \times 3}$
$30 = \boxed{2 \times 3} \times 5$
GCF $= 2 \times 3 = 6$

3.024   proper

3.025   {2, 4}

3.026 $51 = 3 \times \boxed{17}$
$68 = 2 \times 2 \times \boxed{17}$
GCF $= 17$
$\dfrac{51 \div 17}{68 \div 17} = \dfrac{3}{4}$

3.027 $4 = 2 \times 2$
$6 = 2 \times 3$
$8 = 2 \times 2 \times 2$
LCM $= 2 \times 2 \times 2 \times 3 = 24$

3.028   d

3.029   b

$$\begin{array}{cc} 1 & 0 \\ 5\overline{)5} & 1\overline{)0} \\ \underline{5} & \underline{0} \\ 0 & 0 \end{array}$$

$5 = 10_5$

3.030   a
The sum of the digits equals
$2 + 1 + 4 + 1 + 1 = 9$. Since 9 is
divisible by 3, so is 21,411.

3.031   d
$14 = 2 \times 7$
$24 = 2 \times 2 \times 2 \times 3$
LCD $= 2 \times 2 \times 2 \times 3 \times 7 = 168$

3.032   a

3.033   b

3.034   c

3.035   b

SELF TEST 1

1.01 $\dfrac{4}{5}$

1.02 $\dfrac{3}{8}$

1.03 $\dfrac{3}{3} = 1$

1.04 $\dfrac{7}{14} = \dfrac{7 \div 7}{14 \div 7} = \dfrac{1}{2}$

1.05 $9\dfrac{4}{6} = 9\dfrac{4 \div 2}{6 \div 2} = 9\dfrac{2}{3}$

1.06 $\dfrac{1}{5} = \dfrac{1}{5}$

       $\dfrac{1}{15} = \dfrac{1}{3 \times 5}$

       LCD = 3 × 5 = 15

1.07   a. $\dfrac{2}{3} = \dfrac{2 \times 2}{3 \times 2} = \dfrac{4}{6}$

       b. $\dfrac{2}{3} = \dfrac{2 \times 3}{3 \times 3} = \dfrac{6}{9}$

       c. $\dfrac{2}{3} = \dfrac{2 \times 4}{3 \times 4} = \dfrac{8}{12}$

1.08   LCD = 12

$$170\dfrac{2}{3} = 170\dfrac{8}{12}$$
$$+\ 25\dfrac{5}{12} = \ \ 25\dfrac{5}{12}$$
$$\overline{\qquad\qquad 195\dfrac{13}{12}}$$

$$195\dfrac{13}{12} = 195\dfrac{13}{12}$$
$$-\ 10\dfrac{1}{4} = \ \ 10\dfrac{3}{12}$$
$$\overline{\qquad\qquad 185\dfrac{10}{12}}$$

$$185\dfrac{10}{12} = 185\dfrac{10}{12}$$
$$-\ 3\dfrac{1}{2} = \ \ 3\dfrac{6}{12}$$
$$\overline{\qquad 182\dfrac{4}{12} = 182\dfrac{4 \div 4}{12 \div 4} = 182\dfrac{1}{3}\ \text{lbs.}}$$

1.09 $\dfrac{8}{7} = 1\dfrac{1}{7}$

1.010 $\dfrac{10}{8} = \dfrac{10 \div 2}{8 \div 2} = \dfrac{5}{4} = 1\dfrac{1}{4}$

1.011 $\dfrac{5}{4} = 1\dfrac{1}{4}$

1.012 $2\dfrac{6}{12} = 2\dfrac{6 \div 6}{12 \div 6} = 2\dfrac{1}{2}$

1.013 $23\dfrac{8}{11}$

1.014   LCD = 10

$$8\dfrac{4}{5} = \ 8\dfrac{8}{10}$$
$$+\ 40\dfrac{3}{10} = 40\dfrac{3}{10}$$
$$\overline{\qquad 48\dfrac{11}{10} = 48 + 1\dfrac{1}{10} = 49\dfrac{1}{10}}$$

1.015 $17\dfrac{2}{9} = 16 + 1\dfrac{2}{9} = 16\dfrac{11}{9}$

$$3\dfrac{7}{9} = \qquad\qquad 3\dfrac{7}{9}$$
$$\overline{\qquad\qquad\qquad 13\dfrac{4}{9}}$$

1.016   LCD = 9

$$10\dfrac{2}{3} = 10\dfrac{6}{9}$$
$$-\ 6\dfrac{1}{9} = \ 6\dfrac{1}{9}$$
$$\overline{\qquad\qquad 4\dfrac{5}{9}}$$

1.017   LCD = 15

$$\dfrac{2}{5} = \dfrac{6}{15}$$
$$+\ \dfrac{7}{15} = \dfrac{7}{15}$$
$$\overline{\qquad\quad \dfrac{13}{15}}$$

1.018   LCD = 21

$$\dfrac{1}{7} = \dfrac{3}{21}$$
$$+\ \dfrac{11}{21} = \dfrac{11}{21}$$
$$\overline{\qquad \dfrac{14}{21} = \dfrac{14 \div 7}{21 \div 7} = \dfrac{2}{3}}$$

1.019 LCD = 16

$$31\frac{14}{16} = 31\frac{14}{16}$$

$$-11\frac{3}{4} = 11\frac{12}{16}$$

$$20\frac{2}{16} = 20\frac{2 \div 2}{16 \div 2} = 20\frac{1}{8}$$

1.020 LCD = 18

$$6\frac{1}{6} = 6\frac{3}{18} = 5 + 1\frac{3}{18} = 5\frac{21}{18}$$

$$-\frac{7}{18} = \qquad\qquad\qquad \frac{7}{18}$$

$$5\frac{14}{18} =$$

$$5\frac{14 \div 2}{18 \div 2} = 5\frac{7}{9}$$

SELF TEST 2

2.01 $\dfrac{6}{8} = \dfrac{6 \div 2}{8 \div 2} = \dfrac{3}{4}$

2.02 LCD = 6

$$\frac{2}{3} = \frac{4}{6}$$

$$-\frac{1}{6} = \frac{1}{6}$$

$$\frac{3}{6} = \frac{3 \div 3}{6 \div 3} = \frac{1}{2}$$

2.03 $\dfrac{78}{100} = \dfrac{78 \div 2}{100 \div 2} = \dfrac{39}{50}$

2.04 three and fourteen-hundredths

2.05 3.25

2.06

$$\frac{5}{8} = 8\overline{)5.000}$$

$$\begin{array}{r} 0.625 \\ \underline{48} \\ 20 \\ \underline{16} \\ 40 \\ \underline{40} \\ 0 \end{array}$$

$$\frac{5}{8} = \frac{5 \times 5}{8 \times 5} = \frac{25}{40}$$

2.07 $\dfrac{2}{6} = \dfrac{2 \div 2}{6 \div 2} = \dfrac{1}{3}$

$$\frac{1}{3} = 3\overline{)1.000} \quad 0.333\ldots = 0.\overline{3}$$

$$\begin{array}{r} \underline{9} \\ 10 \\ \underline{9} \\ 10 \\ \underline{9} \\ 1 \end{array}$$

2.08

$$\begin{array}{r} 0.32 \\ + 9.50 \\ \hline 9.82 \end{array}$$

2.09

$$\begin{array}{r} \$11.05 \\ - 0.45 \\ \hline \$10.60 \end{array}$$

2.010

$$3.4 = 3.4$$

$$-\frac{3}{5} = 0.6$$

$$2.8 \text{ or}$$

$$3.4 = 3\frac{4}{10} = 3\frac{2}{5} = 2 + 1\frac{2}{5} = 2\frac{7}{5}$$

$$-\frac{3}{5} = \qquad\qquad\qquad \frac{3}{5}$$

$$2\frac{4}{5}$$

2.011 $\dfrac{30}{25} = \dfrac{30 \div 5}{25 \div 5} = \dfrac{6}{5} = 1\dfrac{1}{5}$

2.012 LCD = 8

$$31\frac{6}{8} = 31\frac{6}{8}$$

$$+75\frac{1}{4} = 75\frac{2}{8}$$

$$106\frac{8}{8} = 106 + 1 = 107$$

2.013 LCD = 12

$$21\frac{11}{12} = 21\frac{11}{12}$$

$$-21\frac{2}{3} = 21\frac{8}{12}$$

$$\frac{3}{12} = \frac{3 \div 3}{12 \div 3} = \frac{1}{4}$$

2.014   LCD = 9
$$6\frac{1}{3} = 6\frac{3}{9} = 5 + 1\frac{3}{9} = 5\frac{12}{9}$$
$$-2\frac{5}{9} = \qquad\qquad 2\frac{5}{9}$$
$$\overline{\qquad\qquad\qquad 3\frac{7}{9}}$$

2.015 $43.41

2.016 $101.22

2.017 $297.17

2.018 $0.49

2.019   68.4738

2.020   9,362.5082

2.021   59.889

2.022   75.0833

2.023   31.700
         3.710
         0.963
       + 225.000
         261.373

2.024   13.41200
        − 3.20367
         10.20833

2.025 $21\frac{1}{2} = 21.50$
         $16.4 = 16.40$
         $9.21 = 9.21$
         $\frac{2}{8} = 0.25$
         $\overline{\qquad 47.36}$ or

$$21\frac{1}{2} = \qquad 21\frac{50}{100}$$
$$16.4 = 16\frac{4}{10} = 16\frac{40}{100}$$
$$9.21 = 9\frac{21}{100} = 9\frac{21}{100}$$
$$+\frac{2}{8} = \frac{1}{4} = \frac{25}{100}$$
$$\overline{\qquad\qquad\qquad\qquad\qquad}$$
$$46\frac{136}{100} = 46\frac{136 \div 4}{100 \div 4} = 46\frac{34}{25}$$
$$= 46 + 1\frac{9}{25} = 47\frac{9}{25}$$

2.026 $51.6 = 51\frac{6}{10} = 51\frac{18}{30}$
$$-49\frac{2}{15} = \qquad\qquad 49\frac{4}{30}$$
$$\overline{\qquad\qquad\qquad\qquad}$$
$$2\frac{14}{30} = 2\frac{14 \div 2}{30 \div 2} = 2\frac{7}{15}$$

2.027   25.31
       + 16.90
         42.210
       − 30.478
         11.732

2.028 $7\frac{3}{4} = 7.75$
         7.75
       − 3.65
         4.10
       + 16.81
         20.91

2.029 $2.50
       + 6.37
         $8.87

         $10.00
          8.87
         $ 1.13 change

2.030 $37,497.29
        − 15,028.60
         $22,468.69

SELF TEST 1

1.01 a symbol consisting of three parts: a horizontal bar, a whole number above the bar, and a nonzero whole number below the bar

1.02 the number above the bar in a fraction

1.03 the number below the bar in a fraction

1.04 a symbol with a whole number part and a fraction part

1.05 a fraction with a numerator less than its denominator

1.06 a fraction with a numerator greater than or equal to its denominator

1.07 to put the numerator where the denominator was and vice versa

1.08 $\dfrac{19}{6} \times \dfrac{1}{7} = \dfrac{19}{42}$

1.09 $\dfrac{4}{3} \times \dfrac{81}{7} = \dfrac{108}{7} = 15\dfrac{3}{7}$

1.010 $\dfrac{2}{3} \times \dfrac{3}{4} = \dfrac{1}{2}$

1.011 $\dfrac{17}{8} \times \dfrac{7}{2} = \dfrac{119}{16} = 7\dfrac{7}{16}$

1.012 $\dfrac{2}{3} \times \dfrac{3}{5} = \dfrac{2}{5}$

1.013 $\dfrac{13}{4} \times \dfrac{4}{3} = \dfrac{13}{3} = 4\dfrac{1}{3}$

1.014 $\dfrac{11}{5} \times \dfrac{27}{4} = \dfrac{297}{20} = 14\dfrac{17}{20}$

1.015 $\dfrac{17}{3} \times \dfrac{18}{17} = \dfrac{6}{1} = 6$

1.016 $\dfrac{4}{5} \times \dfrac{15}{16} = \dfrac{3}{4}$

1.017 $\dfrac{3}{8} \times \dfrac{7}{1} = \dfrac{21}{8} = 2\dfrac{5}{8}$

1.018 $\dfrac{6}{7} \times \dfrac{35}{48} = \dfrac{5}{8}$

1.019 $6 \times \dfrac{5}{4} = \dfrac{6}{1} \times \dfrac{5}{4} = \dfrac{15}{2} = 7\dfrac{1}{2}$

1.020 $\dfrac{3}{4} \times \dfrac{16}{21} = \dfrac{4}{7}$

1.021 $\dfrac{7}{1} \times \dfrac{25}{8} = \dfrac{175}{8} = 21\dfrac{7}{8}$

1.022 $\dfrac{19}{9} \div \dfrac{10}{3} = \dfrac{19}{9} \times \dfrac{3}{10} = \dfrac{19}{30}$

1.023 $\dfrac{14 \times 5 + 3}{5} = \dfrac{70 + 3}{5} = \dfrac{73}{5}$

1.024 $\dfrac{6 \times 8 + 1}{8} = \dfrac{48 + 1}{8} = \dfrac{49}{8}$

1.025 $\dfrac{4 \times 4 + 3}{4} = \dfrac{16 + 3}{4} = \dfrac{19}{4}$

1.026 $\dfrac{3 \times 2 + 1}{2} = \dfrac{6 + 1}{2} = \dfrac{7}{2}$

1.027
$$4\overline{)117} = 29\dfrac{1}{4}$$
$$\begin{array}{r} 29 \\ \hline 117 \\ 8 \\ \hline 37 \\ 36 \\ \hline 1 \end{array}$$

1.028 $6\overline{)19}\;\overset{3}{}$ = $3\frac{1}{6}$

$\underline{18}$

1

1.029 $13\overline{)119}\;\overset{9}{}$ = $9\frac{2}{13}$

$\underline{117}$

2

1.030 $23\overline{)3{,}642}\;\overset{158}{}$ = $158\frac{8}{23}$

$\underline{23}$

134

$\underline{115}$

192

$\underline{184}$

8

1.031 $\frac{3}{5} \times \frac{25}{8} = \frac{15}{8} = 1\frac{7}{8}$

1.032 $\frac{67}{10} \div \frac{1}{14} = \frac{67}{10} \times \frac{14}{1} = \frac{469}{5} = 93\frac{4}{5}$

1.033 $3\frac{1}{2} \times \frac{2}{3} = \frac{7}{2} \times \frac{2}{3} = \frac{7}{3} = 2\frac{1}{3}$ cups

1.034 $\frac{3}{4} \div 8 = \frac{3}{4} \times \frac{1}{8} = \frac{3}{32}$ ton

1.035 $12\frac{1}{2} \div 5 = \frac{25}{2} \times \frac{1}{5} = \frac{5}{2} = 2\frac{1}{2}$ min.

1.036 $39\frac{1}{4} \div 14 = \frac{157}{4} \times \frac{1}{14} = \frac{157}{56} = 2\frac{45}{56}$

pounds

1.037 $30 \div 6\frac{2}{3} = 30 \div \frac{20}{3} = \frac{30}{1} \times \frac{3}{20} = \frac{9}{2} = 4\frac{1}{2}$

times

1.038 $5\frac{1}{2} \times 14 = \frac{11}{2} \times \frac{14}{1} = \frac{77}{1} = 77$ hours

1.039 $210 \div 14 = \frac{210}{1} \times \frac{1}{14} = \frac{15}{1} = 15$ yards

1.040 $9\frac{3}{4} \div \frac{1}{4} = \frac{39}{4} \div \frac{1}{4} = \frac{39}{4} \times \frac{4}{1} = \frac{39}{1}$

39 plots

SELF TEST 2

2.01   a horizontal bar with a whole number above it and a nonzero whole number below it

2.02   the number below the bar in a fraction

2.03   a fraction with a value less than 1

2.04   a system based on ten, or a number written in a base-ten system

2.05   a number with a value less than 1 written with a decimal point

2.06   a digit to the right of the decimal point

2.07 $\frac{9}{2} \times \frac{10}{3} = \frac{15}{1} = 15$

2.08 $\frac{6}{11} \times \frac{4}{1} = \frac{24}{11} = 2\frac{2}{11}$

2.09 $\frac{3}{4} \times \frac{5}{6} = \frac{5}{8}$

2.010 $\frac{4}{7} \times \frac{55}{9} = \frac{220}{63} = 3\frac{31}{63}$

2.011 $\frac{25}{8} \times \frac{24}{5} = \frac{15}{1} = 15$

2.012 $\frac{12}{13} \times \frac{39}{48} = \frac{3}{4}$

2.013 $\frac{7}{8} \times \frac{6}{10} = \frac{21}{40}$ or $\frac{7}{8} \times \frac{6}{10} = \frac{21}{40}$

2.014 $\frac{19}{20} \times \frac{5}{16} = \frac{19}{64}$

2.015 $\frac{4}{3} \div \frac{6}{7} = \frac{^2\cancel{4}}{3} \times \frac{7}{\cancel{6}_3} = \frac{14}{9} = 1\frac{5}{9}$

$1\frac{1}{3} = 1.\overline{3}$

$\frac{6}{7} = 0.8571429$

$1.3333333 \div 0.8571429 = 1.5555554$

$1\frac{5}{9} = 1.5555556$

2.016 $\frac{7}{_1\cancel{2}} \times \frac{\cancel{8}^4}{9} = \frac{28}{9} = 3\frac{1}{9}$

$3\frac{1}{2} = 3.5$

$\frac{8}{9} = 0.\overline{8}$

$3.5 \times 0.88888889 = 3.1111112$

$3\frac{1}{9} = 3.1111111$

2.017 $\frac{1}{_4\cancel{16}} \times \frac{\cancel{4}^1}{1} = \frac{1}{4}$

$\frac{1}{16} = 0.0625$

$\frac{1}{4} = 0.25$

$0.0625 \div 0.25 = 0.25$

$\frac{1}{4} = 0.25$

2.018 $\frac{33}{4} \times \frac{3}{2} = \frac{99}{8} = 12\frac{3}{8}$

$8\frac{1}{4} = 8.25$

$1\frac{1}{2} = 1.5$

$8.25 \times 1.5 = 12.375$

$12\frac{3}{8} = 12.375$

2.019 $\frac{^2\cancel{6}}{1} \times \frac{14}{\cancel{3}_1} = \frac{28}{1} = 28$

$6 = 6$

$4\frac{2}{3} = 4.\overline{6}$

$6 \times 4.6666667 = 28$

$28 = 28$

2.020 $\frac{26}{5} \div 3 = \frac{26}{5} \times \frac{1}{3} = \frac{26}{15} = 1\frac{11}{15}$

$5.2 \div 3 = 1.7333333$

$1\frac{11}{15} = 1.7333333$

2.021 4,600

2.022 0.038

2.023 0.041368

2.024 6.3

2.025 0.008143

2.026 153

2.027 $\frac{4}{_1\cancel{9}} \times \frac{\cancel{27}^3}{1} = \frac{12}{1} = 12$

2.028 $\frac{5}{8} \times \frac{3}{4} = \frac{15}{32}$ cup

2.029
$$\begin{array}{r} 210 \text{ packages} \\ 1.5\overline{)315.0.} \\ \underline{30} \\ 15 \\ \underline{15} \\ 0 \end{array}$$

2.030 $7 \times 2\frac{3}{4} = \frac{7}{1} \times \frac{11}{4} = \frac{77}{4} = 19\frac{1}{4}$ feet

2.031
$$\begin{array}{r} 22.059 \rightarrow 22.06 \text{ miles} \\ 8.4\overline{)185.3.000} \quad \text{per gallon} \\ \underline{168} \\ 173 \\ \underline{168} \\ 500 \\ \underline{420} \\ 800 \\ \underline{756} \\ 44 \end{array}$$

2.032
$$\begin{array}{r} \$575 \\ 373\overline{)214475} \\ \underline{1865} \\ 2797 \\ \underline{2611} \\ 1865 \\ \underline{1865} \\ 0 \end{array}$$

2.033
$$
\begin{array}{r}
2.895 \\
\times\ 6.25 \\
\hline
14475 \\
5790 \\
17370 \\
\hline
18.09375 \rightarrow \$18.09
\end{array}
$$

2.034
$$
\begin{array}{r}
0.3008 \rightarrow 0.301 \\
123\overline{)37.0000} \\
\underline{369} \\
1000 \\
\underline{984} \\
16
\end{array}
$$

2.035
$$
\begin{array}{r}
0.001 \\
\times\ 347 \\
\hline
007 \\
004 \\
\underline{003} \\
00.347 = 0.347\ \text{ounces}
\end{array}
$$

2.036
$$
\begin{array}{r}
12.5 \\
\times\ 0.53 \\
\hline
375 \\
\underline{625} \\
6.625 \rightarrow \$6.63
\end{array}
$$

2.037 $\dfrac{1 \times 17 + 16}{17} = \dfrac{17 + 16}{17} = \dfrac{33}{17}$

2.038 $\dfrac{3 \times 8 + 5}{8} = \dfrac{24 + 5}{8} = \dfrac{29}{8}$

2.039 $\dfrac{15 \times 2 + 1}{2} = \dfrac{30 + 1}{2} = \dfrac{31}{2}$

2.040 $\dfrac{103 \times 4 + 3}{4} = \dfrac{412 + 3}{4} = \dfrac{415}{4}$

2.041 $\dfrac{105 \div 105}{315 \div 105} = \dfrac{1}{3}$

2.042 $\dfrac{300 \div 150}{450 \div 150} = \dfrac{2}{3}$

2.043 $\dfrac{504 \div 168}{672 \div 168} = \dfrac{3}{4}$

2.044 $\dfrac{1,512 \div 1,512}{4,536 \div 1,512} = \dfrac{1}{3}$

2.045
$$
\begin{array}{r}
0.81 \\
\times\ 0.003 \\
\hline
0.00243
\end{array}
$$

2.046
$$
\begin{array}{r}
48.3 \\
\times\ 6.1 \\
\hline
483 \\
\underline{2898} \\
294.63
\end{array}
$$

2.047
$$
\begin{array}{r}
41 \\
\times\ 0.3 \\
\hline
12.3
\end{array}
$$

2.048
$$
\begin{array}{r}
22466.666 \rightarrow 22,466.67 \\
0.03\overline{)674.00.000} \\
\underline{6} \\
07 \\
\underline{6} \\
14 \\
\underline{12} \\
20 \\
\underline{18} \\
20 \\
\underline{18} \\
20 \\
\underline{18} \\
20 \\
\underline{18} \\
20 \\
\underline{18} \\
20 \\
\underline{18}
\end{array}
$$

2.049
$$
\begin{array}{r}
882 \\
9.90 \\
1.6\overline{)15.8.40} \\
\underline{144} \\
144 \\
\underline{144} \\
0
\end{array}
$$

2.050
$$
\begin{array}{r}
112.133 \rightarrow 112.13 \\
0.075\overline{)8.410.000} \\
\underline{75} \\
91 \\
\underline{75} \\
160 \\
\underline{150} \\
100 \\
\underline{75} \\
250 \\
\underline{225} \\
250 \\
\underline{225} \\
25
\end{array}
$$

SELF TEST 3

3.01   in a fraction, the number above the bar

3.02   a symbol with a whole number part and a fraction part

3.03   a dot used in decimal place-value notation to separate whole number values from fraction values

3.04   a digit to the right of the decimal point

3.05   per hundred, out of every hundred, or hundredths

3.06 $\frac{1}{3} \times \frac{6^2}{7} = \frac{2}{7}$

3.07 $\frac{248^{124}}{11} \times \frac{3}{2_1} = \frac{372}{11} = 33\frac{9}{11}$

3.08 $\frac{3}{4} \times \frac{3}{1} = \frac{9}{4} = 2\frac{1}{4}$

3.09 $\frac{4^1}{3} \times \frac{15^5}{4_1} = \frac{5}{1} = 5$

3.010 $\frac{8}{1} \times \frac{14^2}{5} = \frac{16}{5} = 3\frac{1}{5}$

3.011 $\frac{9^3}{10_2} \times \frac{5^1}{6_2} = \frac{3}{4}$

3.012 $\frac{12^1}{14_2} \times \frac{21^{3^1}}{36_{2_1}} = \frac{1}{2}$

3.013 $\frac{17^1}{3} \times \frac{8}{17_1} = \frac{8}{3} = 2\frac{2}{3}$

3.014 $\frac{147}{13} = 13\overline{)147}^{\ 11} = 11\frac{4}{13}$

                 13
                 17
                 13
                  4

3.015 $\frac{69}{5} = 5\overline{)69}^{\ 13} = 13\frac{4}{5}$

                 5
                19
                15
                 4

3.016 $\frac{427}{3} = 3\overline{)427}^{\ 142} = 142\frac{1}{3}$

                 3
                12
                12
                07
                 6
                 1

3.017 $\frac{1{,}674}{7} = 7\overline{)1{,}674}^{\ 239} = 239\frac{1}{7}$

                 14
                27
                21
                64
                63
                 1

3.018
```
    68.3
 x 0.87
   4781
  5464
 59.421
```

3.019
```
   0.071
 x 14.38
     568
     213
     284
      71
 1.02098
```

3.020
```
            19.875
 0.16)3.18.000
       16
       158
       144
        140
        128
         120
         112
          80
          80
           0
```

3.021

$$6.41 \overline{)0.01.300} \quad \frac{0.002}{}$$
$$\underline{1\ 282}$$
$$18$$

3.022 $\frac{9}{5} \div \frac{5}{6} = \frac{9}{5} \times \frac{6}{5} = \frac{54}{25} = 2\frac{4}{25}$

$1\frac{4}{5} = 1.8$

$1.8 \div 0.8333333 = 2.168 = 2.17$

$2\frac{4}{25} = 2.16$

3.023 $\frac{{}^1\cancel{11}}{{}_1\cancel{4}} \times \frac{\cancel{8}^2}{\cancel{11}_1} = \frac{2}{1} = 2$

$2\frac{3}{4} = \qquad 2.75$

$\times \quad \frac{8}{11} = \quad \underline{0.7272727}$
$\qquad\qquad\qquad 1925$
$\qquad\qquad\qquad 550$
$\qquad\qquad\qquad 1925$
$\qquad\qquad\qquad 550$
$\qquad\qquad\qquad 1925$
$\qquad\qquad\qquad 550$
$\qquad\qquad\qquad \underline{1925}$
$\qquad\qquad 1.999999925$

$2 =$ approximately 1.999999925

3.024 $\frac{2}{{}_1\cancel{3}} \times \frac{\cancel{9}^3}{7} = \frac{6}{7}$

$\frac{2}{3} = 0.\overline{6}$

$\frac{7}{9} = 0.\overline{7}$

$0.6666667 \div 0.7777778 = 0.8571429$

$\frac{6}{7} = 0.8571429$

3.025 $\frac{{}^7\cancel{14}}{1} \times \frac{7}{\cancel{2}_1} = \frac{49}{1} = 49$

$14 = 1\ 4$

$\times \quad 3\ 1/2 = \underline{3.5}$
$\qquad\qquad\quad 7\ 0$
$\qquad\qquad\quad \underline{4\ 2}$
$\qquad\quad 49 = 4\ 9$

3.026 a. 50%

b. $0.5 = \frac{5}{10} = \frac{5 \div 5}{10 \div 5} = \frac{1}{2}$

3.027 a. 0.16

b. $0.16 = \frac{16}{100} = \frac{16 \div 4}{100 \div 4} = \frac{4}{25}$

3.028 a. $\frac{1}{4} = 1 \div 4 = 0.25 = 25\%$

b. 0.25

3.029 a. $\frac{5}{9} = 5 \div 9 = 0.\overline{5}$ or $0.555... = 55.6\%$

b. 0.556

3.030 a. 38%

b. $0.38 = \frac{38}{100} = \frac{38 \div 2}{100 \div 2} = \frac{19}{50}$

3.031 a. 0.75

b. $75\% = 0.75 = \frac{75}{100} = \frac{75 \div 25}{100 \div 25} = \frac{3}{4}$

3.032

$$\begin{array}{r} 64 \\ \times\ 0.03 \\ \hline 1.92 \end{array}$$

3.033

$$112\overline{)135.000} \quad \frac{1.205}{} = 120.5\%$$
$$\underline{112}$$
$$230$$
$$\underline{224}$$
$$600$$
$$\underline{560}$$
$$40$$

3.034

$$0.12\overline{)90.00} \quad \frac{750}{}$$
$$\underline{84}$$
$$60$$
$$\underline{60}$$
$$0$$

3.035

$$0.75\overline{)147.00} \quad \frac{196}{}$$
$$\underline{75}$$
$$720$$
$$\underline{675}$$
$$450$$
$$\underline{450}$$
$$0$$

3.036
$$\frac{0.2209}{4,721)1,043.0000} = 0.221 = 22.1\%$$
$$\underline{9442}$$
$$9880$$
$$\underline{9442}$$
$$43800$$
$$\underline{42489}$$
$$1311$$

3.037
$$\begin{array}{r} 1.19 \\ \times\ 75 \\ \hline 595 \\ \underline{833} \\ 89.25 \end{array}$$

3.038 $16\frac{1}{2} \div \frac{1}{4} = \frac{33}{2} \times \frac{\cancel{4}^2}{1} = \frac{66}{1} = 66$ pieces

3.039 $15\frac{1}{2} \div 100 = \frac{31}{2} \times \frac{1}{100} = \frac{31}{200}$ of a second

3.040 $13\frac{1}{2} \div 3\frac{1}{4} = \frac{27}{\cancel{2}_1} \times \frac{\cancel{4}^2}{13} = \frac{54}{13}$

$= 4\frac{2}{13}$ degrees

3.041
$$\begin{array}{r} 12.75 \\ \times\ 47.3 \\ \hline 3825 \\ 8925 \\ \underline{5100} \\ 603.075 \text{ miles} \end{array}$$

3.042
$$\begin{array}{r} 3,427 \\ \times\ 0.012 \\ \hline 6854 \\ \underline{3427} \\ 41.124 \text{ inches} \end{array}$$

3.043
$$\begin{array}{r} 14.7 \\ \times\ 59.9 \\ \hline 1323 \\ 1323 \\ \underline{735} \\ 880.53 \text{ cents or } \$8.81 \end{array}$$

3.044
$$\frac{0.05 \text{ or } 5¢}{150)7.50}$$
$$\underline{750}$$
$$0$$

3.045 The base is $4,375.
The rate is 6.5% = 0.065.
The percentage is $4,375 x 0.065 =
$$\begin{array}{r} 4,375 \\ \times\ 0.065 \\ \hline 21875 \\ \underline{26250} \\ 284375 = \$284.38 \end{array}$$

3.046 The base is $12.80.
The rate is 10% = 0.1.
The percentage is $12.80 x 0.1 =
$$\begin{array}{r} 12.8 \\ \times\ 0.1 \\ \hline \$1.28 \end{array}$$

3.047 The base is 1,247.
The rate is 93% = 0.93.
The percentage is 1,247 x 0.93 =
$$\begin{array}{r} 1,247 \\ \times\ 0.93 \\ \hline 3741 \\ \underline{11223} \\ 1,159.71 \rightarrow 1,160 \end{array}$$

3.048 The base is 142.
The rate is 86.3% = 0.863.
The percentage s 142 x 0.863 =
$$\begin{array}{r} 142 \\ \times\ 0.863 \\ \hline 426 \\ 852 \\ \underline{1136} \\ 122.546 \end{array}$$

3.049 The rate is 16% = 0.16.
The percentage is 784.
The base is 784 ÷ 0.16 =
$$\frac{4,900}{0.16)784.00}$$
$$\underline{64}$$
$$144$$
$$\underline{144}$$
$$0$$

3.050 The base is 93.
The percentage is 47.5.
The rate is 47.5 ÷ 93 =
$$\frac{0.5107}{93)47.5000} \rightarrow 0.511 = 51.1\%$$
$$\underline{465}$$
$$100$$
$$\underline{93}$$
$$700$$
$$\underline{651}$$
$$49$$

SELF TEST 1

1.01   a general rule written in mathematical symbols

1.02   the amount of a loan

1.03   the fee paid for the use of borrowed money

1.04   the percent of the principal charged as interest

1.05   put in place of

1.06 $A = L \times W$
$A = 47.3 \times 19.7$

$$
\begin{array}{r}
47.3 \\
\times\ 19.7 \\
\hline
3311 \\
4257 \\
473\ \ \ \\
\hline
\end{array}
$$
$A = \quad 931.81$

1.07 $A = L \times W$
$48 = 8 \times W$
$W = 6$

1.08 $A = L \times W$
$50 = L \times 5$
$L = 10$

1.09 $A = L \times W$
$A = 6\frac{1}{2} \times 2\frac{1}{4} = \frac{13}{2} \times \frac{9}{4} = \frac{117}{8}$
$A = 14\frac{5}{8}$

1.010 $S = C + P$
$\$100 = \$75 + P$
$P = \$25$

1.011 $S = C + P$
$S = \$14.95 + \$9.25$
$S = \$24.20$

1.012 $S = C + P$
$S = \$1.50 + \$0.97$
$S = \$2.47$

1.013   Example:
$I = R \times P$

1.014 $I = R \times P$
$I = 0.18 \times \$350$

$$
\begin{array}{r}
0.18 \\
\times\ 350 \\
\hline
900 \\
54\ \ \ \\
\hline
\end{array}
$$
$I = \ \$63.00$

1.015 $I = R \times P$
$I = 0.06 \times \$6{,}000$

$$
\begin{array}{r}
0.06 \\
\times\ 6000 \\
\hline
\end{array}
$$
$I = \$360.00$

1.016 $I = R \times P$
$I = 0.125 \times \$875$

$$
\begin{array}{r}
0.125 \\
\times\ 875 \\
\hline
625 \\
875\ \ \\
1000\ \ \ \\
\hline
109375
\end{array}
$$
$I = \$109.38$

1.017 $I = R \times P$
$I = 0.1 \times 470$

$$
\begin{array}{r}
0.1 \\
\times\ 470 \\
\hline
070 \\
04\ \ \\
\hline
047.0
\end{array}
$$
$I = \$47$

1.018   Example:
$P = W \times H$

1.019 $P = W \times H$
$P = \$2.75 \times 14$

$$
\begin{array}{r}
\$2.75 \\
\times\ 14 \\
\hline
1100 \\
275\ \ \\
\hline
\end{array}
$$
$P = \quad \$38.50$

1.020 $P = W \times H$
$\$200 = \$5.00 \times H$
$H = 40$

1.021 $P = W \times H$
$\$150 = \$10.00 \times H$
$H = 15$

1.022 $P = W \times H$
$P = \$6.50 \times 37$

$$\begin{array}{r} \$6.50 \\ \times\ \ 37 \\ \hline 4550 \\ 1950 \\ \hline \end{array}$$

$P = \$240.50$

1.023 $d = r \times t$

$d = 60 \times 4\frac{1}{2}$

$\overset{30}{\underset{1}{\cancel{60}}} \times \underset{1}{\frac{9}{\cancel{2}}} = \frac{270}{1}$

$d = 270$ miles

1.024 $P = 4 \times S$
$P = 4 \times 6$
$P = 24$ feet

1.025 $N = L - D$
$N = \$49.50 - \$12.50$
$N = \$37.00$

1.026 $d = r \times t$
$d = 3 \times 0.6$
$d = 1.8$ miles

1.027 $I = R \times P$
$I = 0.18 \times \$250$

$$\begin{array}{r} 0.18 \\ \times\ 250 \\ \hline 900 \\ 36\ \ \\ \hline \end{array}$$

$I = \ \ \$45.00$

1.028 $mpg = m \div g$
$mpg = 250 \div 15$

$$\begin{array}{r} 16.666 \\ 15\overline{)250.000} \\ \underline{15}\ \ \ \ \ \ \ \\ 100 \\ \underline{90} \\ 100 \\ \underline{90} \\ 100 \\ \underline{90} \\ 100 \\ \underline{90} \\ \end{array}$$

$mpg = 16.67$

1.029 $2 \times 5 = 10$

1.030 $11 \times 5 = 55$

1.031 $20 \times 5 = 100$

1.032 $15 \div 2 = 7\frac{1}{2}$ or 7.5

1.033 $14{,}400 \div 2 = 7{,}200$

1.034 $\frac{1}{2} \div 2 = \frac{1}{2} \times \frac{1}{2} = \frac{1}{4}$

SELF TEST 2

2.01   a mathematical sentence that expresses equality between two quantities

2.02   a number that, when substituted for an unknown in an open equation, makes the equation true

2.03   an equation with one or more symbols for which we can substitute numbers

2.04   a general rule stated in symbols

2.05   put in place of

2.06 $\$16.50 = R \cdot \$300$
$\$16.50 \div \$300 = R$

$$\begin{array}{r} 0.055 \\ 300\overline{)16.500} \\ \underline{1500} \\ 1500 \\ \underline{1500} \\ 0 \end{array}$$

$0.055 = R$
$5.5\% = R$

2.07 $I = 0.09 \cdot \$1{,}475$

$$\begin{array}{r} \$1{,}475 \\ \times\ \ 0.09 \\ \hline \end{array}$$

$I = \ \ \$132.75$

2.08 $\$47.00 = 0.125 \times P$
$\$47.00 \div 0.125 = P$

$$
\begin{array}{r}
376 \\
125\overline{)47{,}000} \\
\underline{375} \\
950 \\
\underline{875} \\
750 \\
\underline{750} \\
0
\end{array}
$$

$\$376 = P$

2.09 $x = 509$

2.010 $16 + 493 = x$
$509 = x$

2.011 $x = 6.1 \times 47.3$

$$
\begin{array}{r}
47.3 \\
\times\ \ 6.1 \\
\hline
473 \\
2838 \\
\hline
\end{array}
$$
$x = 288.53$

2.012 $4.3 \div 0.7 = T$

$$
\begin{array}{r}
6.1428 \\
7\overline{)43.0000} \\
\underline{42} \\
10 \\
\underline{7} \\
30 \\
\underline{28} \\
20 \\
\underline{14} \\
60 \\
\underline{56} \\
4
\end{array}
$$

$6.143 = T$

2.013 $\dfrac{13}{2} \times \dfrac{7}{2} = x$

$\dfrac{91}{4} = x$

$22\dfrac{3}{4} = x$

2.014   open

2.015   true; $4 = 4$

2.016   true

2.017   false

2.018   open

2.019 $N = L - D$
$\$8.00 = L - \$3.50$
$\$8.00 + \$3.50 = L$
$\$11.50 = L$

2.020 $A = L \times W$
$A = 5 \times 8$
$A = 40$

2.021 $S = C + P$
$\$9.00 = C + \$3.00$
$\$9.00 - \$3.00 = C$
$\$6.00 = C$

2.022 $I = R \times P$
$\$15.00 = 0.04 \times P$
$\$15.00 \div 0.04 = P$

$$
\begin{array}{r}
375 \\
4\overline{)1500} \\
\underline{12} \\
30 \\
\underline{28} \\
20 \\
\underline{20} \\
0
\end{array}
$$

$\$375 = P$

2.023 $d = r \times t$
$180 = 60 \times t$
$180 \div 60 = t$
$3 = t$
$t = 3$ hours

2.024 $I = R \times P$
$I = 0.18 \times \$1{,}200$

$$
\begin{array}{r}
1{,}200 \\
\times\ 0.18 \\
\hline
9600 \\
1200 \\
\hline
\end{array}
$$
$I = \ \$216.00$

2.025 $1\dfrac{1}{2} + 4\dfrac{1}{2} = 6$

2.026 $7 + 4\dfrac{1}{2} = 11\dfrac{1}{2}$

2.027 $0.5 + 4\frac{1}{2} =$

$\frac{1}{2} + 4\frac{1}{2} = 5$ or

$0.5 + 4\frac{1}{2} =$

$0.5 + 4.5 = 5.0$

2.028 $100 - 90 = 10$

2.029 $15 - 8.3 = 6.7$

2.030 $25 - 25 = 0$

SELF TEST 3

3.01   the amount of a loan

3.02   the fee paid for the use of borrowed money

3.03   a mathematical sentence that expresses equality between two quantities

3.04   a number that, when substituted for an unknown in an open equation, makes the equation true

3.05   a comparison of one quantity with another

3.06   a statement that two ratios are equal

3.07   the inner two terms in a proportion

3.08   the outer two terms in a proportion

3.09
$$S = C + P$$
$$\$49.95 = \$30.00 + P$$
$$\$49.95 - \$30.00 = P$$
$$\$19.95 = P$$

3.010
$$S = C + P$$
$$\$1.50 = C + \$0.60$$
$$\$1.50 - \$0.60 = C$$
$$\$0.90 = C$$

3.011
$$S = C + P$$
$$S = \$23.50 + \$14.95$$
$$S = \$38.45$$

3.012   true

3.013   true

3.014   open

3.015   false

3.016   open

3.017   false

3.018 $I = R \times P$

3.019 $d = r \times t$

3.020 $A = L \times W$

3.021 $S = C + P$

3.022 $P = W \times H$

3.023 $N = L - D$

3.024 $mpg = m \div g$

3.025 $P = 4 \times S$

3.026 $10:25 = 2:5$

3.027 $15:10 = 3:2$

3.028 $10:15 = 2:3$

3.029 $15:25 = 3:5$

3.030   a. $7 \times 16 = 112$
       b. $4 \times 29 = 116$
       c.   no

3.031   a. $2 \times 6 = 12$
       b. $1 \times 12 = 12$
       c.   yes

3.032   a. $44 \times 14 = 616$
       b. $77 \times 8 = 616$
       c.   yes

3.033   a. $5 \times 4 = 20$
       b. $3 \times 6 = 18$
       c.   no

3.034 $mpg = m \div g$
       $mpg = 350 \div 15$

$$15\overline{)35000}\;\;\frac{2333}{}$$
$$\underline{30}$$
$$50$$
$$\underline{45}$$
$$50$$
$$\underline{45}$$
$$50$$
$$\underline{45}$$
$$5$$

       $mpg = 23.33$

3.035 $7:9 = 14:x$
        $9 \cdot 14 = 7 \cdot x$
          $126 = 7 \cdot x$
       $126 \div 7 = x$

$$7\overline{)126}\;\;\frac{18}{}$$
$$\underline{7}$$
$$56$$
$$\underline{56}$$
$$0$$

              $x = 18$ inches

3.036 $d = r \times t$
       $d = 55 \times 12.5$

          12.5
        $\times\;55$
          625
        $\underline{625}$
          687.5
       $d = 687.5$ miles

3.037 $P = 4 \times S$
       $P = 4 \times 15$
       $P = 60$ inches

3.038 $I = R \times P$
       $\$60 = 0.15 \times P$
       $\$60 \div 0.15 = P$

$$15\overline{)6000}\;\;\frac{400}{}$$
$$\underline{60}$$
$$0$$

          $P = \$400$

3.039 $A = L \times W$
       $250 = 25 \times W$
       $250 \div 25 = W$
          $W = 10$ yards

3.040 $7 - 7 = 0$

3.041 $94 - 7 = 87$

3.042 $22 - 7 = 15$

3.043 $2 \times 5 = 10$

3.044   1.5
        $\underline{\times\;3.9}$
         135
        $\underline{45}$
        5.85

3.045 $18 \times \frac{1}{2} = 9$

SELF TEST 1

1.01   A sample that is not random.

1.02   Information from which conclusions can be drawn.

1.03   Percent of one element in the distribution as a part of the whole distribution.

1.04   A sample in which each element of the large group has an equal chance of being chosen.

1.05   Numerical information.

1.06   186, 679, 947, 724, 219

1.07   18,667; 99,477; 24, 219; (0) 1,608; 15,047

1.08   two (for two-digit numerals)

1.09   three (for three-digit numerals)

1.010   six (for six-digit numerals)

1.011

Number	Tally
1	|||
2	||||
3	|||
4	||||
5	⊦⊦⊦⊦ |

1.012   5

1.013

Number	Frequency
1	3
2	4
3	3
4	4
5	6

1.014   20

1.015

Number	Frequency
1	3
2	4
3	3
4	4
5	6
	20

Relative	Frequency		
$\frac{3}{20}$ =		0.15	= 15%
$\frac{4}{20}$ = $\frac{1}{5}$	= 0.2	= 20%	
$\frac{3}{20}$ =		0.15	= 15%
$\frac{4}{20}$ = $\frac{1}{5}$	= 0.2	= 20%	
$\frac{6}{20}$ = $\frac{3}{10}$	= 0.3	= 30%	
			100%

SELF TEST 2

2.01   The range of a set of numbers divided by the number of elements in the set.

2.02   The sum of a set of numbers divided by the number of elements in the set.

2.03   The number in a set that has as many data points below it as it does above it.

2.04   The number that appears most frequently in a set of numbers.

2.05   The difference between the smallest number and the largest number in a set.

2.06

Weight	Tally
3,000	||
3,015	|
3,016	|||
3,017	|
3,033	|
3,098	|

2.07

Weight	Frequency
3,000	2
3,015	1
3,016	3
3,017	1
3,033	1
3,098	1

2.08

Weight	Frequency
3,000	2
3,015	1
3,016	3
3,017	1
3,033	1
3,098	1
	9

Relative	Frequency
$\frac{2}{9} = 0.222 = 22.2\%$	
$\frac{1}{9} = 0.111 = 11.1\%$	
$\frac{3}{9} = \frac{1}{3} = 0.333 = 33.3\%$	
$\frac{1}{9} = 0.111 = 11.1\%$	
$\frac{1}{9} = 0.111 = 11.1\%$	
$\frac{1}{9} = 0.111 = 11.1\%$	
	99.9%

2.09 Mode 3,016

2.010 The set contains 9 numbers.
Median: $9 \div 2 = 4$ remainder
1; $4 + 1 = $ 5th number.
Median = 3,016.

2.011

Number	Frequency
3,000	2
3,015	1
3,016	3
3,017	1
3,033	1
3,098	1
	9

Number x
Frequency
$3,000 \times 2 = 6,000$
$3,015 \times 1 = 3,015$
$3,016 \times 3 = 9,048$
$3,017 \times 1 = 3,017$
$3,033 \times 1 = 3,033$
$3,098 \times 1 = 3,098$
     27,211
Mean = $27,211 \div 9 = 3,023.4$

2.012 $3,098 - 3,000 = 98$

2.013 Range = 98
Average deviation =
$98 \div 9 = 10.9$

2.014

Number	Tally
70	|||
71	|||
72	|||
73	|||
74	||||
75	|||
76	
77	┼┼┼ ||
78	||||
79	|||

2.015

Number	Frequency
70	3
71	3
72	3
73	3
74	4
75	3
76	0
77	7
78	4
79	3

2.016

Number	Frequency
70	3
71	3
72	3
73	3
74	4
75	3
76	0
77	7
78	4
79	3
	33

Relative Frequency

$\frac{3}{33} = \frac{1}{11} =$ 0.091 = 9.1%

$\frac{3}{33} = \frac{1}{11} =$ 0.091 = 9.1%

$\frac{3}{33} = \frac{1}{11} =$ 0.091 = 9.1%

$\frac{3}{33} = \frac{1}{11} =$ 0.091 = 9.1%

$\frac{4}{33} =$ 0.121 = 12.1%

$\frac{3}{33} = \frac{1}{11} =$ 0.091 = 9.1%

$\frac{0}{33} =$ 0.000 = 0.0%

$\frac{7}{33} =$ 0.212 = 21.2%

$\frac{4}{33} =$ 0.121 = 12.1%

$\frac{3}{33} = \frac{1}{11} =$ 0.091 = 9.1%

100.0%

2.017   Mode = 77

2.018   The set contains 33 numbers. Median: $33 \div 2 = 16$ remainder 1; $16 + 1 = 17$th number. Median = 75.

2.019

Number	Frequency
70	3
71	3
72	3
73	3
74	4
75	3
76	0
77	7
78	4
79	3
	33

Number x Frequency

70 x 3 =	210
71 x 3 =	213
72 x 3 =	216
73 x 3 =	219
74 x 4 =	296
75 x 3 =	225
76 x 0 =	0
77 x 7 =	539
78 x 4 =	312
79 x 3 =	237
	2,467

Mean, $2,467 \div 33 = 74.8$

2.020 $79 - 70 = 9$

2.021   Range = 9
Average deviation
$9 \div 33 = 0.3$

SELF TEST 3

3.01   A graph in which data are represented by bars of different heights.

3.02 A sample that is not random.

3.03 A bar graph comparing two or more sets of data.

3.04 A graph in which data are represented by dots connected with lines.

3.05 A graph in which data are represented by pictures.

3.06 A sample in which every member of the large group has an equal chance of being chosen.

3.07 052, 693, 706, 022, 358

3.08 05, 269, 37,060; 22,358; 51,513; 92, 035

3.09

Number	Tally
20	| | | |
21	| |
22	| | |
23	| | | |
24	︱︱︱︱
25	|
26	
27	︱︱︱︱ |
28	| | |
29	| |

3.010

Number	Frequency
10	1
20	3
30	5
40	7
50	5
60	3
70	1

3.011 Mode = 47

3.012 The set contains 30 numbers. Median: $30 \div 2 = 15$; the 15th and 16th numbers are both $9.00. Median = $9.00.

3.013

Number	Frequency	Number x Frequency
2	6	2 x 6 = 12
4	12	4 x 12 = 48
6	10	6 x 10 = 60
8	6	8 x 6 = 48
10	3	10 x 3 = 30
12	3	12 x 3 = 36
	40	234

Mean = $234 \div 40 = 5.85 = 5.9$

3.014 $111 - 33 = 78$

3.015 Range = $34 - 24 = 10$
The set contains 11 numbers.
Average deviation = $10 \div 11 = 0.9$

3.016

3.017

316

3.018

Jan. Feb. Mar. Apr. May June

3.019

□ passing
■ rushing

3.020

Number	Tally
1	|
2	||
3	|||
4	╫╫
5	|||
6	||
7	|

3.021

Number	Frequency
1	1
2	2
3	3
4	5
5	3
6	2
7	1

3.022

Number	Frequency
1	1
2	2
3	3
4	5
5	3
6	2
7	1
	17

Relative	Frequency
$\frac{1}{17} = 0.059$	$= 5.9\%$
$\frac{2}{17} = 0.118$	$= 11.8\%$
$\frac{3}{17} = 0.176$	$= 17.6\%$
$\frac{5}{17} = 0.294$	$= 29.4\%$
$\frac{3}{17} = 0.176$	$= 17.6\%$
$\frac{2}{17} = 0.118$	$= 11.8\%$
$\frac{1}{17} = 0.059$	$= 5.9\%$
	100.0%

3.023 Mode = 4

3.024 The set contains 17 numbers.
Median: $17 \div 2 = 8$ remainder
1; $8 + 1 = $ 9th number.
Median = 4.

3.025

Number	Frequency	Number x Frequency
1	1	1 x 1 = 1
2	2	2 x 2 = 4
3	3	3 x 3 = 9
4	5	4 x 5 = 20
5	3	5 x 3 = 15
6	2	6 x 2 = 12
7	1	7 x 1 = 7
	17	68

Mean = $68 \div 17 = 4$

3.026 $7 - 1 = 6$

3.027   Range = 6

Average deviation =

$6 \div 17 = 0.4$

3.028

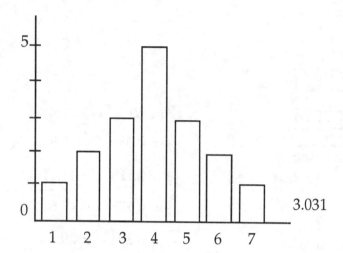

3.029

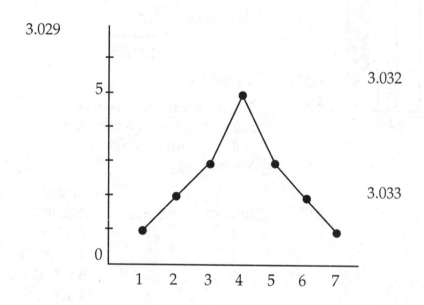

3.030

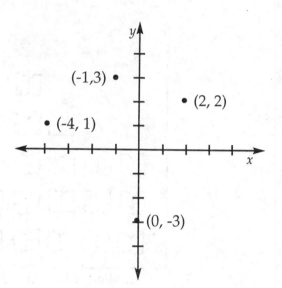

3.031

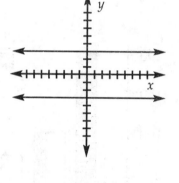

3.032

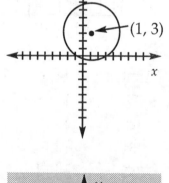

3.033

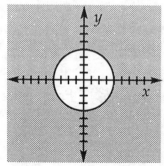

SELF TEST 1

1.01   a. 80,000; 10 + 6
        b. 80,000 > 16
        c. 80,000 ÷ 16 = 5,000

1.02   a. 50 + 2; 40 + 6
        b. 46 < 52
        c. 52 + 46 = 98

1.03   a. 10 + 5; 100 + 90 + 7
        b. 197 > 15
        c. 15 x 197 = 2,955

1.04   a. 800 + 20 + 3;
           900 + 30 + 2
        b. 823 < 932
        c. 932 − 823 = 109

1.05   a. 30 + 2; 40 + 8
        b. 48 > 32
        c. $48 ÷ 32 = 1\frac{1}{2}$ pounds

1.06   a. 30 + 9; 20 + 7
        b. 27 < 39
        c. 39 + 27 = 66

1.07   a. 400 + 50 + 7;
           50 + 9
        b. 457 > 59
        c. 457 x 59 = 26,963

1.08   a. 7,000 + 60 + 2;
           6,000 + 900 + 70 + 4
        b. 6,974 < 7,062
        c. 7,062 − 6,974 = 88

SELF TEST 2

2.01   d

2.02   a

2.03   e

2.04   c

2.05–2.08 Student answers may vary slightly.

2.05 $4\frac{3}{4}$ − 5 in; 12.3 − 12.4 cm;
        64°; 82°; 34°

2.06   5 inches; 12.3 cm; 54°; 36°; 90°

2.07 $5\frac{3}{4}$ inches; 14.6 cm; 49°; 81°; 50°

2.08 $3\frac{1}{2}$ inches; 8.8 cm; 80°; 31°; 69°

2.09   a. 15 = 3 · ⑤
          55 = ⑤ · 11
      GCF = 5

      b.

4's	2's	1's
1	0	1

      $5 = 101_2$

2.010   a. 16 = ⟨2 · 2 · 2 · 2⟩
          48 = ⟨2 · 2 · 2 · 2⟩ · 3
      GCF = 2 · 2 · 2 · 2 = 16

      b.

16's	8's	4's	2's	1's
1	0	0	0	0

      $16 = 10000_2$

2.011   a. 4 = 2 · 2
          6 = 2 · 3
      LCM = 2 · 2 · 3 = 12

      b.

8's	4's	2's	1's
1	1	0	0

      $12 = 1100_2$

2.012   a. 8 = 2 · 2 · 2
          12 = 2 · 2 · 3
      LCM = 2 · 2 · 2 · 3 = 24

      b.

16's	8's	4's	2's	1's
1	1	0	0	0

      $24 = 11000_2$

2.013   ∈

2.014   ∉

2.015   ∈

2.016   ∉

2.017   ∈

2.018   ∉

2.019 $C = \pi \times d$
$C = 3.14 \times 18$
$C = 56.52$ inches

2.020 $C = 2 \times \pi \times r$
$100 = 2 \times 3.14 \times r$
$100 = 6.28 \times r$
$\dfrac{100}{6.28} = r$
$r = 15.9$ meters

2.021 $34 \times 20 = 680$ sq. ft.

2.022   a. $94 \times 50 = 4{,}700$ sq. ft.
b. $74 \times 42 = 3{,}108$ sq. ft.
c. $4{,}700 - 3{,}108 = 1{,}592$ sq. ft.
d. $1{,}000 + 500 + 90 + 2$

2.023   a. $36 \times 78 = 2{,}808$ sq. ft.
b. $27 \times 78 = 2{,}106$ sq. ft.
c. Length of singles back-court:
$4'\,6'' = 4\frac{1}{2}$ ft.

$2 \times 4\frac{1}{2} = 2 \times \frac{9}{2} = 9$ ft.
$36 - 9 = 27$ ft.

Width of singles back-court:
$21 + 21 = 42$
$78 - 42 = 36$
$36 \div 2 = 18$ ft.
Area $= 27 \times 18 = 486$ sq. ft.
d. Length of singles fore-court:

$4'\,6'' = 4\frac{1}{2}$ ft.

$2 \times 4\frac{1}{2} = 2 \times \frac{9}{2} = 9$ ft.

$36 - 9 = 27$ ft.

Width of singles fore-court:
21 ft.
Area $= 27 \times 21 = 567$ sq. ft.
e. Area of singles service court $=$
$2 \times$ area of singles forecourt $=$
$2 \times 567 = 1{,}134$ sq. ft.
f. $2{,}808 > 2{,}106 > 1{,}134 > 567 > 486$

SELF TEST 3

3.01   a. $\frac{3}{4} + \frac{1}{2} = \frac{3}{4} + \frac{2}{4} = \frac{5}{4}$ or $1\frac{1}{4}$

b. $\frac{3}{4} - \frac{1}{2} = \frac{3}{4} - \frac{2}{4} = \frac{1}{4}$

c. $\frac{3}{4} \times \frac{1}{2} = \frac{3}{8}$

d. $\frac{3}{4} \div \frac{1}{2} = \frac{3}{\underset{2}{\cancel{4}}} \times \frac{\overset{1}{\cancel{2}}}{1} = \frac{3}{2}$ or $1\frac{1}{2}$

3.02 $2\frac{2}{3} = \frac{(2 \times 3) + 2}{3} = \frac{8}{3}$

a. $\frac{3}{4} + \frac{8}{3} = \frac{9}{12} + \frac{32}{12} = \frac{41}{12}$ or $3\frac{5}{12}$

b. $\frac{8}{3} - \frac{3}{4} = \frac{32}{12} - \frac{9}{12} = \frac{23}{12}$ or $1\frac{11}{12}$

c. $\frac{\overset{1}{\cancel{3}}}{\underset{1}{\cancel{4}}} \times \frac{\overset{2}{\cancel{8}}}{\underset{1}{\cancel{3}}} = \frac{2}{1} = 2$

d. $\frac{3}{4} \div \frac{8}{3} = \frac{3}{4} \times \frac{3}{8} = \frac{9}{32}$

3.03   a. $\frac{3}{4} + 5 = 5\frac{3}{4}$

b. $5 - \frac{3}{4} = \frac{20}{4} - \frac{3}{4} = \frac{17}{4}$ or $4\frac{1}{4}$

c. $\frac{3}{4} \times 5 = \frac{3}{4} \times \frac{5}{1} = \frac{15}{4}$ or $3\frac{3}{4}$

d. $\frac{3}{4} \div 5 = \frac{3}{4} \times \frac{1}{5} = \frac{3}{20}$

3.04   a. $\frac{3}{4} + \frac{4}{5} = \frac{15}{20} + \frac{16}{20} = \frac{31}{20}$ or $1\frac{11}{20}$
b. $\frac{4}{5} - \frac{3}{4} = \frac{16}{20} - \frac{15}{20} = \frac{1}{20}$
c. $\frac{3}{\underset{1}{\cancel{4}}} \times \frac{\overset{1}{\cancel{4}}}{5} = \frac{3}{5}$
d. $\frac{3}{4} \div \frac{4}{5} = \frac{3}{4} \times \frac{5}{4} = \frac{15}{16}$

3.05   a. $\frac{3}{4} + \frac{1}{6} = \frac{9}{12} + \frac{2}{12} = \frac{11}{12}$
b. $\frac{3}{4} - \frac{1}{6} = \frac{9}{12} - \frac{2}{12} = \frac{7}{12}$
c. $\frac{\overset{1}{\cancel{3}}}{4} \times \frac{1}{\underset{2}{\cancel{6}}} = \frac{1}{8}$
d. $\frac{3}{4} \div \frac{1}{6} = \frac{3}{\underset{2}{\cancel{4}}} \times \frac{\overset{3}{\cancel{6}}}{1} = \frac{9}{2}$ or $4\frac{1}{2}$

3.06   a. 0.33 + 0.75 = 1.08

       b. 0.75 − 0.33 = 0.42

       c.
```
      0.33
    x 0.75
      165
     231
    0.2475
```

       d.
```
              0.44
   0.75)0.33,00
            300
            300
            300
              0
```

3.07   a. 0.33 + 7.0 = 7.33

       b. 7.0 − 0.33 = 6.67

       c.
```
      0.33
    x 7.0
    2.310 = 2.31
```

       d.
```
        0.04714
    7)0.33000
      28
      50
      49
      10
       7
      30
      20
       2
```

3.08   a. 0.33 + 1.59 = 1.92

       b. 1.59 − 0.33 = 1.26

       c.
```
      0.33
    x 1.59
      297
      165
      33
    0.5247
```

       d.
```
            0.20754
    1.59)0.33,00000
           318
          1200
          1113
           870
           795
           750
           636
            14
```

3.09   a. 0.33 + 0.86 = 1.19

       b. 0.86 − 0.33 = 0.53

       c.
```
      0.33
    x 0.86
      198
      264
    0.2838
```

3.09 cont.

       d.
```
              0.38372
   0.86)0.33,00000
            258
            720
            688
            320
            258
            620
            602
            180
            172
              8
```

3.010 a. 0.33 + 0.12 = 0.45

       b. 0.33 − 0.12 = 0.21

       c.
```
      0.33
    x 0.12
      066
      033
    0.0396
```

       d.
```
             2.75
   0.12)0.33,00
           24
           90
           84
           60
           60
            0
```

3.011 {Willie, John, Kevin, Greg, Jeff, Fred, Richard, Mike, Sam, Cheryl, Nancy, Vicki, Dawn, Amy, Sandy}

3.012 φ

3.013 {John, Kevin, Richard, Sam}

3.014 {Cheryl, Nancy}

3.015

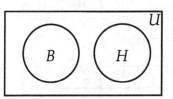

3.016

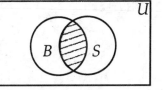

3.017

1's
1

$1 = 1_2$

3.018

2's	1's
1	0

$2 = 10_2$

3.019

2's	1's
1	1

$3 = 11_2$

3.020

4's	2's	1's
1	0	0

$4 = 100_2$

3.021

4's	2's	1's
1	0	1

$5 = 101_2$

3.022

4's	2's	1's
1	1	0

$6 = 110_2$

3.023

4's	2's	1's
1	1	1

$7 = 111_2$

3.024

8's	4's	2's	1's
1	0	0	0

$8 = 1000_2$

3.025

8's	4's	2's	1's
1	0	0	1

$9 = 1001_2$

3.026

8's	4's	2's	1's
1	0	1	0

$10 = 1010_2$

SELF TEST 4

4.01 a. $\frac{1}{2} + \frac{3}{4} = \frac{2}{4} + \frac{3}{4} = \frac{5}{4}$ or $1\frac{1}{4}$

b. $\frac{3}{4} - \frac{1}{2} = \frac{3}{4} - \frac{2}{4} = \frac{1}{4}$

c. $\frac{3}{4} \times \frac{1}{2} = \frac{3}{8}$

d. $\frac{3}{4} \div \frac{1}{2} = \frac{3}{\underset{2}{\cancel{4}}} \times \frac{\overset{1}{\cancel{2}}}{1} = \frac{3}{2}$ or $1\frac{1}{2}$

4.02 $1\frac{1}{5} = \frac{(1 \times 5) + 1}{5} = \frac{6}{5}$

a. $\frac{3}{4} + \frac{6}{5} = \frac{15}{20} + \frac{24}{20} = \frac{39}{20}$ or $1\frac{19}{20}$

b. $\frac{6}{5} - \frac{3}{4} = \frac{24}{20} - \frac{15}{20} = \frac{9}{20}$

c. $\frac{3}{\underset{2}{\cancel{4}}} \times \frac{\overset{3}{\cancel{6}}}{5} = \frac{9}{10}$

d. $\frac{3}{4} \div \frac{6}{5} = \frac{3}{4} \times \frac{5}{\underset{2}{\cancel{6}}} = \frac{5}{8}$

4.03 a. $\frac{3}{4} + \frac{5}{6} = \frac{9}{12} + \frac{10}{12} = \frac{19}{12}$ or $1\frac{7}{12}$

b. $\frac{5}{6} - \frac{3}{4} = \frac{10}{12} - \frac{9}{12} = \frac{1}{12}$

c. $\frac{\overset{1}{\cancel{3}}}{4} \times \frac{5}{\underset{2}{\cancel{6}}} = \frac{5}{8}$

d. $\frac{3}{4} \div \frac{5}{6} = \frac{3}{\underset{2}{\cancel{4}}} \times \frac{\overset{3}{\cancel{6}}}{5} = \frac{9}{10}$

4.04 a. $\frac{3}{4} + \frac{1}{3} = \frac{9}{12} + \frac{4}{12} = \frac{13}{12}$ or $1\frac{1}{12}$

b. $\frac{3}{4} - \frac{1}{3} = \frac{9}{12} - \frac{4}{12} = \frac{5}{12}$

c. $\frac{\overset{1}{\cancel{3}}}{4} \times \frac{1}{\underset{1}{\cancel{3}}} = \frac{1}{4}$

d. $\frac{3}{4} \div \frac{1}{3} = \frac{3}{4} \times \frac{3}{1} = \frac{9}{4}$ or $2\frac{1}{4}$

4.05 a. $0.29 + 0.36 = 0.65$
b. $0.36 - 0.29 = 0.07$
c.

$$
\begin{array}{r}
0.29 \\
\times\, 0.36 \\
\hline
174 \\
87 \\
\hline
0.1044
\end{array}
$$

4.05 (cont.)

d.
$$0.36\overline{)0.29{,}00000} = 0.80555$$

```
        0.80555
0.36)0.29,00000
     288
      200
      180
      200
      180
      200
      180
       20
```

0.80555 = 0.8056 rounded
to four decimal places

4.06
a. 0.29 + 1.58 = 1.87
b. 1.58 − 0.29 = 1.29
c.
```
    1.58
  x 0.29
   1422
    316
  0.4582
```

d.
```
         0.18354
1.58)0.29,00000
      158
     1320
     1264
      560
      474
      860
      790
      700
      632
       68
```

4.07
a. 0.29 + 0.77 = 1.06
b. 0.77 − 0.29 = 0.48
c.
```
    0.29
  x 0.77
    203
    203
  0.2233
```

d.
```
         0.37662
0.77)0.29,00000
      231
      590
      539
      510
      462
      480
      462
      180
      154
       26
```

4.08
a. 0.29 + 2.02 = 2.31
b. 2.02 − 0.29 = 1.73
c.
```
    0.29
  x 2.02
    058
    58
  0.5858
```

d.
```
         0.14356
2.02)0.29,00000  = 0.1436
      202             rounded
      880             to the
      808             nearest
      720             tenth
      606
     1140
     1010
     1300
     1212
       88
```

4.09 {2, 4}

4.010 {5, 7}

4.011 φ

4.012 {0, 2, 4, 5, 6, 7, 8}

4.013 {1, 3, 5, 6, 7, 8, 9}

4.014 {1, 2, 3, 4, 5, 7, 9}

4.015

1's
1

$1 = 1_2$

4.016

2's	1's
1	1

$3 = 11_2$

4.017

4's	2's	1's
1	0	1

$5 = 101_2$

4.018

4's	2's	1's
1	1	1

$7 = 111_2$

4.019

8's	4s	2's	1's
1	0	0	1

$9 = 1001_2$

4.020 $5\% = 0.05$

9,000
$\times\ 0.05$
$\overline{450.00}$ = \$450

4.021 $I = R \times P$
$I \div P = R$
$\quad R = I \div P$
$\quad\quad 0.07\% = 7\%$
$500\overline{)35.00}$
$\quad\quad \underline{3500}$
$\quad\quad\quad 0$

4.022 $I = R \times P$
$I \div R = P$
$\quad P = I \div R$
$6\% = 0.06$
$\quad\quad 1200.00 = \$1,200$
$0{,}06\overline{)72{.}00{,}00}$
$\quad\quad \underline{6}$
$\quad\quad 12$
$\quad\quad \underline{12}$
$\quad\quad\ 0$

4.023 $\quad\quad 12.13 =$ 12.1 rounded
$8{,}3\overline{)100.7{,}00}$ to the near-
$\quad\ \underline{83}$ est tenth
$\quad 177$
$\quad \underline{166}$
$\quad 110$
$\quad\ \underline{83}$
$\quad 270$
$\quad \underline{249}$
$\quad\ 21$

4.024 $mpg = m \div g$
$\quad\quad g = m \div mpg$
$\quad\quad 28.37 = 28.4$ rounded
$25.2\overline{)715.1{,}00}$ to the
$\quad \underline{504}$ nearest
$\quad 2111$ tenth
$\quad \underline{2016}$
$\quad\ 950$
$\quad\ \underline{756}$
$\quad 1940$
$\quad \underline{1764}$
$\quad\ 176$

4.025 $mpg = m \div g$
$mpg \times g = m$
$\quad\quad m = mpg \times g$

15.9
$\underline{\times\ 2.5}$
795
$\underline{318}$
39.75 miles = 39.8 rounded to nearest tenth

4.026 10
$\underline{\times\ 20}$
200

4.027 $A = L \times W$
$A \div W = L$
$\quad L = A \div W$
$13\frac{3}{4} \div 5\frac{1}{2} =$

$\overset{5}{\cancel{}}\frac{55}{4} \div \frac{11}{\cancel{2}}^1 =$

$\frac{2\cancel{5}5}{4} \times \frac{\cancel{2}1}{11} = \frac{5}{2}$ or $2\frac{1}{2}$

4.028 $A = L \times W$
$A \div L = W$
$\quad W = A \div L$

$\quad\ 15$
$15\overline{)225}$
$\quad \underline{15}$
$\quad\ 75$
$\quad \underline{75}$
$\quad\ \ 0$

4.029

Score	Tally
6	|
7	|||
8	|
9	|||
10	||
11	|
12	||||
13	|

4.030

Score	Frequency
6	1
7	3
8	1
9	3
10	2
11	1
12	4
13	1
	16

Relative Frequency

$\frac{1}{16}$ = 0.0625 = 6.25%

$\frac{3}{16}$ = 0.1875 = 18.75%

$\frac{1}{16}$ = 0.0625 = 6.25%

$\frac{3}{16}$ = 0.1875 = 18.75%

$\frac{2}{16} = \frac{1}{8}$ = 0.125 = 12.50%

$\frac{1}{16}$ = 0.0625 = 6.25%

$\frac{4}{16} = \frac{1}{4}$ = 0.25 = 25.00%

$\frac{1}{16}$ = 0.0625 = 6.25%

100.00%

4.031   The set contains 16 numbers. Median: $16 \div 2 = 8$; the 8th and 9th numbers are 9 and 10. Median = $\frac{9+10}{2} = \frac{19}{2} = 9\frac{1}{2}$ or 9.5.

4.032   Mode = 12

4.033

Score	Frequency	Score x Frequency
6	1	6 x 1 = 6
7	3	7 x 3 = 21
8	1	8 x 1 = 8
9	3	9 x 3 = 27
10	2	10 x 2 = 20
11	1	11 x 1 = 11
12	4	12 x 4 = 48
13	1	13 x 1 = 13
	16	154

Mean = $154 \div 16 = 9.625$

4.034 $13 - 6 = 7$

4.035   Range = 7
Average deviation =
$7 \div 16 = 0.4375$

4.036

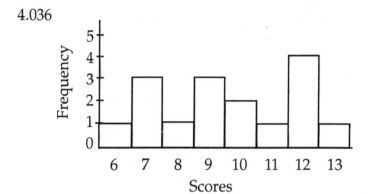

1. $6,526 = 6(1,000) + 5(100)$
    $+ 2(10) + 6(1)$
    $= 6,000 + 500 + 20 + 6$

2. $91,047 = 9(10,000) + 1(1,000)$
    $+ 0(100) + 4(10) + 7(1)$
    $= 90,000 + 1,000 + 0 + 40$
    $+ 7 = 90,000 + 1,000 +$
    $40 + 7$

3. $742,060 = 7(100,000) +$
    $4(10,000) + 2(1,000)$
    $+ 0(100) + 6(10)$
    $+ 0(1) = 700,000$
    $+ 40,000 + 2,000 + 0$
    $+ 60 + 0 = 700,000$
    $+ 40,000 + 2,000 + 60$

4. $3(100,000) + 2(1,000) + 2(100)$
    $+ 3 = 300,000 + 2,000 + 200$
    $+ 3 = 302,203$

5. $4(10,000) + 7(1,000)$
    $+ 9(100) + 6(10) = 40,000$
    $+ 7,000 + 900 + 60 = 47,960$

6. $5(1,000,000) + 5(10,000)$
    $+ 5(10) = 5,000,000$
    $+ 50,000 + 50 = 5,050,050$

7. $6 < 7$

8. $15 = 15$

9. $23 > 20$

10. $6,733 + 1,500 + 7,446$
    $= 7,000 + 2,000 + 7,000$
    $= 16,000$

11. $8,949 - 5,324 =$
    $9,000 - 5,000 = 4,000$

12. $3,443 - 2,476 =$
    $3,000 - 2,000 = 1,000$

13. $$\begin{array}{r} \overset{1\,1}{3}21 \\ 543 \\ +\,768 \\ \hline 1,632 \end{array}$$

14. $$\begin{array}{r} \overset{8}{9},\overset{12}{3}\overset{11}{2}\overset{1}{5} \\ -\,4,888 \\ \hline 4,437 \end{array}$$

15. $$\begin{array}{r} \overset{4}{5}\overset{1}{4},623 \\ -\,7,020 \\ \hline 47,603 \end{array}$$

16. $5,280 + 321 + 72,561$
    $= 78,162$

17. $6,499 - 4,132 = 2,367$

18. $53 + 64 + 75 + 86 + 97$
    $= 375$

19. $a - b = 23$

   a. $40 - b = 23$
            $b = 17$
   b. $63 - b = 23$
            $b = 40$
   c. $32 - b = 23$
            $b = 9$
   d. $a - 10 = 23$
            $a = 33$
   e. $a - 17 = 23$
            $a = 40$

The table for $a - b = 23$ is

a	40	63	32	d. 33	e. 40
b	a. 17	b. 40	c. 9	10	17

20.

a.
$$a + 6 = b$$
$$10 + 6 = b$$
$$16 = b$$

b.
$$8 + 6 = b$$
$$14 = b$$

c.
$$15 + 6 = b$$
$$21 = b$$

d.
$$a + 6 = 19$$
$$a = 13$$

e.
$$a + 6 = 26$$
$$a = 20$$

The table for $a + 6 = b$ is

a	10	8	15	d. 13	e. 20
b	a. 16	b. 14	c. 21	19	26

21. { (3, 8), (10, 15), (47, 52) }
Each number pair has a difference of 5. Since the first number is smaller than the second number in each number pair, the first number plus 5 equals the second number. Therefore, the sentence is $a + 5 = b$ (or $b - 5 = a$, or $b - a = 5$)

22. { (21, 18), (4, 1), (50, 47) }
Each number pair has a difference of 3. Since the first number is larger than the second number in each number pair, the first number minus 3 equals the second number, or the first number minus the second number equals 3. Therefore, the sentence is $a - 3 = b$ (or $a - b = 3$, or $b + 3 = a$)

23. { (10, 20), (15, 25), (50, 60) }
Each number pair has a difference of 10. Since the first number is smaller than the second number in each number pair, the first number plus 10 equals the second number. Therefore, the sentence is $a + 10 = b$ (or $b - a = 10$, or $b - 10 = a$)

24.

$$\begin{array}{r} 979 \\ - 423 \\ \hline 556 \end{array} \qquad \begin{array}{r} {}^{1\,1} \\ 556 \\ + 364 \\ \hline 920 \end{array} \qquad \begin{array}{r} {}^{8\ 11}920 \\ - 189 \\ \hline 731 \end{array}$$

25.

$$\begin{array}{r} 24{,}024 \\ + 300 \\ \hline 24{,}324 \end{array} \qquad \begin{array}{r} {}^{1\ 13\ 12} 24{,}324 \\ - 5{,}791 \\ \hline 18{,}533 \end{array}$$

$$\begin{array}{r} {}^{1\,1} \\ 18{,}533 \\ + 97 \\ \hline 18{,}630 \end{array}$$

26.

$$\begin{array}{r} {}^{8\,1} 179{,}179 \\ - 4{,}343 \\ \hline 174{,}836 \end{array} \qquad \begin{array}{r} {}^{1} \\ 174{,}836 \\ + 202 \\ \hline 175{,}038 \end{array}$$

$$\begin{array}{r} {}^{6\,1}\ \ {}^{2\,1} 175{,}038 \\ - 9{,}009 \\ \hline 166{,}029 \end{array}$$

27.

$$\begin{array}{r} 37 \\ - 12 \\ \hline 25 \end{array}$$

When Cary was born, his father was 25 years old.

28. The word "total" is a clue to add.

$$\begin{array}{r} {}^{1}\ \ {}^{2}\ {}^{1} \\ \$3{,}725 \\ 300 \\ 75 \\ 150 \\ + 250 \\ \hline \$4{,}500 \end{array}$$

total cost of the car with extras.

29. The word "total" is a hint to add.

$$\begin{array}{r} {}^{1} \\ 19 \\ 25 \\ + 31 \\ \hline 75 \end{array}$$

yrs. is the total of their ages.

1.
$$\begin{array}{r} 15 \\ 5\overline{)75} \end{array}$$
   a.   5 is the divisor.
   b.   15 is the quotient.
   c.   75 is the dividend.

2.
$$\begin{array}{r} 67 \\ \times\ 7 \\ \hline 469 \end{array}$$
   a.   469 is the product.
   b.   7 is the multiplier.
   c.   67 is the multiplicand.

3.   Function rule: Multiply by 7, then add 4

Number	Function		
1	1 x 7	=	7
	7 + 4	=	⑪
4	4 x 7	=	28
	28 + 4	=	㉜
7	7 x 7	=	49
	49 + 4	=	㊿㉝

4.   Function rule: Divide by 6, then add 7

Number	Function		
12	12 ÷ 6	=	2
	2 + 7	=	⑨
36	36 ÷ 6	=	6
	6 + 7	=	⑬
54	54 ÷ 6	=	9
	9 + 7	=	⑯

5. $6^3 = 6 \times 6 \times 6 = 216$

6. $5^4 = 5 \times 5 \times 5 \times 5 = 625$

7. $2^5 = 2 \times 2 \times 2 \times 2 \times 2 = 32$

8. $4^3 = 4 \times 4 \times 4 = 64$

9.   1, 4, 16, a. ___ , b. ___ , c. ___
   Multiplying each member by 4 results in the member following it.
   1 x 4 = 4
   4 x 4 = 16

   The next three members are
   a.   16 x 4 = 64
   b.   64 x 4 = 256
   c.   256 x 4 = 1,024

10.   128, 64, 32, 16, a. ___ ,
   b. ___ , c. ___
   To find the divisor of the pattern, divide any member by the one following it.
   32 ÷ 16 = 2
   The divisor is 2. The next three members are
   a.   16 ÷ 2 = 8
   b.   8 ÷ 2 = 4
   c.   4 ÷ 2 = 2

11.   0, 6, 12, 18, a. ___ , b. ___ ,
   c. ___
   Each member is a multiple of 6.
   6 x 0 = 0
   6 x 1 = 6
   6 x 2 = 12
   6 x 3 = 18
   The next three members are
   a.   6 x 4 = 24
   b.   6 x 5 = 30
   c.   6 x 6 = 36

12.   729, 243, 81, a. ___ , b. ___ ,
   c. ___
   To find the divisor of the pattern, divide any member by the one following it.
   243 ÷ 81 = 3
   The divisor is 3.
   The next three members are
   a.   81 ÷ 3 = 27
   b.   27 ÷ 3 = 9
   c.   9 ÷ 3 = 3

13. The word "per" means to divide.

Mr. Jones makes $755 per month.

$$\begin{array}{r} 755 \\ 12\overline{)9{,}060} \\ \underline{84} \\ 66 \\ \underline{60} \\ 60 \\ \underline{60} \\ 0 \end{array}$$

14. "Kilometers per liter" means "kilometers divided by liters."

$$\begin{array}{r} 6 \text{ kilometers per liter} \\ 90\overline{)540} \\ \underline{540} \\ 0 \end{array}$$

15.

Day		Day	
Monday	25	Monday	33
Tuesday	28	Tuesday	30
Wednesday	30	Wednesday	30
Thursday	32	Thursday	28
Friday	33	Friday	31
Saturday	33	Saturday	33
Sunday	34	Sunday	34
	215		219

Add the totals for both weeks together.

$$\begin{array}{r} 215 \\ + 219 \\ \hline 434 \end{array}$$

Divide the total by the number of days in 2 weeks.

$$\begin{array}{r} 31 \\ 14\overline{)434} \\ \underline{42} \\ 14 \\ \underline{14} \\ 0 \end{array}$$

The average maximum temperature was 31° Celsius.

16.

$$\begin{array}{r} 475 \\ 67 \\ \hline 3325 \\ 2850 \\ \hline 31{,}825 \end{array}$$

17.

$$\begin{array}{r} 368 \\ 58 \\ \hline 2944 \\ 1840 \\ \hline 21{,}344 \end{array}$$

18.

$$\begin{array}{r} 219 \\ 29 \\ \hline 1971 \\ 438 \\ \hline 6{,}351 \end{array}$$

19.

$$\begin{array}{r} 354 \\ 91 \\ \hline 354 \\ 3186 \\ \hline 32{,}214 \end{array}$$

20.

$$\begin{array}{r} 786 \\ 438 \\ \hline 6288 \\ 2358 \\ 3144 \\ \hline 344{,}268 \end{array}$$

21.

$$\begin{array}{r} 548 \text{ R}41 \\ 58\overline{)31{,}825} \\ \underline{290} \\ 282 \\ \underline{232} \\ 505 \\ \underline{464} \\ 41 \end{array}$$

22.

$$\begin{array}{r} 318 \text{ R}38 \\ 67\overline{)21{,}344} \\ \underline{201} \\ 124 \\ \underline{67} \\ 574 \\ \underline{536} \\ 38 \end{array}$$

23.
```
       1,110  R24
29)32,214
    29
    32
    29
    31
    29
    24
     0
    24
```

24.
```
          6,543  (R0)
47)307,521
   282
   255
   235
   202
   188
   141
   141
     0
```

25.
```
        6,199  R9
98)607,511
   588
   195
    98
   971
   882
   891
   882
     9
```

26.
```
        7,055  R19
77)543,254
   539
   425
   385
   404
   385
    19
```

27.
```
           1,507  (R0)
765)1,152,855
    765
    3878
    3825
    5355
    5355
       0
```

28.
```
           7,067  R369
489)3,456,132
    3423
    3313
    2934
    3792
    3423
     369
```

29.
```
        29,557 (R0)
6)177,342
  12
  57
  54
  33
  30
  34
  30
  42
  42
   0
```

30.
```
        35,130  R1
7)245,911
  21
  35
  35
  09
   7
  21
  21
  01
   0
   1
```

1.   true

2.   false

3.   true

4.   false

5.   false

6.   true

7.   false

8.   false

9.   true

10.   false

11.   e

12.   j

13.   g

14.   h

15.   a

16.   f

17.   i

18.   c

19.   b

20.   d

21.

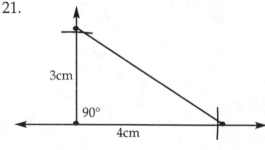

22.

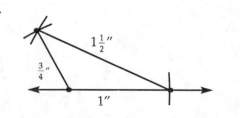

23.

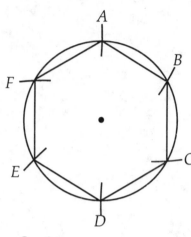

24. $C = 2\pi r$
$C = 2 \times 3.14 \times 3$
$C = 18.84$ inches

25. $A = \pi r^2$
$A = 3\frac{1}{7} \times 7 \times 7$
$A = \frac{22}{7} \times 49$
$A = \frac{22}{{}_1 7} 1 \times \frac{\cancel{49}^{7}}{1}$
$A = 154$ sq. inches

26.   Each ∠ of the hexagon in Problem 23 measures 120°. The sum of the ∠'s = 6 x 120° = 720°.

27.   f

28.   e

29.   c

30.   k

31.   h

32.   a

33.   i

34.   d

35.   g

36.   b

1. $\dfrac{21}{44}$

2. $\dfrac{45}{72} = \dfrac{45 \div 9}{72 \div 9} = \dfrac{5}{8}$

3. $54 \div 9 = 6$
   $\dfrac{5}{9} = \dfrac{5 \times 6}{9 \times 6} = \dfrac{30}{54}$

4. $\dfrac{45}{7} = 45 \div 7 = 6\dfrac{3}{7}$

5. Multiply $7 \times 5 = 35$ and add 2; $5\dfrac{2}{7} = \dfrac{37}{7}$.

6. $\dfrac{7}{8}$ is larger than $\dfrac{7}{9}$ because eighths are larger than ninths.

7. $\dfrac{1}{5} = 0.2$ from the table, or
   $\dfrac{1}{5} = \dfrac{1 \times 20}{5 \times 20} = \dfrac{20}{100} = 0.2$
   $2\dfrac{1}{5} = 2.2$

8. $0.89 = \dfrac{89}{100}$

9. $\dfrac{71}{6} = 71 \div 6 = 11\dfrac{5}{6}$

10. 5.6 is larger than 0.56

11. seventy-two and forty-eight hundredths

12. Move the decimal point two places to the right and add the % sign.
    $0.87 = 87\%$

13. Convert 7 nickels to pennies.
    $7 \times 5 = 35$ pennies
    $\dfrac{35}{50} = \dfrac{35 \div 5}{50 \div 5} = \dfrac{7}{10}$ or $7:10$

14. $56\% = \dfrac{56}{100} = \dfrac{56 \div 4}{100 \div 4} = \dfrac{14}{25}$

15. Move the decimal point two places to the right and add the % sign.
    $0.00034 = 0.034\%$

16. Move the decimal point two places to the right and add the % sign.
    $5.18 = 518\%$

17. $\dfrac{25}{2} = 25 \div 2 = 12.5$ or 12.50

18. Move the decimal point two places to the left and drop the % sign.
    $0.13\% = 0.0013$

19. b
    $\dfrac{29}{3} = 29 \div 3 = 9\dfrac{2}{3}$

20. a

21. c

22. c
    $\dfrac{5}{8} = \dfrac{5 \times 2}{8 \times 2} = \dfrac{10}{16}$

23. a
    $1,000:1 = 2,500:?$
    $1 \times 2,500 = 1,000 \times ?$
    $2,500 = 1,000 \times ?$
    $? = 2.5$

24. b

25. b
    $7¢$ of $\$1.00 = \dfrac{7}{100} = 7\%$

26. f

27. d

28. g

29. a

30.   h

31.   b

32.   e

33.   c

34.   j

35.   i

36.   Examples:

a. $\dfrac{5}{12} = \dfrac{5 \times 2}{12 \times 2} = \dfrac{10}{24}$

b. $\dfrac{5}{12} = \dfrac{5 \times 3}{12 \times 3} = \dfrac{15}{36}$

c. $\dfrac{5}{12} = \dfrac{5 \times 4}{12 \times 4} = \dfrac{20}{48}$

d. $\dfrac{5}{12} = \dfrac{5 \times 5}{12 \times 5} = \dfrac{25}{60}$

e. $\dfrac{5}{12} = \dfrac{5 \times 6}{12 \times 6} = \dfrac{30}{72}$

1. i

2. h

3. f

4. e

5. l

6. j

7. k

8. g

9. b

10. a

11. c

12. d

13. ⊂

14. infinite

15. {a, e, i, o, u}

16. $5 \times 10^4 = 5 \times 10,000 = 50,000$

17. $29 = 20 + 9 = XXIX$

18. $1000_2 = 1 \times 2^3$
    $= 1 \times 8$
    $= 8$

19. {1, 2, 3, 4, 6, 8, 12, 24}

20. The factors of 6 are 2 x 3; therefore, 6 has factors other than 1 and itself.

21. 36 is divisible by 2:
    $36 = 2 \times 18$
    18 is divisible by 2:
    $36 = 2 \times 2 \times 9$
    9 is divisible by 3:
    $36 = 2 \times 2 \times 3 \times 3 \text{ or } 2^2 \times 3^2$

22. $14 = 2 \times \boxed{7}$
    $21 = 3 \times \boxed{7}$
    GCF = 7

23. The elements are r, s, t, u, and v.

24. {4}

25. 30 is 1,000 times greater than 0.03.

26. empty set (ϕ)

27. 6×10^{-4}

28. $2 = 2$
    $7 = 7$
    $12 = 2 \times 2 \times 3$
    $20 = 2 \times 2 \times 5$
    $LCM = 2 \times 2 \times 3 \times 5 \times 7 = 420$

29.
    $$25\overline{)27} \quad 5\overline{)2} \quad 1\overline{)2}$$
    with quotients 1, 0, 2
    $$\frac{25}{2} \quad \frac{0}{2} \quad \frac{2}{0}$$
    $27 = 102_5$

30. {1, 2, 3, 4, 6, 8}

31. c

32. d

33. b   Since the units digit is even, the number is divisible by 2.

34. c
    $39 = 3 \times \boxed{13}$
    $65 = 5 \times \boxed{13}$
    GCF = 13

    $\dfrac{39 \div 13}{65 \div 13} = \dfrac{3}{5}$

35. a
    $111_2 = 1 \times 2^2 + 1 \times 2^1 + 1 \times 1$
    $= 4 + 2 + 1$
    $= 7$

36. b
    $CXIV = 100 + 10 + 4 = 114$

1. $\frac{7}{5} = 1\frac{2}{5}$

2. LCD = 10

$$3\frac{1}{2} = 3\frac{5}{10} = 2 + 1\frac{5}{10} = 2\frac{15}{10}$$
$$-1\frac{7}{10} = \qquad\qquad 1\frac{7}{10}$$
$$\rule{6cm}{0.4pt}$$
$$1\frac{8}{10} = 1\frac{8 \div 2}{10 \div 2} = 1\frac{4}{5}$$

3. $4.4 = 4\frac{4}{10} = 4\frac{4 \div 2}{10 \div 2} = 4\frac{2}{5}$

4.
$$\begin{array}{r} 16.348 \\ +\ 3.720 \\ \hline 20.068 \end{array}$$

5.
$$\begin{array}{r} 116.400 \\ -\ 21.368 \\ \hline 95.032 \end{array}$$

6. a. Examples:
$$\frac{3}{8} = \frac{3 \times 2}{8 \times 2} = \frac{6}{16} \quad or$$
$$\frac{3}{8} = \frac{3 \times 3}{8 \times 3} = \frac{9}{24}$$

   b.
$$\begin{array}{r} 0.375 \\ 8\overline{)3.000} \\ \underline{24} \\ 60 \\ \underline{56} \\ 40 \\ \underline{40} \\ 0 \end{array}$$

7. a. Examples:
$$6\frac{5}{10} = 6\frac{5 \div 5}{10 \div 5} = 6\frac{1}{2} \quad or$$
$$6\frac{5}{10} = 6\frac{5 \times 2}{10 \times 2} = 6\frac{10}{20}$$

   b.
$$\begin{array}{r} 0.5 \\ 20\overline{)10.0} \\ \underline{100} \\ 0 \end{array}$$
$$6\frac{5}{10} = 6.5$$

8. 72.0348

9. 10.1

10. LCD = 24
$$\frac{3}{4} = \frac{18}{24}$$
$$\frac{1}{8} = \frac{3}{24}$$
$$+\ \frac{5}{12} = \frac{10}{24}$$
$$\rule{3cm}{0.4pt}$$
$$\frac{31}{24} = 1\frac{7}{24}$$

11. LCD = 6
$$\frac{1}{3} = \frac{2}{6}$$
$$\frac{1}{2} = \frac{3}{6}$$
$$+\ \frac{1}{6} = \frac{1}{6}$$
$$\rule{3cm}{0.4pt}$$
$$\frac{6}{6} = 1$$

12. $23\frac{18}{13} = 23 + 1\frac{5}{13} = 24\frac{5}{13}$

13. LCD = 14
$$\frac{1}{2} = \frac{7}{14}$$
$$-\ \frac{5}{14} = \frac{5}{14}$$
$$\rule{3cm}{0.4pt}$$
$$\frac{2}{14} = \frac{2 \div 2}{14 \div 2} = \frac{1}{7}$$

14. LCD = 36
$$38\frac{3}{4} = 38\frac{27}{36}$$
$$-\ 14\frac{5}{9} = 14\frac{20}{36}$$
$$\rule{3cm}{0.4pt}$$
$$24\frac{7}{36}$$

15. $4\frac{2}{7} = 3 + 1\frac{2}{7} = 3\frac{9}{7}$
$$-3\frac{4}{7} = \qquad\qquad 3\frac{4}{7}$$
$$\rule{3cm}{0.4pt}$$
$$\frac{5}{7}$$

16.   455.091

17.   3.0801

18.   3.682
     + 37.650
        41.332
     − 0.230
        41.102

19. $ 6.05
            11.55
     + 100.00
        $117.60

20.   a. $18.00
              1.05
              2.25
          + 4.80
             $26.10

         b. $35.00
          − 26.10
             $ 8.90

1.   d

2.   c

3.   a

4.   e

5.   b

6.
```
  1 0 5
x   0.5
   52.5
```

7.
```
           0.4405  →   0.441 =
34.5 )15.2.0000       44.1%
      1380
      1400
      1380
      2000
      1725
       275
```

8.
```
             8,591.304  →   8,591.30
0.23 )1976.00.000
      184
      136
      115
      210
      207
       30
       23
       70
       69
      100
       92
        8
```

9.
```
     754
x   0.06
   45.24
```

10.
```
            64.233   →   64.23
0.685 )44.000.000
       4110
       2900
       2740
       1600
       1370
       2300
       2055
       2450
       2055
        395
```

11.
```
          0.8236   →   0.824 =
14.57 )12.00.0000      82.4%
       11656
        3440
        2914
        5260
        4371
        8890
        8742
         148
```

12.
$$\overset{6}{\cancel{18}} \times \frac{19}{\cancel{3}_1} = \frac{114}{1} = 114$$

$18 = 18$

$6\frac{1}{3} = 6.\overline{3}$

$18 \times 6.3333333 = 113.99999$

$114 =$ approximately 113.99999

13.
$$\frac{5}{3} \div \frac{3}{4} = \frac{5}{3} \times \frac{4}{3} = \frac{20}{9} = 2\frac{2}{9}$$

$1\frac{2}{3} = 1.\overline{6}$

$\frac{3}{4} = 0.75$

$1.666667 \div 0.75 = 2.2222226$

$2\frac{2}{9} = 2.2222222$

14.
$$\overset{5}{\underset{1}{\cancel{25}}} \times \overset{3}{\underset{1}{\cancel{24}}} = \frac{15}{1} = 15$$

$3\frac{1}{8} = 3.125$

$4\frac{4}{5} = 4.8$

3.125 x 4.8 = 15

15 = 15

15. $\frac{8}{9} \times \frac{5}{3} = \frac{40}{27} = 1\frac{13}{27}$

$\frac{8}{9} = 0.\overline{8}$

$\frac{3}{5} = 0.6$

0.8888889 ÷ 0.6 = 1.4814815

$1\frac{13}{27} = 1.4814815$

16. The rate is 75% = 0.75.
The percentage is $48.
The base is 48 ÷ 0.75.

```
        $64
0.75 )48.00
      450
      300
      300
        0
```

17.
```
     14.6
   x 59.9
    1314
    1314
    730
  874.54 = 875 cents or $8.75
```

18. The base is $83.40.
The rate is 10% = 0.1.
The percentage is $83.40 x 0.1 =

```
   83.4
 x  0.1
  $8.34
```

19.
```
     1.7357      →     1.736 grams
14 )24.3000
    14
    103
     98
     50
     42
     80
     70
    100
     98
      2
```

20.
```
      0.3185      →      0.319
135 )43.0000
     405
     250
     135
    1150
    1080
     700
     675
      25
```

21.
```
     12.5
   x 0.69
   1125
   750
  8.625 = $8.63
```

22. The base is $4,000.
The percentage is $3,600.
The rate is $3,600 ÷ $4,000. =
0.9 = 90%

```
4,000 )3,600.0
       36000
           0
```

23. The base is $120.
The percentage is $18.
The rate is $18 ÷ $120 =
0.15 = 15%

```
120 )18.00
     120
     600
     600
       0
```

24. The rate is 38% = 0.38.
The percentage is 450.
The base is 450 ÷ 0.38 =

```
       1,184
0.38 )450.00
      38
      70
      38
     320
     304
     160
     152
       8
```

1. the amount of a loan

2. put in place of

3. a mathematical sentence expressing equality between two quantities

4. a comparison of one quantity with another

5. in a proportion, the two outermost terms

6.
$$d = r \times t$$
$$12 = r \times 3$$
$$12 \div 3 = r$$
$$4 = r$$

7.
$$d = r \times t$$
$$6 = r \times 1$$
$$6 \div 1 = r$$
$$6 = r$$

8.
$$d = r \times t$$
$$16 = 8 \times t$$
$$16 \div 8 = t$$
$$2 = t$$

9. true; 13 = 13

10. open

11. false

12. 4:10 = 2:5

13. 4:6 = 2:3

14. 6:10 = 3:5

15. 6:4 = 3:2

16.
$$6 \times 2 = 12$$
$$4 \times 3 = 12$$
$$12 = 12; \text{true}$$

17.
$$3 \times 39 = 117$$
$$13 \times 9 = 117$$
$$117 = 117; \text{true}$$

18.
$$200 \times 4 = 800$$
$$100 \times 7 = 700$$
$$800 = 700; \text{not a true equation}$$

19.
$$2{:}3 = x{:}4.5$$
$$3 \cdot x = 2 \cdot 4.5$$
$$3 \cdot x = 9$$
$$x = 9 \div 3$$
$$x = 3 \text{ inches}$$

20.
$$P = W \times H$$
$$P = \$2.95 \times 40$$
$$\begin{array}{r} \$2.95 \\ \times \quad 40 \\ \hline \end{array}$$
$$P = \quad \$118.00$$

21. $9 \div 3 = 3$

22. $27 \div 3 = 9$

23. $135 \div 3 = 45$

24. $15 + 16 = 31$

25. $1 + 17 = 18$

26. $2\frac{1}{2} + 6\frac{3}{8} = 2\frac{4}{8} + 6\frac{3}{8} = 8\frac{7}{8}$

1. A sample that is not random.

2. A sample in which each member of the large group has an equal chance of being chosen.

3.

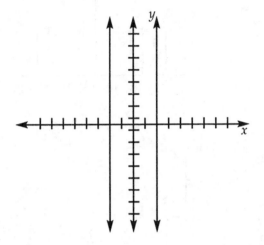

4.

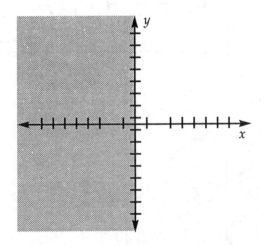

5.

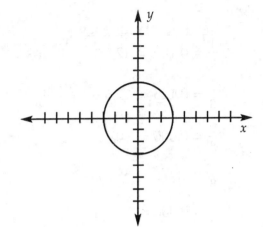

6.

Score	Tally
12	|
13	|
14	
15	|||
16	
17	|
18	
19	
20	||

7.

Score	Frequency
12	1
13	1
14	0
15	3
16	0
17	1
18	0
19	0
20	2

8.

Score	Frequency
12	1
13	1
14	0
15	3
16	0
17	1
18	0
19	0
20	2
	8

8.   cont.

Relative   Frequency

$\frac{1}{8} = 0.125 = 12.5\%$

$\frac{1}{8} = 0.125 = 12.5\%$

$\frac{0}{8} = 0.000 = 0.0\%$

$\frac{3}{8} = 0.375 = 37.5\%$

$\frac{0}{8} = 0.000 = 0.0\%$

$\frac{1}{8} = 0.125 = 12.5\%$

$\frac{0}{8} = 0.000 = 0.0\%$

$\frac{0}{8} = 0.000 = 0.0\%$

$\frac{2}{8} = 0.25 = \underline{25.0\%}$

100.0%

9.   Mode = 15

10   The set contains 8 numbers.
Median: $8 \div 2 = 4$, the 4th
and 5th numbers are both 15.
Median = 15.

11.

Number	Frequency	Number x Frequency
12	1	12 x 1 = 12
13	1	13 x 1 = 13
14	0	14 x 0 = 0
15	3	15 x 3 = 45
16	0	16 x 0 = 0
17	1	17 x 1 = 17
18	0	18 x 0 = 0
19	0	19 x 0 = 0
20	2	20 x 2 = 40
	8	127

Mean = $127 \div 8 = 15.875 = 15.9$

12. $20 - 12 = 8$

13.   Range = 8
Average deviation = $8 \div 8 = 1$

14.

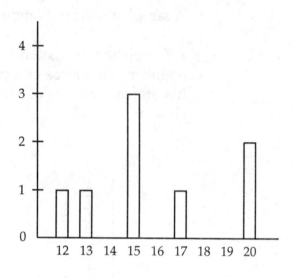

15.

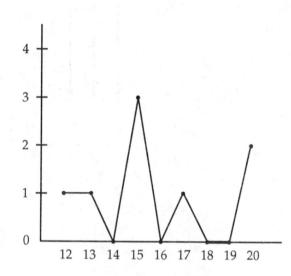

16.

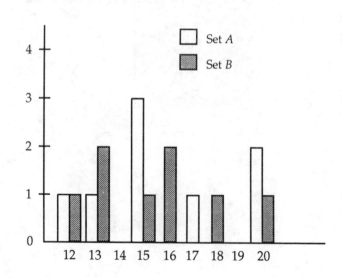

1. a. $\frac{1}{3} + \frac{3}{8} = \frac{8}{24} + \frac{9}{24} = \frac{17}{24}$

    b. $\frac{3}{8} - \frac{1}{3} = \frac{9}{24} - \frac{8}{24} = \frac{1}{24}$

    c. $\frac{1}{\underset{1}{\cancel{3}}} \times \frac{\cancel{3}^{1}}{8} = \frac{1}{8}$

    d. $\frac{1}{3} \div \frac{3}{8} = \frac{1}{3} \times \frac{8}{3} = \frac{8}{9}$

2. $1\frac{1}{2} = \frac{(1 \times 2) + 1}{2} = \frac{3}{2}$

    a. $\frac{1}{3} + \frac{3}{2} = \frac{2}{6} + \frac{9}{6} = \frac{11}{6}$ or $1\frac{5}{6}$

    b. $\frac{3}{2} - \frac{1}{3} = \frac{9}{6} - \frac{2}{6} = \frac{7}{6}$ or $1\frac{1}{6}$

    c. $\frac{1}{\underset{1}{\cancel{3}}} \times \frac{\cancel{3}^{1}}{2} = \frac{1}{2}$

    d. $\frac{1}{3} \div \frac{3}{2} = \frac{1}{3} \times \frac{2}{3} = \frac{2}{9}$

3. a. $0.52 + 1.76 = 2.28$
    b. $1.76 - 0.52 = 1.24$
    c.
```
      1.76
    x 0.52
      352
     880
   0.9152
```

    d.
```
         0.29545 = 0.2955
    1.76 )0.52.00000      rounded
           352            to the
           1680           nearest
           1584           ten-thousandth
            960
            880
            800
            704
            960
            880
             80
```

4. a. $0.52 + 0.18 = 0.70$
    b. $0.52 - 0.18 = 0.34$
    c.
```
      0.52
    x 0.18
      416
     052
   0.0936
```

4. cont.
```
            2.88888 = 2.8889
    0.18 )0.52.00000      rounded
           36             to the
           160            nearest
           144            ten-thousandth
           160
           144
           160
           144
           160
           144
           160
           144
            16
```

5. $\in$

6. $\notin$

7. $\in$

8. $\notin$

9.
4's	2's	1's
1	0	1

$5 = 101_2$

10.
4's	2's	1's
1	0	1

$6 = 110_2$

11.
4's	2's	1's
1	1	1

$7 = 111_2$

12.
8's	4's	2's	1's
1	0	0	0

$8 = 1000_2$

13. $A \cap B$

14. $A \cup B$

15. b

16. c

17. a

18. d

19. e

20. g

21.
$$3:18 = \underline{\quad}:36$$
$$18 \times \underline{\quad} = 3 \times 36$$
$$18 \times \underline{\quad} = 108$$
$$\underline{6} = 108 \div 18$$

22.
$$2:\underline{\quad} = 5:60$$
$$\underline{\quad} \times 5 = 2 \times 60$$
$$\underline{\quad} \times 5 = 120$$
$$\underline{24} = 120 \div 5$$

23.
$$\underline{\quad}:13 = 12:4$$
$$13 \times 12 = \underline{\quad} \times 4$$
$$156 = \underline{\quad} \times 4$$
$$156 \div 4 = \underline{39}$$

24.
$$15:75 = 1:\underline{\quad}$$
$$75 \times 1 = 15 \times \underline{\quad}$$
$$75 = 15 \times \underline{\quad}$$
$$75 \div 15 = \underline{5}$$

25.
$$18:72 = \underline{\quad}:64$$
$$72 \times \underline{\quad} = 18 \times 64$$
$$72 \times \underline{\quad} = 1,152$$
$$\underline{16} = 1,152 \div 72$$

26. c

27. b

28. g

29. d

30. a

31. e

32. true; 5 = 5

33. open

34. false; 5 ≠ 6

35. open

36. false; 15 ≠ 5

37.

Score	Tally
1	|
2	++++ ++++ |
3	| |
4	++++ | | | |
5	| |
6	++++ |
7	|
8	| | | |
9	|
10	| | |

38.

Score	Frequency
1	1
2	11
3	2
4	9
5	2
6	6
7	1
8	4
9	1
10	3
	40

Relative Frequency

$$\frac{1}{40} = 0.025 \qquad = 2.5\%$$

$$\frac{11}{40} = 0.275 \qquad = 27.5\%$$

$$\frac{2}{40} = \frac{1}{20} = 0.05 = 5.0\%$$

$$\frac{9}{40} = 0.225 \qquad = 22.5\%$$

$$\frac{2}{40} = \frac{1}{20} = 0.05 = 5.0\%$$

$$\frac{6}{40} = \frac{3}{20} = 0.15 = 15.0\%$$

$$\frac{1}{40} = 0.025 \qquad = 2.5\%$$

$$\frac{4}{40} = \frac{1}{10} = 0.1 \quad = 10.0\%$$

$$\frac{1}{40} = 0.025 \qquad = 2.5\%$$

$$\frac{3}{40} = 0.075 \qquad = \underline{7.5\%}$$
$$100.0\%$$

39. The set contains 40 numbers.
Median: $40 \div 2 = 20$; the
20th and 21st numbers are both
4. Median = 4.

40. Mode = 2

41.

Score	Frequency
1	1
2	11
3	2
4	9
5	2
6	6
7	1
8	4
9	1
10	3
	40

Score x Frequency
$1 \times 1 = 1$
$2 \times 11 = 22$
$3 \times 2 = 6$
$4 \times 9 = 36$
$5 \times 2 = 10$
$6 \times 6 = 36$
$7 \times 1 = 7$
$8 \times 4 = 32$
$9 \times 1 = 9$
$10 \times 3 = 30$
189

Mean = $189 \div 40 = 4.725$

42. $10 - 1 = 9$

43. Range = 9
Average deviation = $9 \div 40 = 0.225$

44.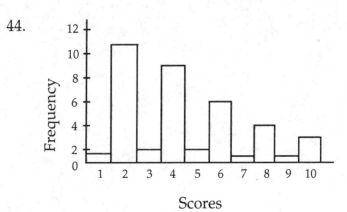

1.
$$\begin{array}{r} {}^{2}76 \\ 94 \\ 38 \\ +\ 22 \\ \hline 230 \end{array}$$

2.
$$\begin{array}{r} {}^{1\,1}582 \\ 763 \\ +\ 948 \\ \hline 2{,}293 \end{array}$$

3.
$$\begin{array}{r} {}^{8\,1}5{,}694 \\ -\ 309 \\ \hline 5{,}385 \end{array}$$

4. $6 + 23 + 342 + 5{,}106 = 5{,}477$

5. $692 - 317 = 375$

6. $70{,}005 - 6{,}947 = 63{,}058$

7. $6{,}457 =$
   a. 6,460 rounded to the nearest ten
   b. 6,500 rounded to the nearest hundred
   c. 6,000 rounded to the nearest thousand

8.
$$\begin{aligned} a + b &= 20 \\ \text{a.}\quad a + 9 &= 20 \\ a &= 11 \\ \text{b.}\quad a + 6 &= 20 \\ a &= 14 \\ \text{c.}\quad 2 + b &= 20 \\ b &= 18 \\ \text{d.}\quad 17 + b &= 20 \\ b &= 3 \\ \text{e.}\quad 13 + b &= 20 \\ b &= 7 \end{aligned}$$

The table for $a + b = 20$ is

a.	2	17	13	a. 11	b. 14
b.	c. 18	d. 3	e. 7	9	6

9.
$$\begin{aligned} a - 5 &= b \\ \text{a.}\quad a - 5 &= 5 \\ a &= 10 \\ \text{b.}\quad a - 5 &= 8 \\ a &= 13 \\ \text{c.}\quad 7 - 5 &= b \\ 2 &= b \\ \text{d.}\quad 16 - 5 &= b \\ 11 &= b \\ \text{e.}\quad 5 - 5 &= b \\ 0 &= b \end{aligned}$$

The table for $a - 5 = b$ is

a	7	16	5	a. 10	b. 13
b	c. 2	d. 11	e. 0	5	8

10. $\{(17, 8), (9, 0), (29, 20)\}$
Each number pair has a difference of 9. Since the first number is larger than the second number in each number pair, the first number minus 9 equals the second number, or the first number minus the second number equals 9. Therefore, the sentence is $a - 9 = b$, or $a - b = 9$, or $b + 9 = a$.

11. $\{(8, 12), (3, 17), (15, 5)\}$
Each number pair has a sum of 20. Therefore, the sentence is $a + b = 20$, or $20 - b = a$, or $20 - a = b$.

12. $7{,}010 = 7{,}000 + 10$

13. $23{,}698 = 20{,}000 + 3{,}000 + 600 + 90 + 8$

14. $540{,}006 = 500{,}000 + 40{,}000 + 6$

15. $8(10{,}000) + 2(1{,}000) + 4(10) =$
$80{,}000 + 2{,}000 + 40 = 82{,}040$

16. $3(1{,}000{,}000) + 7(1{,}000) + 6(100) + 9 =$
$3{,}000{,}000 + 7{,}000 + 600 + 9 = 3{,}007{,}609$

17. $1(100{,}000) + 5(10{,}000) + 1(100) =$
$100{,}000 + 50{,}000 + 100 = 150{,}100$

18. $8 + 15 > 22 - 8$
$23 > 14$; true

19. $13 + 10 - 7 < 42 - 23$
$16 < 19$; true

20. $48 + 75 = 162 - 49$
$123 = 113$; false

21. $652 + 1{,}320 + 75$
$= 700 + 1{,}300 + 100 = 2{,}100$

22. $6{,}872 - 2{,}927 = 6{,}900 - 2{,}900 = 4{,}000$

23. $350 + 1{,}649 + 800$
$= 400 + 1{,}600 + 800 = 2{,}800$

24.
$$\begin{array}{r} {}^{1}622 \\ +508 \\ \hline 1{,}130 \end{array}\qquad\begin{array}{r} {}^{0}{}^{10}{}^{12}{}^{1}\\ \cancel{1}{,}\cancel{1}\cancel{3}0 \\ -193 \\ \hline 937 \end{array}\qquad\begin{array}{r} {}^{8}{}^{1}\\ 9\cancel{3}7 \\ -356 \\ \hline \boxed{581} \end{array}$$

25.
$$\begin{array}{r} {}^{6}{}^{11}{}^{1}\\ 3\cancel{7}{,}2\cancel{1}2 \\ -4{,}321 \\ \hline 32{,}891 \end{array}$$

$$\begin{array}{r} {}^{1}\\ 32{,}891 \\ +22 \\ \hline 32{,}913 \end{array}$$

$$\begin{array}{r} {}^{1}{}^{18}{}^{10}{}^{1}\\ 3\cancel{2}{,}9\cancel{1}3 \\ -999 \\ \hline \boxed{31{,}914} \end{array}$$

26.
$$\begin{array}{r} 8{,}214 \\ +8{,}213 \\ \hline 16{,}427 \end{array}\qquad\begin{array}{r} {}^{0}{}^{1}\\ \cancel{1}6{,}427 \\ -8{,}215 \\ \hline 8{,}212 \end{array}\qquad\begin{array}{r} 8{,}212 \\ -8{,}212 \\ \hline \boxed{0} \end{array}$$

27.
$$\begin{array}{r} {}^{1}22\cancel{c} \\ 17\cancel{c} \\ +13\cancel{c} \\ \hline 52\cancel{c} \end{array}\qquad\begin{array}{r} {}^{0}{}^{9}{}^{1}\\ \$\cancel{1}.00 \\ -0.52 \\ \hline \$0.48 \text{ or } 48\text{¢ change} \end{array}$$

28.
$$\begin{array}{r} 1952 \\ +4 \\ \hline \end{array}$$ Bill was born in 1956 when his parents celebrated their fourth wedding anniversary.

$$\begin{array}{r} 1952 \\ +20 \\ \hline 1972 \end{array}$$ Mr. and Mrs. Smith celebrated their twentieth wedding anniversary in 1972.

$$\begin{array}{r} {}^{6}{}^{1}\\ 19\cancel{7}2 \\ -1956 \\ \hline 16 \end{array}$$ Bill was 16 years old when his parents celebrated their twenthieth wedding anniversary.

29.
$$\begin{array}{r} {}^{1}\\ \$42 \\ +69 \\ \hline \$111 \end{array}$$ is the amount the group earned in the two days.

$$\begin{array}{r} {}^{0}{}^{10}{}^{1}\\ \$\cancel{1}\cancel{1}1 \\ -26 \\ \hline \$85 \text{ profit} \end{array}$$

350

1. a. 9 is the quotient.
    b. 63 is the dividend. $7\overline{)63}^{\,9}$
    c. 7 is the divisor.

2. a. 5 is the multiplier.   75
    b. 75 is the multiplicand. $\times\,5$
    c. 375 is the product.   375

3. Function rule: Multiply by 6, then subtract 3

Number	Function
1	1 x 6 = 6
	6 − 3 = ③
7	7 x 6 = 42
	42 − 3 = ㊴
11	11 x 6 = 66
	66 − 3 = �63

4. Function rule: Divide by 7, then add 4

Number	Function
7	7 ÷ 7 = 1
	1 + 4 = ⑤
21	21 ÷ 7 = 3
	3 + 4 = ⑦
49	49 ÷ 7 = 7
	7 + 4 = ⑪

5. $2^4 = 2 \times 2 \times 2 \times 2 = 16$

6. $5^3 = 5 \times 5 \times 5 = 125$

7. $7^3 = 7 \times 7 \times 7 = 343$

8. $3^4 = 3 \times 3 \times 3 \times 3 = 81$

9. 0, 5, 10, 15, a. ____, b. ____, c. ____
    Each member is a multiple of 5.
    5 x 0 = 0
    5 x 1 = 5
    5 x 2 = 10
    5 x 3 = 15

    The next three members are
    a. 5 x 4 = 20
    b. 5 x 5 = 25
    c. 5 x 6 = 30

10. 243, 81, 27, a. ____, b. ____, c. ____
     To find the divisor of the pattern, divide any member by the one following it.
     81 ÷ 27 = 3
     The divisor is 3.
     The next three members are
     a. 27 ÷ 3 = 9
     b. 9 ÷ 3 = 3
     c. 3 ÷ 3 = 1

11. 1, 5, 25, a. ____, b. ____, c. ____
     Multiplying each member by 5 results in the member following it.
     1 x 5 = 5
     5 x 5 = 25
     The next three members are
     a. 25 x 5 = 125
     b. 125 x 5 = 625
     c. 625 x 5 = 3,125

12. 256, 128, 64, 32, a. ____, b. ____, c. ____
     To find the divisor of the pattern, divide any member by the one following it.
     64 ÷ 32 = 2
     The divisor is 2.
     The next three members are
     a. 32 ÷ 2 = 16
     b. 16 ÷ 2 = 8
     c. 8 ÷ 2 = 4

13. $\overset{3}{4}5$
     $\times\,7$
     315 kilometers

14. Add the scores. $\overset{3}{9}5$
                           85
                           88
                           86
                           92
                           94
                          $+\,90$
                           630

351

Divide the total by the number of scores.

```
      90
7)630        students average
  63         score was 90
  ──
  00
```

15.
```
      6
40)240
   240
   ───
     0
```

His pay per hour is $6.

16.
```
    1 4
    1 3
    327
     75
  ──────
  1,635
  2,289
 ───────
 24,525
```

17.
```
     6 4
     681
      86
  ──────
  4,086
  5448
 ───────
 58,566
```

18.
```
     8 7
     0 0
     598
      91
  ──────
  ¹598
  5382
 ───────
 54,418
```

19.
```
     6 4
     0 0
     1 1
     275
     902
  ──────
   550
  ⁰000
  2,475
 ───────
 248,050
```

20.
```
      5
      4
      1
     608
     752
  ──────
 ¹1216
  3040
  4256
 ───────
 457,216
```

21.
```
      518 R 6
47)24,352
   2 35
  ─────
     85
     47
    ───
    382
    376
    ───
      6
```

22.
```
       418 R57
76)31,825
   304
   ───
   142
    76
    ──
    665
    608
    ───
     57
```

23.
```
       850 R1
92)78,201
   736
   ───
   4 60
   4 60
   ─────
     01
```

24.
```
       8,189 R51
58)475,013
   464
   ───
   110
    58
    ──
    521
    464
    ───
    573
    522
    ───
     51
```

25.
```
       6,198 (R 0)
89)551,622
   534
   ───
   176
    89
    ──
    872
    801
    ───
    712
    712
    ───
      0
```

26.
```
       8,075 R20
67)541,045
   536
   ───
   504
   469
   ───
   355
   335
   ───
    20
```

27.
```
         2,358 R16
948)2,235,400
    1 896
    ─────
    3394
    2844
    ────
    5500
    4740
    ────
    7600
    7584
    ────
      16
```

28.
```
            7,075 (R 0)
   576)4,075,200
       4 032
          4320
          4032
          2880
          2880
             0
```

29.
```
       19,301 R 5
   9)173,714
    9
    83
    81
     27
     27
      014
        9
        5
```

30.
```
       53,438 R 3
   8)427,507
    40
    27
    24
     35
     32
      30
      24
       67
       64
        3
```

1. true

2. false

3. true

4. false

5. true

6. true

7. false

8. false

9. true

10. false

11. e

12. b

13. c

14. a

15. h

16. g

17. d

18. j

19. f

20. i

21.

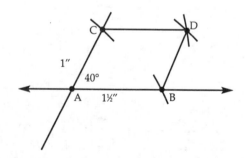

22.

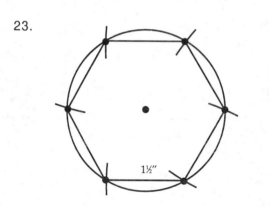

23.

24. C = 2πr
    C = 2 x 3.14 x 2
    C = 12.56"

25. A = πr²
    A = $3\frac{1}{7}$ x 7 x 7
    A = $\frac{22}{1\!\!/7}$ x $\frac{\overset{7}{\cancel{49}}}{1}$
    A = 154 sq. cm

26. sum of ∠'s = 40° + 140° + 40° + 140°
    sum of ∠'s = 360°

27. e

28. i

29. a

30. g

31. k

32. b

33. h

34. d

35. j

36. f

1. $\frac{34}{43}$

2. $\frac{39}{65} = \frac{3 \times 13}{5 \times 13} = \frac{3}{5}$

3. $56 = 7 \times 8$

   $\frac{3}{7} = \frac{3 \times 8}{7 \times 8} = \frac{24}{56}$

4. $51 \div 9 = 5\frac{6}{9} = 5\frac{2}{3}$

5. Multiply $8 \times 4 = 32$ and add 3; the answer is $\frac{35}{8}$.

6. Since sixths are larger than sevenths, $\frac{5}{6}$ is larger than $\frac{5}{7}$.

7. $\frac{3}{5} = \frac{60}{100}$, so $4\frac{3}{5} = 4.6$.

8. $0.57 = \frac{57}{100}$

9. $65 \div 4 = 16\frac{1}{4}$

10. 4.7 is larger

11. fifty-six and thirty-nine hundredths

12. Move the decimal point two places to the right and add the % sign: $0.79 = 79\%$.

13. The units must be the same, so convert nickels to pennies: $9 \times 5 = 45$ pennies. The ratio is $\frac{34}{45}$ or $34{:}45$.

14. 76% means $\frac{76}{100} = \frac{19 \times 4}{25 \times 4} = \frac{19}{25}$

15. Move the decimal point two places to the right and add the % sign: $0.00082 = 0.082\%$

16. Move the decimal point two places to the right and add the % sign: $4.65 = 465\%$

17. $33 \div 4 = 8\frac{1}{4} = 8.250$ or 8.25

18. Move the decimal point two places to the left and drop the % sign: $0.31\% = 0.0031$.

19. $33 \div 4 = 8\frac{1}{4}$ (a)

20. c

21. $9 \div 10,000 = 0.0009$ (d)

22. $\frac{5}{7} = \frac{5 \times 2}{7 \times 2} = \frac{10}{14}$ (b)

23. $1,000{:}1 = 3,400{:}?$

    $1 \times 3,400 = 1,000 \times ?$

    $\frac{3,400}{1,000} = ?$

    $? = 3.4$ liters (a)

24. $9 \div 8 = 1\frac{1}{8} = 1.125$ (b)

25. c

26. f

27. i

28. k

29. e

30. g

31. j

32. d

33. h

34. c

35. b

36. Examples; any order:

    $\frac{7}{12} = \frac{7 \times 2}{12 \times 2} = \frac{14}{24}$; $\frac{7}{12}$; $\frac{7 \times 3}{12 \times 3} = \frac{21}{36}$

    $\frac{7}{12} = \frac{7 \times 4}{12 \times 4} = \frac{28}{48}$; $\frac{7}{12} = \frac{7 \times 5}{12 \times 5} = \frac{35}{60}$;

    $\frac{7}{12} = \frac{7 \times 6}{12 \times 6} = \frac{42}{72}$

1. f

2. a

3. g

4. l

5. j

6. h

7. k

8. d

9. c

10. b

11. e

12. i

13. $\in$

14. finite

15. {the vowels}

16. $5 \times 10^{-4} = 5 \times 0.0001 = 0.0005$

17. XIV

18. $100_2 = 1 \times 2^2 = 4$

19. {1, 2, 4, 8, 16, 32}

20. 7 has only itself and 1 as factors.

21. $48 = 2 \times 2 \times 2 \times 2 \times 3$ or $2^4 \times 3$

22. $15 = 3 \times 5$
$25 = 5 \times 5$
GCF = 5 (5 is the common factor)

23. The elements are +, −, x, and ÷.

24. $7x = 35$
$x = \{5\}$

25. 2 is 1,000 times greater than 0.002.

26. {*B*}

27. $6,000 = 6 \times 1,000 = 6 \times 10^3$

28. $3 = 1 \times 3$
$6 = 2 \times 3$
$12 = 2^2 \times 3$
$18 = 2 \times 3^2$
LCM = $2 \times 2 \times 3 \times 3 = 36$ (Choose each factor the greatest number of times that it appears)

29. $30 = 25 + 5$
$30 = 5^2 + 5$
$30 = 110_5$

Twenty-fives	Fives	Ones
1	1	0

30. {2, 4}

31. $0.001 \times 8 = 0.008$ (d)

32. a

33. (d) Since the units' digit is not even, the number is not divisible by 2 (a).

The sum of the digits = $1 + 4 + 6 + 1 + 5 + 1 + 5 = 23$. 23 is not divisible by 3, so the number is not divisible by 3 (b).

Since the number is not divisible by 2, it is not divisible by 4 (c).

Since the units' digit is 5, the number is divisible by 5 (d).

34. $\frac{46}{69} = \frac{2 \times 23}{3 \times 23} = \frac{2}{3}$ (c)

35. $1011_2 = (1 \times 2^3) + (1 \times 2^1) + (1 \times 2^0) = 8 + 2 + 1 = 11$ (b)

36. LXIV = 50 + 10 + 4 = 64 (b)

1. $\frac{5}{8} + \frac{7}{8} = \frac{12}{8} = 1\frac{4}{8} = 1\frac{1}{2}$

2. $\quad 4\frac{1}{3} = \overset{3}{\cancel{4}}\,\overset{16}{\frac{4}{12}}$

$\quad -1\frac{7}{12} = 1\frac{7}{12}$

$\qquad\qquad 2\frac{9}{12} = 2\frac{3}{4}$

3. $3.25 = 3\frac{1}{4}$

4. $\quad\;\; 11.632$
   $\quad +\;6.691$
   $\quad\;\; 18.323$

5. $\quad\;\; 36.180$
   $\quad -17.863$
   $\quad\;\; 18.317$

6. Examples:

   a. $\frac{2}{5} = \frac{2\times2}{5\times2} = \frac{4}{10}$ or

   $\quad \frac{2}{5} = \frac{2\times3}{5\times3} = \frac{6}{15}$

   b. $\frac{2}{5} = 0.4$

7. Examples:

   a. $3\frac{10}{16} = 3\frac{10\div2}{16\div2} = 3\frac{5}{8}$ or

   $\quad 3\frac{10}{16} = 3\frac{5}{8} = 3\frac{5\times3}{8\times3} = 3\frac{15}{24}$

   b. $3\frac{10}{16} = 3\frac{5}{8} = 3.625$

8. 35.0106

9. 70.1

10. $\quad \frac{1}{2} = \frac{5}{10}$

$\quad\;\; \frac{2}{5} = \frac{4}{10}$

$\quad +\frac{1}{10} = \frac{1}{10}$

$\qquad\quad \frac{10}{10} = 1$

11. $\quad \frac{2}{3} = \frac{12}{18}$

$\quad\;\; \frac{5}{9} = \frac{10}{18}$

$\quad +\frac{5}{6} = \frac{15}{18}$

$\qquad\quad \frac{37}{18} = 2\frac{1}{18}$

12. $\qquad\quad \frac{1}{9}$

$\qquad\;\; 3\frac{8}{9}$

$\quad +19\frac{5}{9}$

$\quad\;\; 22\frac{14}{9} = 23\frac{5}{9}$

13. $\quad \frac{2}{3} = \frac{10}{15}$

$\quad -\frac{1}{15} = \frac{1}{15}$

$\qquad\quad \frac{9}{15} = \frac{3}{5}$

14. $25\frac{2}{3} = 25\frac{16}{24}$

$\;-13\frac{5}{8} = 13\frac{15}{24}$

$\qquad\qquad 12\frac{1}{24}$

15. $\quad \overset{6}{\cancel{7}}\,\overset{11}{\frac{2}{9}}$

$\quad -4\frac{7}{9}$

$\qquad 2\frac{4}{9}$

16. 778.246

17. 5.6713

18. $\quad 25.782 \qquad\quad 30.372$
    $\;+\;4.59 \qquad\quad -\;0.85$
    $\quad 30.372 \qquad\quad 29.522$

19. $\quad\quad \$8.05$
    $\quad\quad\; 16.40$
    $\;+\;300.00$
    $\quad \$324.45$

20. a. $\quad \$25.00$
    $\qquad\;\; 1.65$
    $\qquad\;\; 3.50$
    $\quad +\;0.80$
    $\quad \$30.95$

    b. $\quad \$40.00$
    $\quad -30.95$
    $\quad\;\; \$9.05$

1. d

2. b

3. a

4. e

5. c

6. 0.8% = 0.008

```
      741
    x 0.008
    ───────
     5.928
```

7.
```
          0.569 = 56.9%
    52.0.)29.6.000
          2600
          ────
          3600
          3120
          ────
          4800
          4680
          ────
           120
```

8. 103% = 1.03

```
              250.485
    1.03.)258.00.000
          206
          ───
          520
          515
          ───
           500
           412
           ───
            880
            824
            ───
             560
             515
             ───
              45
```

9. 19% = 0.19

```
               252.6315 or 252.632
    0.19.)48.00.0000
          38
          ──
          100
           95
           ──
            50
            38
            ──
           120
           114
           ───
             60
             57
             ──
             30
             19
             ──
            110
             95
             ──
             15
```

10. 23% = 0.23

```
       0.78
     x 0.23
     ──────
       234
       156
     ──────
     0.1794
```

11.
```
              0.4146 = 41.5%
    0.369,)0.153.0000
           1476
           ────
            540
            369
            ───
           1710
           1476
           ────
           2340
           2214
           ────
            126
```

12. $16 \times 8\frac{1}{3} = 16 \times \frac{25}{3} = \frac{400}{3} = 133\frac{1}{3}$

$16 \times 8.333333 = 133.3332$

$133\frac{1}{3} = 133.333333$

13. $1\frac{3}{4} \div \frac{2}{3} = \frac{7}{4} \times \frac{3}{2} = \frac{21}{8} = 2\frac{5}{8}$

$1.75 \div 0.6666666 = 2.6250002$

$2\frac{5}{8} = 2.625$

14. $8\frac{1}{3} \times 5\frac{4}{5} = \frac{\overset{5}{25}}{3} \times \frac{29}{\underset{1}{5}} = \frac{145}{3} = 48\frac{1}{3}$

$8.3333333 \times 5.8 = 48.333333$

$48\frac{1}{3} = 48.333333$

15. $\frac{7}{8} \div \frac{5}{3} = \frac{7}{8} \times \frac{3}{5} = \frac{21}{40}$

$0.875 \div 1.6666666 = 0.525$

$\frac{21}{40} = 0.525$

16. 21% = 0.21

```
              1,119.0 = 1,119 students
    0.21.)235.00.0
          21
          ──
          25
          21
          ──
          40
          21
          ──
          190
          189
          ───
           10
```

17.
$$\begin{array}{r} 0.08 = 8\% \\ 150\overline{)12.00} \\ \underline{12\,00} \\ 0 \end{array}$$

24.
$$\begin{array}{r} 24.0 = \$24 \\ 0.75.\overline{)18.00.0} \\ \underline{15\,0} \\ 3\,00 \\ \underline{3\,00} \\ 0 \end{array}$$

18.
$$\begin{array}{r} 0.9 = 90\% \\ 300\overline{)270.0} \\ \underline{2700} \\ 0 \end{array}$$

or

$$\frac{270}{300} = \frac{\overset{9}{\cancel{270}}}{\underset{10}{\cancel{300}}} = 90\%$$

19.
$$\begin{array}{r} 21.5 \\ \times\ 0.43 \\ \hline 645 \\ 860\ \ \\ \hline 9.245\ = \$9.25 \end{array}$$

or

$$21\frac{1}{2} \times \frac{43}{100} = \frac{43}{2} \times \frac{43}{100} = \frac{1,849}{200}$$

$$= 9\frac{49}{200} = 9.245 = \$9.25$$

20.
$$\begin{array}{r} 0.324 \\ 145\overline{)47.000} \\ \underline{43\,5}\ \ \\ 3\,50\ \ \\ \underline{2\,90}\ \ \\ 600 \\ \underline{580} \\ 20 \end{array}$$

21.
$$\begin{array}{r} 2.078\ \text{grams} \\ 23\overline{)47.800} \\ \underline{46}\ \ \\ 1\,80 \\ \underline{1\,61} \\ 190 \\ \underline{184} \\ 6 \end{array}$$

22. A tithe is 10% = 0.1.

$$\begin{array}{r} 97.30 \\ \times\ 0.1 \\ \hline 9.730\ = \$9.73 \end{array}$$

23.
$$\begin{array}{r} 0.639 \\ \times\ 7.3 \\ \hline 1917 \\ 4473\ \ \\ \hline 4.6647\ = \$4.66 \end{array}$$

1. a general rule expressed in mathematical symbols

2. the fee paid by the borrower for the use of borrowed money

3. a number that, when substituted for a symbol in an open equation, makes the equation true

4. a statement that two ratios are equivalent

5. in a proportion, the two innermost terms

6. $S = C + P$
$\$49.95 = \$30.00 + P$
$\$49.00 - \$30.00 = P$
$\$19.95 = P$

7. $S = C + P$
$\$1.50 = C + \$0.60$
$\$1.50 - \$0.60 = C$
$\$0.90 = C$

8. $S = C + P$
$S = \$23.50 + \$14.95$
$S = \$38.45$

9. true

10. false

11. open

12. 12:33 or 4:11

13. 21:33 or 7:11

14. 12:21 or 4:7

15. 21:12 or 7:4

16. 9 x 16 = 144
4 x 36 = 144
true

17. 1 x 18 = 18
6 x 4 = 24
not a true proportion

18. 10 x 19 = 190
5 x 37 = 185
not a true proportion

19. $S = C + P$
$S = \$89.90 + \$55.00$
$S = \$144.90$

20. 3:7 = 9:x
63 = 3 • x
63 ÷ 3 = x
21 = x

21. 1.5
    x 1.5
    ─────
     75
     15
    ─────
    2.25

22. 20
    x 1.5
    ─────
    100
    20
    ─────
    30.0 = 30

23. 1.5
    x 4
    ─────
    6.0 = 6

24. 18
    − 11
    ─────
    7

25. 1.5
    −0.8
    ─────
    0.7

26. 125
    − 30
    ─────
    95

1. A sample that is not random.

2. A sample in which each member of the large group has an equal chance of being chosen.

3.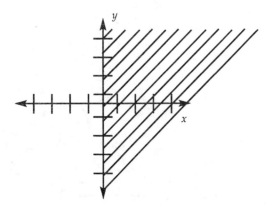

4.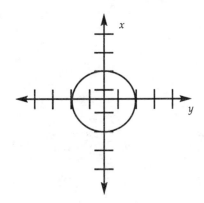

5.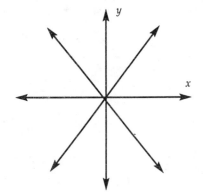

6.

Score	Tally
12	I
13	II
14	
15	I
16	II
17	
18	I
19	
20	I

7.

Score	Frequency
12	1
13	2
14	0
15	1
16	2
17	0
18	1
19	0
20	1

8.

Score		Relative Frequency
12	$\frac{1}{8}$	= 12.5%
13 $\frac{2}{8}$ =	$\frac{1}{4}$	= 25.0%
14	0	= 0%
15	$\frac{1}{8}$	= 12.5%
16 $\frac{2}{8}$ =	$\frac{1}{4}$	= 25.0%
17	0	= 0%
18	$\frac{1}{8}$	= 12.5%
19	0	= 0%
20	$\frac{1}{8}$	= 12.5%
		100.0%

9. 13 and 16

10. $\frac{15+16}{2} = \frac{31}{2} = 15\frac{1}{2} = 15.5$

11.

```
13            15.375  =  15.4
20          8)123.000
16            8
16            43
18            40
13             3 0
15             2 4
12              60
123             56
                40
                40
                 0
```

12. $20 - 12 = 8$

13. average deviation

$$= \frac{\text{range}}{\text{total number of data items}}$$

$$= \frac{8}{8} = 1$$

14.

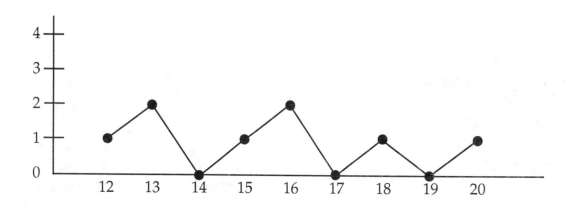

15.

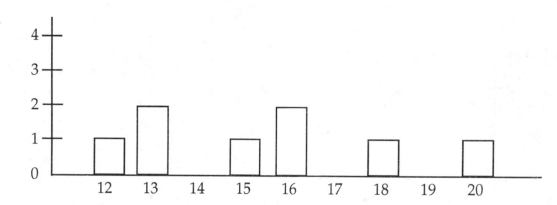

16.

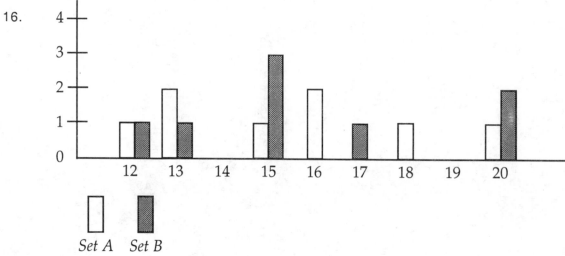

Set A *Set B*

1. a. $\frac{2}{5} + \frac{5}{6} = \frac{12}{30} + \frac{25}{30} = \frac{37}{30} = 1\frac{7}{30}$

    b. LCD = 30

    $\frac{5}{6} - \frac{2}{5} = \frac{5 \times 5}{6 \times 5} - \frac{2 \times 6}{5 \times 6}$

    $= \frac{25}{30} - \frac{12}{30} = \frac{13}{30}$

    c. $\frac{1\!\!\!/2}{1\,5} \times \frac{5\!\!\!/^1}{6\,3} = \frac{1}{3}$

    d. $\frac{2}{5} \div \frac{5}{6} = \frac{2}{5} \times \frac{6}{5} = \frac{12}{25}$

2. a. $\frac{2}{5} + 1\frac{1}{3} = \frac{2}{5} + \frac{4}{3}$

    LCD = 15

    $\frac{2 \times 3}{5 \times 3} + \frac{4 \times 5}{3 \times 5}$

    $= \frac{6}{15} + \frac{20}{15} = \frac{26}{15} = 1\frac{11}{15}$

    b. LCD = 15

    $\frac{4}{3} - \frac{2}{5} = \frac{4 \times 5}{3 \times 5} - \frac{2 \times 3}{5 \times 3}$

    $= \frac{20}{15} - \frac{6}{15} = \frac{14}{15}$

    c. $\frac{2}{5} \times \frac{4}{3} = \frac{8}{15}$

    d. $\frac{2}{5} \div \frac{4}{3} = \frac{2}{5} \times \frac{3}{4\,_2} = \frac{3}{10}$

3. a.
    ```
      0.47
    +1.62
    ─────
      2.09
    ```

    b.
    ```
      1.62
    −0.47
    ─────
      1.15
    ```

    c.
    ```
      0.47
      1.62
    ──────
       094
      282
       47
    ──────
    0.7614
    ```

    d.
    ```
              0.29012 = 0.2901
    1.62.)0.47.00000
              324
             1460
             1458
    ──────────
              200
              162
    ──────────
              380
              324
    ──────────
               56
    ```

4. a.
    ```
      0.47
    +0.23
    ─────
      0.70
    ```

    b.
    ```
      0.47
    −0.23
    ─────
      0.24
    ```

    c.
    ```
       0.47
     x 0.23
    ───────
        141
         94
        000
    ───────
     0.1081
    ```

    d.
    ```
            2.04347 = 2.0435
    0.23.)0.47.00000
            46
           100
            92
            80
            69
           110
            92
           180
           161
            19
    ```

5. $\notin$

6. $\in$

7. $\notin$

8. $\in$

9. 1001_2

8's	4's	2's	1's
1	0	0	1

10. 1010_2

8's	4's	2's	1's
1	0	1	0

11. 1011_2

8's	4's	2's	1's
1	0	1	1

12. 1100_2

8's	4's	2's	1's
1	1	0	0

13. $A \cap B = \phi$

14. $A \cup B = A$ or
    $A \cap B = B$

15. g

16. e

17. d

18. a

19. c

20. b

21. $14 \times ? = 2 \times 35$
    $14 \times ? = 70$
    $? = \frac{70}{14} = 5$

22. $? \times 1 = 25 \times 20$
    $? = 500$

23. $32 \times 6 = ? \times 2$
    $192 = ? \times 2$
    $? = \frac{192}{2} = 96$

24. $30 \times 20 = 5 \times ?$
    $600 = 5 \times ?$
    $? = \frac{600}{5} = 120$

25. $160 \times ? = 80 \times 42$
    $160 \times ? = 3,360$
    $? = \frac{3,360}{160} = 21$

26. d

27. e

28. c

29. a

30. f

31. g

32. true

33. open

34. false

35. open

36. false

37.

Score	Tally
1	IIII
2	IIII
3	IIII
4	IIII II
5	IIII III
6	III
7	III
8	III
9	II
10	I

38.

Score	Frequency	Relative Frequency	
1	4	$\frac{4}{40} = \frac{1}{10}$	= 10.0%
2	4	$\frac{4}{40} = \frac{1}{10}$	= 10.0%
3	5	$\frac{5}{40} = \frac{1}{8}$	= 12.5%
4	7	$\frac{7}{40}$	= 17.5%
5	8	$\frac{8}{40} = \frac{1}{5}$	= 20.0%
6	3	$\frac{3}{40}$	= 7.5%
7	3	$\frac{3}{40}$	= 7.5%
8	3	$\frac{3}{40}$	= 7.5%
9	2	$\frac{2}{40} = \frac{1}{20}$	= 5.0%
10	1	$\frac{1}{40}$	= 2.5%
	40		100.0%

39. Arrange the numbers in order from lowest to highest.

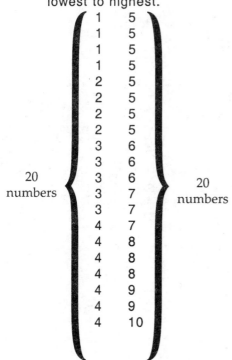

20 numbers

20 numbers

```
1    5
1    5
1    5
1    5
2    5
2    5
2    5
2    5
3    6
3    6
3    6
3    7
3    7
4    7
4    8
4    8
4    8
4    9
4    9
4   10
```

The two numbers in the middle are 4 and 5. Add them and divide by 2: 4 + 5 = 9 and 9 ÷ 2 = 4.5, which is the median.

40. 5

41. Add the numbers.

   1 x 4 =   4
   2 x 4 =   8
   3 x 5 =   15
   4 x 7 =   28
   5 x 8 =   40
   6 x 3 =   18
   7 x 3 =   21
   8 x 3 =   24
   9 x 2 =   18
  10 x 1 =   10
            ____
            186

Divide this sum by the count.

```
        4.65 is the mean
   40)186.00
      160
      ____
      260
      240
      ____
       200
       200
       ____
         0
```

42. range = difference between smallest and largest of the numbers

   10 − 1 = 9

43. average deviation

   $= \dfrac{\text{range}}{\text{total number of data items}}$

   $= \dfrac{9}{40} =$

```
        0.225
   40)9.000
      8 0
      ____
      1 00
        80
      ____
        200
        200
      ____
          0
```

44.

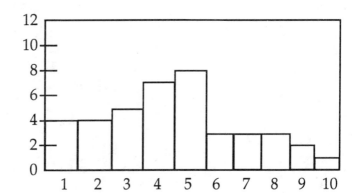